THEORETICAL FUNDAMENTALS OF ATMOSPHERIC OPTICS

THEORETICAL FUNDAMENTALS OF ATMOSPHERIC OPTICS

Yu.M. Timofeyev and A.V. Vasi'lev

Viva Books

New Delhi | Mumbai | Chennai | Kolkata | Bangalore | Hyderabad | Kochi

First Indian Edition 2009

VIVA BOOKS PRIVATE LIMITED

- 4737/23 Ansari Road, Daryaganj, New Delhi 110 002
  E-mail: vivadelhi@vivagroupindia.net, Tel. 42242200
- 76, Service Industries, Shirvane, Sector 1, Nerul, Navi Mumbai 400 706
  E-mail: vivamumbai@vivagroupindia.net, Tel. 27721273, 27721274
- Jamals Fazal Chambers, 26 Greams Road, Chennai 600 006
  E-mail: vivachennai@vivagroupindia.net, Tel. 28290304, 28294241
- B-103, Jindal Towers, 21/1A/3 Darga Road, Kolkata 700 017
  E-mail: vivakolkata@vivagroupindia.net, Tel. 22836381, 22816713
- 7, Sovereign Park Apartments, 56-58, K. R. Road, Basavanagudi, Bangalore 560 004
  E-mail: vivabangalore@vivagroupindia.net, Tel. 26607409, 26607410
- 101-102 Mughal Marc Apartments, 3-4-637 to 641, Narayanguda, Hyderabad 500 029
  E-mail: vivahyderabad@vivagroupindia.net, Tel. 27564481, 27564482
- Beevi Towers, First Floor, SRM Road, Kaloor, Kochi 682 018, Kerala
  E-mail: vivakochi@vivagroupindia.net, Tel: 0484-2403055, 2403056

www.vivagroupindia.com

Published by arrangement with
Cambridge International Science Publishing
7, Meadow Walk, Great Abington
Cambridge CB1 6AZ, UK

ISBN-10: 81-309-0943-X
ISBN-13: 978-81-309-0943-1

Published by Vinod Vasishtha for Viva Books Private Limited, 4737/23, Ansari Road, Daryaganj, New Delhi-110 002. Printed and bound by Anand Sons, Delhi-110 092.

Contents

Preface

Various atmospheric phenomena have been of interest to man from time immemorial. Both of the life of people and of other representatives of the fauna and flora has depended greatly on the weather and climate on our planet and on illumination conditions. The dependence of the mankind on the weather and climate is still very considerable Fluctuations of precipitation, anomalous temperatures and wind have the controlling effect on the life of people. Long-term droughts result in the deaths of tens and hundreds of thousands of people, regardless of the help provided by various international and charitable organisations and funds. Tens and hundreds of people die during flooding, cyclones and storms.

One of the books written by Aristotle is termed 'Meteorology' and is concerned in particular with the description and attempts to explain various atmospheric phenomena. For example, a relatively rational explanation of phenomena such as halo and rainbow is proposed. The subjects, discussed in 'Meteorology', are now the subjects of various sciences, not only meteoroly. The Meteorology book is interesting not only as an attempt to explain natural phenomena but also as experience with the application of unified principles for explaining different phenomena. The book may be regarded as the first book on meteorology available to us. The accurate publishing date is not known, but it may be assumed that it was written more than 2300 years ago.

Many optics laws of were discovered a long time ago. For example, the law of direct propagation of light is found in a report on optics attributed to Euclides (300 years BC) and it is likely that this phenomenon had been known long time prior to this date. The law of light reflection is also mentioned in the book Optics by Euclides. The phenomenon of light refraction was already known to Aristotle (400 years BC).

The sources of current atmospheric optics are found in optical sciences which have been formed to a large extent on the basis of observation of natural optical phenomena. Studies of various atmospheric optical phenomena were carried out by scientists such as Newton, Foucault, Euler, Rehmer, Huygens, Lomonosov and many others. Current atmospheric optics includes information on the physical state of the planet atmosphere, various sections of classic optics, radiation transfer theory, atomic and molecular spectroscopy, and electrodynamics.

If main attention in the initial stages of development of atmospheric optics was paid to investigations in the theory of visibility and radiation energetics of the atmosphere, then at the present time atmospheric optics studies and describes greatly differing optical phenomena, both from the energetics viewpoint and from the viewpoint of angular, spectral and temporal dependences of the characteristics of the radiation field and of the factors which determine these characteristics. A powerful impetus for the further development of atmospheric optics has been the need to perform remote measurements (ground-based, space) of different atmospheric parameters and of the surface. To realise these methods, it is necessary to achieve even deeper understanding of various processes of interaction of radiation with the medium (atmosphere and surface) and even higher accuracy of definition of the quantitative parameters of the interaction.

Atmospheric optics is part of the atmospheric physics – the science of physical processes in the Earth atmosphere (and also other planners). Atmospheric physics includes theoretical description and experimental investigations of all atmospheric phenomena. Atmospheric physics and, in particular, atmospheric optics are connected with greatly differing scientific disciplines, because atmospheric processes influence almost all aspects of the life of the mankind. A suitable example confirming this, is the connection of atmospheric optics with medicine: the number of skin cancer patients is directly connected with the amount of ultraviolet radiation of the Sun reaching the Earth surface.

The interests of the current atmospheric optics include:

–the processes of transformation of radiation energy of the Sun in the atmosphere and on the surface, and the formation of different types of atmospheric radiation as components of the radiation balance of the planet;

–the processes of propagation and transformation of different types of radiation (solar, thermal, non-equilibrium) which determine the temporal, spatial, polarisation and other characteristics of radiation fields, in particular, illumination of the surface;

–the radiation field as a source of information on the optical and physical characteristics of the atmosphere and the surface.

This book has been written on the basis of long-term experience of reading lectures and seminar course by the authors at the Department of Atmospheric Physics of the Faculty of Physics of the St. Petersburg State University.

The first chapter is concerned with a brief examination of the main special features of the structure of the solar system. Information on the planets of the solar system is provided. The main characteristics of Earth and changes of this movement are discussed. The structure of the Sun and special features of formation of solar radiation are investigated. Data are presented on the variability of solar energy reaching the Earth and on the spectral distribution of solar radiation.

The second chapter discusses essential general information on the Earth atmosphere. The latest information on the division of the atmosphere into layers, based on different criteria, is provided. The spatial and temporal variations of the main structural parameters of the atmosphere – temperature, density and pressure, are discussed. Various forms of the equation of state of the gas and the statics equations, barometric relationships, and applications of these relationships are outlined. Special attention is paid to the characteristics of the gas composition of the atmosphere, temporal variability and long-term variations caused by various antropogeneous factors. The data on atmospheric aerosols, clouds and precipitation are briefly analysed. Various characteristics of these parameters – the concentration of particles, the size distribution function of the particles and intensity of precipitation are introduced.

The third chapter discusses the determination of the main optical characteristics – intensity, radiation flux, investigations of the division of the electromagnetic spectrum into individual ranges. The main processes of interaction of radiation with the gas–aerosol medium are analysed. The radiation transfer equation in different forms is introduced. The polarisation characteristics of radiation and the vector form of the transfer equation are presented.

The fourth chapter discusses in detail molecular absorption in the atmosphere and its main special features. Various types of molecular absorption and parameters describing this process are analysed. Attention is also given to the characteristics of selective molecular absorption, such as the position, intensity, half-width of spectral lines, and their dependence on the parameters of the physical state of the medium. The main mechanisms of formation of the absorption line shapes – natural broadening, broadening as a result of collisions, and the Doppler effect, are discussed. The quantum-mechanics form of the transfer equation is presented, and the equation is used to derive expressions for the characteristics of selective molecular absorption.

The fifth chapter describes various types of scattering in the Earth atmosphere. The coefficient and phase function of the molecular scattering of radiation are derived. The polarisation characteristics of molecular scattering are analysed. Scattering both on individual aerosol particles and aerosol scattering in the atmosphere as a whole are investigated. Various optical phenomena, connected with aerosol scattering, are studied. Various types of non-coherent scattering in the atmosphere and its significance in solving atmospheric physics problems are dealt with. The optical phenomena, determined by radiation refraction, are investigated and many atmospheric optics phenomena are explained.

The sixth chapter introduces and analyses various optical characteristics of the reflecting properties of natural surfaces. Surface albedo, the spectral brightness coefficient and reflection factor are defined, and the relationships between various reflection characteristics presented. A large number of examples of the optical characteristics of the underlying surface in different ranges of the spectrum is given.

The seventh chapter is concerned with investigation of various aspects of the theory of atmospheric radiation transfer. The case of the transfer of thermal radiation for a plane-parallel model of the atmosphere is investigated in detail. Various methods of deriving the transmittance functions are outlined. Information is provided on various modelling approaches in calculations of the transmittance function and the approximate methods, used in the calculations of the transmittance functions and characteristics of thermal radiation, are analysed. Information on the non-equilibrium infrared radiation of the Earth atmosphere is provided.

Chapter 8 describes the fundamentals of the theory of solar radiation transfer. Information is given on various methods of solving the radiation transfer equation with multiple light scattering taken into account. The nature and main special features of various types of atmospheric glow are investigated. Special features of the formation of the ozone layer of the Earth and various photochemical processes in the Earth atmosphere are also briefly described.

In chapter 9, attention is given to the study of the problems of radiation energetics of the atmosphere. The relationships for calculating the solar energy flux at the upper boundary of the atmosphere at different moments of time and at different latitudes, are presented. The radiation balance of the underlying surface, the atmosphere and of the planet as a whole, is discussed.

The tenth chapter investigates direct and inverse problems of radiation transfer theory and atmospheric optics. Special attention is given to various remote measurement methods. The remote methods of measuring the parameters of the atmosphere and the Earth surface are classified, examples of various passive remote measurement methods, using the measurement of the radiation, passing through the investigated medium, and radiation of the atmosphere–underlying surface system are given. Active remote measurement methods are briefly characterised.

The appendix contains information on the main measurement units of atmospheric and optical parameters.

The authors are grateful to their colleagues, G.I. Gorchakov, D.I. Naringer and M.V. Tonkov, who reviewed the individual chapters of the book and made a number of important comments. They are especially grateful to G.M. Shved who made many useful critical comments and constructive proposals which greatly increased the quality of presentation of the material.

Finally, the authors would like to express their gratitude to colleagues of the Department of Atmospheric Physics of the Moscow State University (department head V.E. Kunitsyn) who edited the book.

The book could not be written without the help of colleagues of the Department of Atmospheric Physics of the Scientific Research Institute of Physics of the St. Petersburg State University, E.M. Sulgina, T.A. Naumova and L.N. Poberovskaya, who worked very hard on the preparation of the book for publication.

CHAPTER 1

THE SOLAR SYSTEM: PLANETS AND THE SUN

1.1. The planets of the solar system

In examination of the physics and properties of the atmosphere, it is useful to mention the main facts of the solar system, the Earth as a planet, its size and shape, special features of its orbit, the Sun and its radiation, and other planets. This is also important because the concepts such as 'atmospheric physics' and 'atmospheric optics' are applicable not only to the Earth but also to any other planet with an atmosphere.

The planets

The Earth is one of the nine planets of our solar system [16, 50, 94]. Table 1.1 gives the main astronomical parameters of the planets.

Table 1.1. Main astronomical parameters of planets

Planet	R, a.u.	r, km	T	t	φ	g, units of g_0	A
Mercury	0.387	2439	88 d	59 d	0°	0.37	0.056
Venus	0.723	6051	225 d	–243 d	3°	0.88	0.72
Earth	1.000	6378	365 d	23 h 56'	23° 27'	1.00	0.29
Mars	1.524	3379	687 d	24 h 37'	25° 12'	0.38	0.16
Jupiter	5.203	71300	11.86 y	9 h 50'	3°	2.64	0.343
Saturn	9.555	60100	29.46 y	10 h 14'	26° 44'	1.15	0.342
Uranus	19.22	24500	84.01 y	17 h 17'	98°	1.17	0.34
Neptune	30.11	25100	164.8 y	16 h 7'	28° 48'	1.18	0.29
Pluto	39.44	1500(?)	247.7 y	6 d 9 h	?	0.046(?)	0.4

Comment: R – The mean distance of a planet from the sun in astronomic units (1 a.u. = 149500000 km – the mean distance from the Earth to the sun); r – the radius of the planet; T – the period of rotation of the planet around the sun (in d – days, y – years); t period of rotation around its axis (the minus sign indicates rotation in the direction opposite to that of the Earth);φ_1 is the angle between the perpendicular to the plane of the eliptic and axis of rotation of the planet;[1] g – the rotational acceleration of the surface of the planet expressed in the units of acceeleration on the surface of the Earth g_0 = 9.81 m·s^{-1}; A is the integral albedo of the planet – fraction of the reflected energy of the sun arriving on the planet.

[1] The plane of the ecliptic – the plane in which the Earth rotates around the sun

The data presented in Table 1.1 clearly indicate large differences of the planets of the solar system in respect of different characteristics. In the following paragraph, it will be shown that the atmospheres of the planets also greatly differ. This fact is of considerable research and scientific value. The nature has presented to the man unique possibilities of examining very 'contrasting' planets. If it is taken into account that the possibilities of organisation of directional or even full-scale experiments – the main method of examining the physical processes – in the physics of the atmosphere ae greatly restricted, then the examination of, in particular, atmospheres of different planets offers a unique possibility of verifying the physical models, for example, the general circulation of the atmosphere.

Finally, when discussing the planets, it is also important to mention their satellites. This is also important because of the fact that the Earth has a single and very large satellite and reflection of solar radiation from the satellite during the night plays a significant role in the formation of the radiation field in the atmosphere of the Earth (although the radiation from the Moon at the upper boundary of the atmosphere of the Earth is almost 6 orders of magnitude lower than the radiation from the Sun). Mars has two satellites (Phobos and Deimos). The giant planets have many satellites. The largest satellites, with the size only slightly smaller than that of Mercury, are the sixth satellite of Saturn – Titan and the third satellite of Jupiter, Ganymede. Some satellites, for example, Titan, have relatively thick atmospheres.

1.2. Main parameters of the atmosphere of the planets

Main characteristics of planet atmospheres

All the planets of the solar system have atmospheres [50, 94]. Mercury and Pluto have very rarefied atmospheres and, to a certain approximation, it may be assumed that these planets have no atmosphere.

The atmospheres of the planets, like the planets themselves, greatly differ. Table 1.2 shows the data on the gas composition of the atmospheres of the planets and satellites, and the range of gases corresponds to the order of decreasing concentration.

Table 1.3 gives the main physical parameters of the atmospheres of the planets (with the exception of Pluto for which only a few reliable data are available). It should be mentioned that the physical parameters of the atmospheres are linked by the relationships well-

Table 1.2. Gas composition of the atmosphere of planets and satellites [94]

Planet	Main composition	Small gas components
Mercury	CO_2	
Venus	CO_2	Ar, H_2O, CO, HCl, HFr
Earth	N_2,O_2	Ar, H_2O, CO_2, CH_4, O_3 NO and others
Mars	CO_2	N_2, O_2, CO, H_2O, H_2, O_3, NO
Jupiter	H_2, He	NH_3, PH_3, CH_4, C_2H_6, C_2H_4, and others
Saturn	H_2, He	NH_3, PH_3, CH_4, C_2H_6, C_2H_2, CO
Uranus	H_2, He	CH_4
Neptune	H_2, He	CH_4, C_2H_2, C_2H_4, C_2H_6
Pluto	CH_4	H_2, He
Io	SO_2	
Titan	N_2	Ar, H_2, CH_4, C_2H_2, C_2H_4
Triton	N_2	CH_4

Table 1.3. Main physical parameters of planet atmospheres [94]

Planet	M, g/cm²	μ, g/mol	c_p, J/g·K	κ	γ_a, K/km	v_s, m/s	T_{ee}, K
Mercury	<0.1	44	0.85	1.28	4.6	350	500
Venus	10^5	44	0.85	1.28	10.5	240	230
Earth	10^3	29	1.0	1.41	9.8	320	255
Mars	16	44	0.85	1.28	4.4	230	216
Jupiter	10^3	2.6	10	1.45	2.5	790	134
Saturn	2.10^3	2.6	10	1.45	0.93	670	97
Uranus	10^3	2.2	13	1.42	0.76	540	54
Neptune	10^3	2.2	13	1.42	1.0	450	38

Comment: M – The mass of the column of the atmosphere above the unit area of the surface; μ – the mean molecular mass determined by the chemical composition of the atmosphere; c_p – the specific heat capacity per unit mass at constant pressure; $\kappa = c_p/c_v$ – the ratio of the specific heat capacities at constant pressure and volume; γ_a – the adiabatic vertical temperature gradient (decrease of temperature with height at adiabatic lifting of an air particle related to the unit height); v_s –the speed of sound; T_{ee} – the equilibrium temperature of outgoing radiation of the planet.

known in thermodynamics [2, 8, 43, 77, 79]:

$$c_p = c_v + \frac{R}{\mu}, \quad c_p = \frac{\kappa}{\kappa - 1}\frac{R}{\mu}, \quad \gamma_a = \frac{g}{c_p}, \quad v_s = \sqrt{\frac{\kappa R T}{\mu}},$$

where R is the universal gas constant, T is the gas temperature.

It should be mentioned that by analogy with the Earth, the atmosphere refers to the air shell above the solid surface of the planet. However, in giant planets there may not be surfaces in our sense of the word and, according to theoretical models, their hydrogen–helium atmosphere transfers to the solid state at a depth of more than

1000 km because of the very high pressure (in this case, the density of the gas at such a 'surface' reaches 0.1 g/cm^3, and temperature – several thousands of degrees). Therefore, the parameters of the giant planets should be determined using, instead of the surface, the upper edge of the cloud layer completely covering these planets.

Equilibrium temperature of the planet

The most important parameter determining the physical conditions on a planet is the amount of the energy received by the planets from the Sun. The solar constant of the planets S_0 is the amount of solar energy received per unit time on the unit area normal to the solar rays at the mean distance of the planet from the Sun. The planet absorbs and reflects the received radiation. The latter process is described by its integral albedo (see later). In addition to this, the planet generates atmospheric radiation escaping into the cosmos. To simplify considerations, it will be assumed that the outgoing radiation of the planet is governed by the laws of radiation of an absolutely black body (for more details see chapter 3). It is well known that the integral radiation of the absolutely black body (ABB) (radiation integrated in respect of all wavelengths) is governed by the Stefan–Boltzmann law. The total radiation energy of the absolutely black body from the unit area per unit time is determined by the relationship [20, 32, 91, 92]:

$$E_b = \sigma_B T^4,$$

where σ is the Stefan–Boltzmann constant ($\sigma = 5.67032 \cdot 10^{-8}$ W · m^{-2} · K^{-4}); T is the temperature of the absolutely black body. If this equation is used for characterising the outgoing radiation of the planet, equating it to E_b, then the corresponding value of T in the Stephan–Boltzmann law is the effective temperature of radiation of the planet T_{ee}.

If a planet has no internal energy sources or these sources are negligible, radiation temperature T_{ee} can be determined from the energy balance. The latter consists of the equality of the energy of radiation arriving from the Sun on the surface of the planet and the energy reflected from the surface, and the outgoing atmospheric radiation of the planet. In this case, it must be taken into account that the arriving (and reflected) solar energy of radiation falls (and is reflected) on the area (and is reflected by the area) of the cross-section of the planet normal to the solar rays (πr^2), and outgoing radiation is emitted by the entire surface of the planet (the surface of the sphere $4\pi r^2$). Therefore, the balance equation may be written in the following form:

$$\pi r^2 S_0 = A\pi r^2 S_0 + 4\pi r^2 \left(\sigma_B T_{ee}^4\right), \qquad (1.2.1)$$

where A is the integral albedo of the planet (see further chapter 6); T_{ee} is the equilibrium effective temperature of the planets. From (1.2.1) we obtain:

$$T_{ee} = \left(\frac{S_0(1-A)}{4\sigma_B}\right)^{1/4} \qquad (1.2.2)$$

T_{ee} may be regarded as the measure of the energy supplied to the atmosphere because it is determined by the quantity S_0 – the solar constant and albedo of the planet: the value $(1-A)$ is the fraction of the solar energy absorbed by the planet. The equilibrium effective temperature of the planet retains its physical meaning even if the planet has internal heat sources. For these planets, measured values of T_{ee} are such that the energy balance (1.2.1) is not fulfilled even at the albedo $A = 0$, i.e. in the case in which the entire solar energy, falling on the planet, is absorbed. In this situation, the ratio of the energy balances may be written in the form:

$$4\pi r^2 \left(\sigma_B T_{ee}^4\right) = (1-A)\pi r^2 S_0 + E_i, \qquad (1.2.3)$$

where E_i is the energy of internal heat sources of the planet. The estimates of the value of E_i for the Earth give the values of the order of $4.3+0.6\cdot 10^{13}$ W. The solar energy, absorbed by our planet, is $\sim 1.2\cdot 10^{17}$ W, which corresponds to the relative contribution of the internal heat sources of the Earth of $\sim 3.5\cdot 10^{-4}$. The identical situation is observed in the case of Venus, Mars and, probably, Uranus. Large internal heat sources have been discovered in Jupiter, Saturn and Neptune. For these planets, the outgoing radiation energy is greater than the energy required for reaching equilibrium with the solar radiation absorbed by the planet.

We determine the relative radiation balance of the planet R_0 as the ratio of the outgoing energy of atmospheric radiation to the absorbed solar energy. In the absence of internal heat sources

Table 1.4. Energy balances of outer planets of the solar system [94]

Parameter	Jupiter	Saturn	Uranus	Neptune
Albedo	0.343+0.032	0.342+0.030	0.300+0.049	0.31+0.04
Absorbed solar energy in 1 s (10^{16} W)	50.14+2.48	11.14+0.50	0.526+0.037	0.192+0.010
Emitted (outgoing) energy in 1 s (10^{16} W)	83.65+0.84	19.77+0.32	0.560+0.011	0.534+0.036
Internal energy 1 s (10^{16} W)	33.5+2.6	8.36+0.60	0.034+0.038	0.342+0.037
Relative radiation balance R_0	1,67+0.09	1.78+0.09	1.06+0.08	2.78+0.18

internal heat R_0 = 1. For the Earth, the relative radiation balance is very close to this value, R_0 = 1.00035. For Jupiter, Saturn and Neptune, the radiation balance R_0 greatly exceeds unity (R_0~1.7–2.8). Table 1.4 gives the components of radiation balance of a number of planets and also of their albedo [94].

Differences in planet atmospheres

Analysis of the data, presented in Table 1.3, shows that the main parameters of the planet atmospheres greatly differ. For example, the mass of the unit column of the atmosphere of the Mercury M is less than 0.1 g/cm^2, i.e. the atmosphere of the Mercury is 'very thin'. A relatively 'thin' atmosphere is also found on Mars (M ~ 16 g/cm^2). On the other hand, the atmosphere of Venus is very 'thick'. The value of M for this atmosphere is of the order of 10^5 g/cm^2, the pressure on the surface of Venus reaches 90–100 atm.

The molecular weight of air on Mercury, Neptune and Mars is the same, μ = 44. This is associated with the fact that the atmospheres of these planets consists mainly of carbon dioxide (Table 1.2). The atmosphere of the Earth consists mainly of oxygen and nitrogen. The molecular weight of this mixture is μ = 29. It should be mentioned that the Earth is the only planet in which oxygen is one of the main gas components. The low values of the molecular weights of the atmospheres of the giant planets (μ = 2.2–2.6) are determined by their main gas components – hydrogen and helium.

The specific heat capacity at constant pressure c_p for the planets of the Earth group (Mercury, Venus, Mars, Earth) is close to 1 J/g·K, whereas for the giant planets it is of the order of 10 J/g·K. This is associated mainly with the difference in the gas composition of the planet atmospheres. The adiabatic gradient of temperature is small in the giant planets: from 0.76 to 2.5 K/km and the maximum in the case of Venus and Earth (up to 10 K/km).

The effective temperature of the planet T_{ee} is maximum on

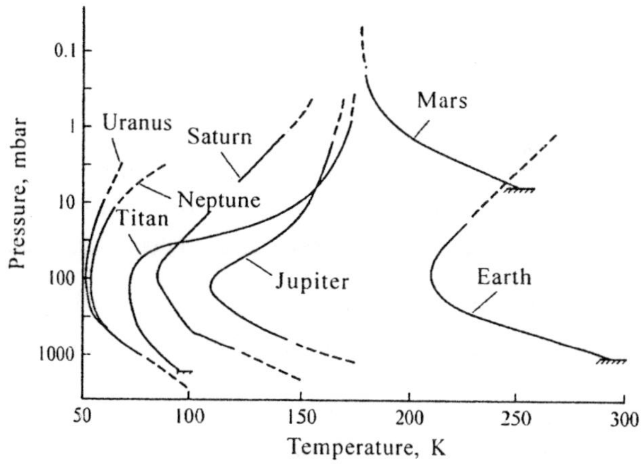

Fig. 1.1. Temperature profiles in planet atmospheres [94].

Mercury (approximately 500 K). This is determined mainly by the proximity of Mercury to the Sun and, consequently, by a high value of S_0, and also by the low value of the albedo (A = 0.06). It is interesting to note that if Mercury had an albedo equal to 0.72, i.e., as on Venus, its equilibrium temperature would be 355 K. The minimum values of the equilibrium temperature are recorded on the giant planets. For example, on Neptune, this temperature is only 38 K. There is almost a monotonic decrease of the values of T_{ee} with an increase of the distance from the Sun and, consequently, a decrease of solar constants S_0. An exception from this rule is detected only in the case of Venus because of the very high value of its albedo (A = 0 .72) determined by intensive reflection of solar energy from its cloud layer.

Figure 1.1 shows the mean temperature profiles in the atmospheres of various planets [94]. The graph shows clearly both differences and identical features of the temperature profiles. For example, almost all planets are characterised by an almost linear decrease of temperature in the lower layers of the atmosphere (this region is referred to as the troposphere). At the same time, on the Earth and giant planets temperature increases in the upper layers of the atmosphere because of the absorption of solar radiation by different atmospheric gases. In the mean, all temperatures of the planet atmospheres decrease with an increase of the distance from the Sun. This is determined by a decrease of the solar constants S_0 of these planets.

1.3. Special features of the orbit of the Earth

Rotation of the Earth

The mean distance from the Earth of the Sun is 149 500 000 km. This value is used in astronomy as one of the units of the distance (1 a.u. – astronomic unit). This is the most important characteristic, but it is not the only characteristic determining many special features of the physical state of the Earth atmosphere. Of special importance are also the periods of rotation of the planets around the Sun and the natural axis of rotation, the angle of inclination of the axis to the ecliptic plane ψ ($\psi = 90°-\varphi$).

The parameters of the orbit of rotation of the Earth around the Sun and around the natural axis [16, 37, 43] are the most important factors determining the amount of radiation energy (radiative energy) of the Sun reaching different areas of the Earth in different times of year, and also as a result, the weather and climate of the planets. Every 24 hours, the Earth carries out a single complete rotation from west to east in relation to the direction on the Sun around the axis passing through the poles. The direction of the axis of natural rotation of the Earth in relatively short periods of time (of the order of thousands of years) may be regarded as constant. This rotation is the the reason for the change of the day and the night on our planet, taking place during the illumination of the different regions of the Earth by the Sun. At the same time, the Earth travels around the Sun carrying out one complete rotation in approximately 365 days.

As a result of rotation around the natural axis, the Earth is not strictly spherical, and a special term, geoid, is introduced for designing its shape. The oblate ellipsoid of revolution (spheroid) describes the shape of our planet more accurately than the sphere. The equatorial and polar radii of the Earth (the major and minor half-axes of the ellipsoid) are equal to 6378.5 and 6356.0 km, respectively, i.e. differ by 21.5 km. The mean radius of the Earth is assumed to be 6370 km. The axis of rotation of the Earth is not perpendicular to the plane of the orbit (ecliptic). The angle between the axis of revolution and the normal to the ecliptic plane is 23.5°. This angle is referred to as the inclination angle. In particular, this special feature determines the presence of seasonal variations of the amount of solar radiation in different latitudes of our planet. In movement of the Earth along the orbit around the Sun the axis of revolution retains (to a first approximation, see later) its direction,

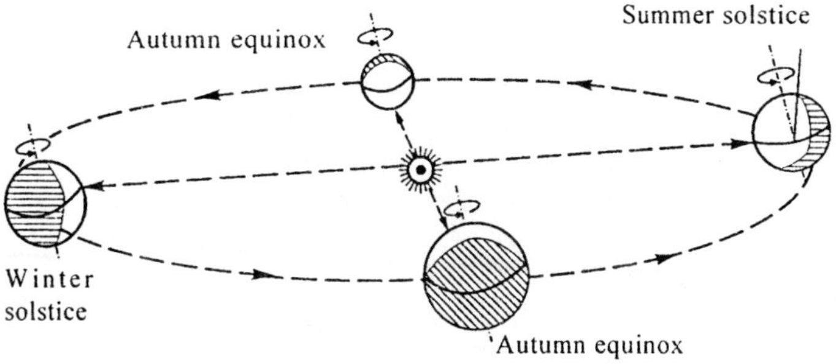

Fig. 1.2. Positions of the Earth in the orbit.

and in different positions on the orbit different 'points' of the planets have different values of radiation energy (Fig. 1.2).

Because of the disturbing effect, exerted on the rotation of the Earth by the bodies of the solar system, the axis of revolution of the Earth carries out a very complicated movement in space [16, 37, 43, 49]. In particular, it slowly describes a cone remaining inclined always under the angle of approximately 66.5° in relation to the plane of movement of the Earth. This movement of the Earth axis is referred to as precession. The precession period is approximately 21 000 years and determines the mean direction of the axis in space in various ages. It should be stressed that precession does not change the amount of solar energy emitted to different latitudes of the Earth, but it does change the time of start of the individual seasons.

In addition to precession motion, the axis of rotation of the Earth carried out different 'fine oscillations' around its mean position, the main of which has a period of ~18.6 years and is referred to as the nutation of the Earth axis (nutation is the consequence of attraction of the Earth's spheroid by the Moon). There is also a component of the variation of the angle of inclination (in the range ~1.5° with a period of ~41 000 years). With time, as a result of the effect of the parameter, the eccentricity of the Earth changes. These changes have a non-periodic nature with a typical period of a approximately 100 000 years.

The current investigations in climatology (or, more accurately, paleoclimatology – the science of the climate of the Earth in the past) based on, for example, analysis of deep water bottom sediments, containing information on the climate of our planet over a period of the last 500 000 years, have confirmed the presence of 100 000-, 40 000- and 23 000-year components of climatic variations.

The sky sphere

The consequence of rotation of the Earth around its axis is the visible movement of the Sun, the Moon and stars in the sky. In astronomy, on the basis of long-term tradition, this movement is described as the movement of heavenly bodies in the sky sphere, i.e. imaginary sphere, with the observer in its centre [16]. The tangent plane to the Earth surface, passing through the examination point, is referred to as the horizon plane. The angle between the horizon plane and the direction to the heavenly body is referred to as the celestial altitude above the horizon. This angle may vary from −90 to +90°. The negative values of the celestial altitude correspond to the position of the body below the horizon, i.e. not visible to the observer.* The upper point of the sky sphere (with an altitude of +90°) is referred to as the zenith, the opposite (lower) point of the sky sphere (with an altitude of −90°) is referred to as nadir.

In atmospheric optics, as in astronomy, the position of the Sun and other heavenly bodies is characterised by their altitude above the horizon. In addition to this, the angles counted from the zenith–nadir axis are often determined. The angle, counted from the zenith, is referred to as the zenith angle, the angle counted from the nadir as the nadir angle. The zenith angle is used most frequently in examination from the Earth surface, and the nadir angle in observation from space. These angles are linked by the obvious relationship $\theta = 90° - \psi = 180 - \theta'$, where θ is the zenith angle, ψ is the altitude above the horizon, θ' is the nadir angle.

Because of the daily rotation of the Earth, the heavenly bodies also carry out circular movements during the day. The maximum altitude of a heavenly body above the horizon during the day is referred to as the upper culmination, the minimum as the lower culmination. The moments of intersection of the plane of the horizon by the heavenly body are referred to as sunrise and sunset. If the lower culmination of the heavenly body has a positive altitude, then the body is non-setting. In this case, if the upper culmination has a negative altitude, this heavenly body is non-rising.

Times of year

The change of the times of year on the Earth is the most

---

*This is not completely accurate if we take into account atmospheric refraction which will be discussed in chapter 5.

characteristic feature of the climate. It is determined by the rotation of the Earth around the Sun and also by the inclination of the Earth's axis to the ecliptic plane under the angle of 66.5°. Figure 1.2 shows the diagram characterising the position of the Earth in different seasons of the year. During the summer solstice which falls on approximately June 22nd, the Sun at midday is in the zenith at 23.5° of the northern latitude (Tropic of Cancer). This day is characterised by the maximum supply of the radiative energy of the Sun to the northern hemisphere (and conversely minimum to the southern hemisphere). During the summer solstice the altitude of the Sun above the horizon and the duration of the day in the northern hemisphere reach maximum values, and in the direction to the north from the northern polar circle (66.5° northern latitude) the Sun remains all day above the horizon. In the southern hemisphere the altitude of the Sun and the duration of day in this time are minimum. In June solstice, the Sun does not appear above the horizon in the south from the southern polar circle (66.5° southern latitude), i.e. a polar night occurs. Naturally, this time corresponds to summer in the northern hemisphere and to winter in the southern one.

The inverse pattern is detected in the course of December (winter for the northern hemisphere) solstice (December 22). On December 22nd, the Sun is in the zenith already at 23.5° southern latitude (Tropic of Capricorn). In this period, in the northern direction of more than 66.5° the northern hemisphere has the polar night, and the southern hemisphere (more southern than 66.5°) a polar day. During the spring (March 21st) and autumn (September 22nd) equinox the duration of the day and night is always the same everywhere and equals 12 h and the Sun is in the zenith at midday on the equator.

Tropics and polar circles

The definitions of the tropics and polar circles of the Earth can easily be generalized to other planets. The tropic (northern or southern) is the maximum distance along the latitude from the equator where the Sun may be in the zenith [16]. The polar circle (northern or southern) is the maximum (in respect of latitude) distance from the pole where the Sun maybe non-rising (or, which is the same, non-setting).

If we return to Table 1.1, it may be seen that the inclination angles differ for different planets. For examples, for Jupiter it is only 3°, the axis of rotation of Jupiter is almost normal to the plane of the orbit. For this reason, Jupiter does not show any large variations of the incoming radiative energy of the Sun at different latitudes, i.e., in our understanding, this planet does not have different times of year. On the other hand, for Uranus, the inclination angle of the planet is close to 90°, i.e. the planet axis of rotation is almost identical with the ecliptic plane. Correspondingly, the tropics on Jupiter are located at the equator and the polar circles at the pole. For Uranus the pattern is reversed and for inhabitants of the Earth completely unusual: the tropics are located at the poles and the polar circles at the equator.

Additional information on the Earth

The Earth rotates around the Sun around an ellipse with a medium eccentricity of $e = 0.0167$. The closest point of the orbit in relation to the Sun is referred to as the perigelium. The distance from the Sun to the perigelium is $a(1-e)$, where a is the major half-axis of the ellipse (for the Earth $a=1.4959787 \cdot 10^{11}$ m), i.e. it is equal to $1.4709958 \cdot 10^{11}$ m (approximately 147 million kilometres). The point of the orbit most remote from the Sun is referred to as the aphelian and is equal to $a(1+e)$, i.e. approximately 152 million kilometres. It is interesting to note that the velocity of motion of the Earth along the orbit differs – maximum at perigelium (30.3 km/s) and minimum at aphelian (29.3 km/c).

Finally, we mention other important information on the planet Earth. The surface of the Earth equals 510 098 million km^2. Of this, 70.8% is occupied by world oceans. In the northern hemisphere, the world ocean occupies 60.7%, and in the southern hemisphere 80.9% of the surface. The latter quantity sometimes indicate that the planet Earth in the southern hemisphere is 'a water planet'. The mass of the Earth is $5.9737 \cdot 10^{24}$ kg. The mass of the atmosphere is only $5.15 \cdot 10^{18}$ kg, and 50% of the mass of the atmosphere is concentrated to an altitude of 5 km, and 99% to 35 km. The mean density of the planet Earth is 5.5 g·cm^{-3}. It approximately coincides with the density of Mercury, Venus and Mars, but is considerably higher than the density of giant planets whose density is close to 1 g·cm^{-3}; in the case of Saturn it is even lower than unity (0.7 g·cm^{-3}). Infact, the low values of the mean density of giant planets are also associated with the fact that their radius is regarded as the upper boundary of the cloud cover (see Table 1.3).

1.4. The Sun and its radiation

The main source of energy in the solar system is the Sun and as regards its mass it is a medium star [16, 32, 37, 42]. Its mass is $1.99 \cdot 10^{30}$ kg, i.e. 99.87% of all mass, concentrated in the solar system. The radius of the Sun is $6.96 \cdot 10^{5}$ km. From these data we can easily obtain the mean density of solar matter, 1.41 $g \cdot cm^{-3}$. When discussing the dimensions of the Sun, it should be mentioned that its radius is 109 times greater than the radius of the Earth. From the surface of the Earth, the Sun is visible in the form of a disk whose angular dimensions are slightly greater than half a degree ($9.3 \cdot 10^{-3}$ rad). These angular dimensions change (±1.7%) during movement of the Earth along the elliptical orbit.

According to the current views, the main elements of the Sun are hydrogen and helium (91 and 9% according to the number of atoms, 73 and 26.5% by mass). In addition to these two elements, the Sun also contains many others. The most widely found elements are magnesium, iron, calcium, silicon, neon, carbon, etc. The reaction of thermonuclear synthesis of four hydrogen atoms in one helium atom is the source of energy of the Sun.

Structure of the Sun

The generation of energy on the Sun takes place in the vicinity of the centre [37, 49]. The propagation of photons to the outer boundary is accompanied by the processes of absorption of radiation (by iron atoms, for example) and re-radiation. Another process of transfer of energy from the centre of the Sun to its periphery is convection. According to current estimates, the temperature of the Sun in the centre reaches 5 million degrees. The temperature of the outer layers is considerably lower, the minimum value is 5000–6000 K.

The part of the Sun visible to us is the photosphere. In particular, the photosphere is characterised by generation of a large part of outgoing solar radiation. The photosphere is a relatively thin layer of the order of 400–500 km. The regions positioned above the photosphere are referred to as the 'solar atmosphere'. They are subdivided into the chromosphere, transition zone (TZ) and solar corona (Fig.1.3). The chromosphere is located above the photosphere and its thickness is 2000–4000 km.

The vertical temperature profile in the photosphere and the solar atmosphere is shown in Fig.1.3. It may be seen that the minimum temperature, T = 4200 K, is detected at the upper boundary of the

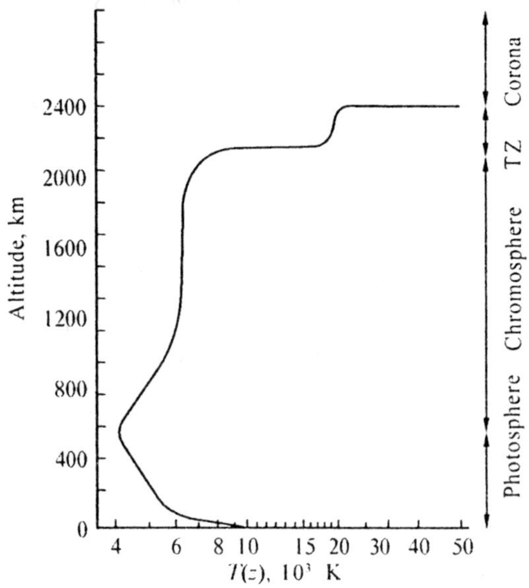

Fig.1.3. Temperature profile $T(z)$ in the photosphere, chromosphere and transition zone (TZ) of the Sun [20].

photosphere. Further, in the chromosphere and the transition zone the temperature increases to 20000–50000 K. In the Sun corona, temperature increases to values of the order of 1 500 000 K at the altitude equal to the Sun radius.

The photosphere is the region of a highly rarefied gas with a characteristic density of $(1–3)\cdot 10^{-7}$ cm^{-3} and a pressure of 5–150 mbar [49, 79]. The radiation of the photosphere has a continuous spectrum. The layer of relatively 'cold' gases in the upper layer of the atmosphere and lower part of the chromosphere absorbs a part of this radiation in the spectral lines of absorption of the atoms of carbon, silicon, cadmium, magnesium, iron, and so on. This absorption results in the formation of the absorption spectrum of the Sun. Therefore, the spectrum of the outgoing radiation of the Sun shows clearly these absorption lines – Fraunhofer's lines (the name originates from a German scientist Josef Fraunhofer who discovered these lines).

If we approximate the spectral dependence of the outgoing radiation of the Sun using Planck's formula for the radiation of an absolutely black body (see Chapter 3), then the best approximation is obtained for the temperatures of 5800–6000 K [91] (Fig.1.4). However, in different spectral ranges similar equivalent temperatures differ. This is caused by the fact that the formation of outgoing radiation of the Sun in different spectral ranges takes place at different altitudes. On the other hand, it may be seen (Fig.1.3) that the temperature in the photosphere differs at different altitudes. The 'normal' (non-perturbed) photosphere has a grainy structure – the so-called granulation. The granules are light 'tiny spots' separated by darker lines, with a diameter from 200 to 1300 km, the brightness is 10–30% higher than the mean surrounding background with the mean lifetime of 8–10 min.

The chromosphere, situated above the photosphere, is evidently highly heterogeneous and has a fibrous structure. In the non-perturbed atmosphere, spicules and supergranules are constantly observed. Spicules are the 'teeth' at the upper boundary of the chromosphere, visible during Sun's eclipse. Their thickness is 500-600 km and on average they have the length of 7500 km at the Sun's equator and 7800 km at the poles. The lifetime of the spicules is 2–5 min. Supergranulation is a large cellular structure of the solar atmosphere also found in the photosphere, and especially clearly visible in the chromosphere. The diameter of the supergranules is approximately 3000 km, their total number on the Sun disk is 2500, lifetime approximately 24 hours.

Sun's corona is located above the chromosphere in the transition zone. To a distance of $z = 2R$ the corona is visible in the spectral lines of natural radiation – the lines of highly ionised iron, nickel and calcium. The corona also emits in x-ray and radiowave ranges.

Solar activity

In addition to the granules, supergranules and spicules, typical of the 'normal', non-perturbed solar atmosphere, the atmosphere also contains a number of perturbations, i.e. heterogeneities with a relatively short lifetime. The main of them are spots, faculas, chromospheric bursts, proturberances, coronal loops. We shall characterise briefly these special features of the structure of the solar atmosphere referring the reader for detailed examination to the monographs concerned with the physics of the Sun, for example (for example, the monograph by A.S. Monin).

Solar spots are the most clearly visible heterogeneity of the Sun surface. The spots form as small dark 'pores', with angular diameters of (2–4)". A typical solar spot consists of a dark central 'shadow' with a mean diameter of approximately 17 500 km and the brightness of (20–30)% of the brightness of the surrounding background of the non-perturbed photosphere. The linear dimensions of the spots vary in a wide range from 10 000 km to 150 000 km. The typical area of a spot is approximately 10^{-4} of the entire area of the visible Sun disk. The duration of existance of individual solar spots varies from several days to several months. The solar spots are found mainly in the latitude zone ±40° in relation to the solar equator. The solar spots appear dark only as regards the contrast with the non-perturbed photosphere. The mean radiation temperature of the 'shadow' of large solar spots in the visible range of the spectrum is 4200 K, i.e., approximately 1600 K lower than the temperature of the non-perturbed photosphere. The number of observed solar spots greatly varies.

There are periods of maximum and minimum number of spots on the Sun disk. The mean period between the appearance of the maximum and minimum number of the solar spots is approxiamtely eleven years (the so-called 11 year solar cycle). In the years of the maximum number of the solar spots (the years of solar actvity) on the Sun surface there are frequent cases of ejection of particles and radiation bursts. In the period of the minimum solar spots (the minimum solar activites) bursts are far less frequent. Usually, the bursts are localised in the vicinity of the complicated groups of the solar spots and are referred to as solar flares. The ejection of the particles and radiation bursts from large flares have a strong effect on the condition in the magnetic field of the Earth (magnetic storms – interference in radio communications) and also of the upper layers of the atmosphere. For example, powerful proton bursts influence the ozone layer.

Faculas are relatively long-life brighter regions in the vicinity of the solar spots. The bright elements of the chromospheric faculas are referred to as floccules. The photospheric faculas have a granular structure. The brightness is on average 10% higher than that of the non-perturbed photosphere. In the brightest facula granules this difference may reach 40–45% and in some cases even 150%.

Chromospheric bursts are the short term increases of the intensity of radiation of limited areas of the chromosphere in the vicinity of the solar spots. On average, the areas of the bursts equal $1.6 \cdot 10^{-4}$ of the area of the visible disk. During the bursts, the brightness of radiation in the short wave part of the spectrum of the Sun may increase by up to 10 times. The duration of the bursts changes from several minutes to several hours (mean duration 20 min). The bursts are a frequent phenomenon on the Sun: on average, a burst occurs every 7 hours of the life of a group of solar spots. In the period of maximum solar activity, bursts occur on the Sun on average every two hours. Radiation bursts in the visible part of the spectrum are accompanied by strong radiation in the x-ray and ultraviolet ranges, the so-called 'large bursts of radiowave radiation' and also by the emission of corpuscular fluxes and cosmic rays.

Protuberances are strip-shaped 'cold' areas of densening of the gas in the internal part of the Sun corona. A typical 'still' protuberance is up to 200 000 km long (sometimes the length may even reach 1.9 million km), thickness 50 000 km, width not more than 6000 km. The Sun corona also contains rays whose thickness at the base is approximately 7000 km and the electron density is five times greater than in the surrounding matter; coronal holes are the regions of rarefaction of the corona.

This brief survey of different formations on the Sun shows the complicated nature of phenomena in the Sun atmosphere and its non-stationary nature and heterogeneity which leads to variations of the intensity of radiation over the Sun disk. Therefore, it is important to examine the problem of the constancy of the spectral composition and integral (in respect of spectrum) radiation of the entire Sun disk and, as a consequence, the constancy of solar – integral and spectral – constants.

The entire complex of the non-stationary phenomena in the solar atmosphere is referred to as solar activity. Various parameters – indices – are used for quantitative characterisation of solar activities [49]. In most cases, the index of solar activity is used, i.e. the Wolf number W proportional to the sum of the total number of the spots f and the tenfold number of their groups g:

$$W=k(10g+f),$$

where k is the empirical coefficient of reduction of the observations to the unit conditions, and depends on the quality of astronomical instruments and experience of the observer.

Radiation spectrum of the Sun

The distribution of the electromagnetic radiation emitted by the Sun and arriving at the upper boundary of the Earth's atmosphere depending on the wavelength λ is referred to as the Sun spectrum. The definition of the Sun spectrum should also include requirements from the definition of the solar constant (section 1.2) as the

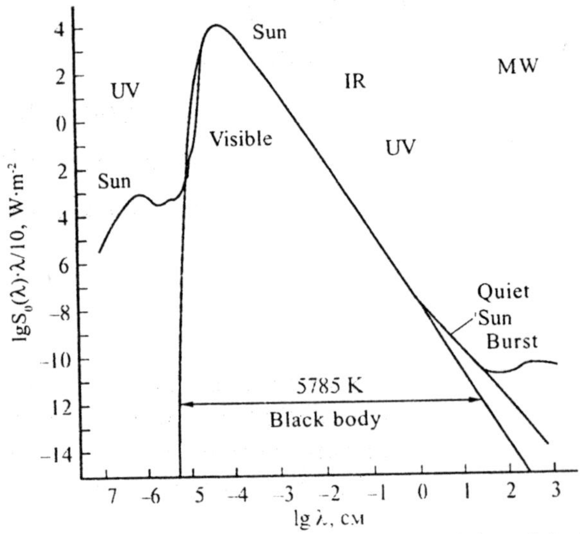

Fig.1.4. Solar radiation spectrum in comparison with the radiation of the black body at T = 5785 K. UV, Visible, IR, MW – ultraviolet, visible, infrared, microwave radiation of the Sun, λ is the wavelength, cm [91].

incoming solar energy per unit term on unit area normal to the rays at the mean distance between the Earth and the Sun. This quantity is often referred to as the spectral solar constant $S_0(\lambda)$, and for the solar constant shown in section 1.2, we use the more accurate term 'the integral solar constant S_0':

$$S_0 = \int_0^\infty S_0(\lambda)d\lambda. \quad (1.4.1)$$

Solar radiation is the main source of heating the atmosphere and the underlying surface. Therefore, the measurements of the Sun spectrum and the solar constant have already been continuing for a long time. Initially, these measurements were taken on Earth's surface, later using altitude aerostats. In analysis of these measurements (especially measurements from the Earth's surface) it is necessary to take into account the attenuation of radiation by the atmosphere. In recent years, measurements of $S_0(\lambda)$ and S_0 have been taken using equipment installed on satellites. According to the latest data, the mean value of the spectral solar constant of the Earth is 1366.1 W·m^{-2} [http://spacewx.com/wavelengthdescript.html].

The standard spectrum of the Sun with a low spectral resolution is shown in Fig.1.4 [91]. For comparison, the figure gives the curve

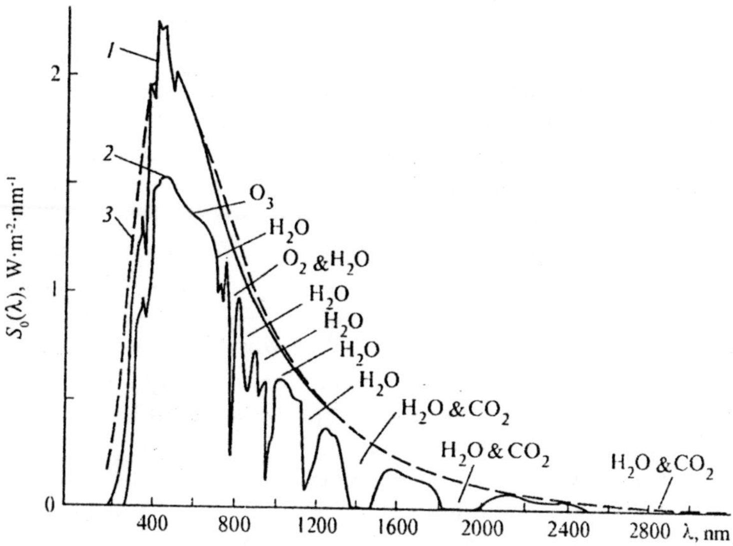

Fig. 1.5. Solar radiation at the upper boundary of the atmosphere (1) and surface of the Earth (2); black body radiation (3) – at T = 5900 K. Absorption bands of ozone (O_3), water vapour (H_2O), oxygen (O_2) and carbon dioxide (CO_2) are shown [44].

of radiation of a black body (Planck's function) at a temperature T = 5785 K. This curve efficiently approximates the Sun spectrum in its middle part – in the wavelength range 0.2 μm–1 cm. At the edges of the solar spectrum in ultraviolet and radiowave range the differences from the radiation of the black body at T=5785 K are very large, especially at surges on the Sun. If we examine the Sun spectrum at a higher spectral resolution, we can see the presence of many Fraunhofer lines caused, as already mentioned, by the absorption of different elements in the solar photosphere and chromosphere. Of the entire solar energy arriving on the Earth approximately 40% is the visible range (0.4–0.7 μm), 10% shorter wavelength, 50% longer wavelength of radiation.

Figure 1.5. also shows these solar radiation spectra arriving at the upper boundary of the atmosphere and on the Earth's surface, i.e. the radiation attenuated by the atmosphere. It may be seen that a large fraction of the solar energy does not reach the surface, it is absorbed by the atmosphere or scattered or reflected back into the cosmic space. Consequently, the absorption and scattering of solar radiation by atmospheric gases plays a significant role in the atmospheric processes. As indicated by Fig. 1.5, the spectrum of the Sun on the Earth's surface contains bands and lines of absorption of the Earth's atmosphere (absorption bands of H_2O,

CO_2 and O_3). In contrast to Fraunhofer lines, these lines are referred to as telluric. Analysis of the absorption in the atmosphere is presented in Chapter 4. Although the Sun spectrum is often approximated by the radiation of an absolutely black body, giving it different effective (equivalent) radiation temperature, this approach is very approximate (Figs. 1.4 and 1.5). In particular, because of this approach, as already mentioned previously, the effective temperature of solar radiation is not a constant and differs for different wavelengths. For example, in the spectrum range from 0.1 to 0.3 μm the equivalent temperature of the Sun changes in a wide range from 4600 K to more than 6000 K. The curve at T = 5785 K, as shown in Fig.1.4, efficiently approximates solar radiation in the region of the maximum, but results in large errors in shortwave and radiowave ranges. Correspondingly, the effective temperatures of radiation at the edges of the Sun spectrum greatly exceed 6000 K. The Sun radiation spectrum contains a large number of absorption lines (Fraunhofer) but in its shortwave part (shorter than a wavelength of 130 nm) there are also radiation lines. It is important to note a series of Layman radiation lines determined by transitions between the excited and ground states of the hydrogen atoms. The Layman-α radiation line with a wavelength of 121.5 nm is especially strong.

Variability of solar constant and Sun spectrum

A fundamental problem for many Earth sciences is the variability of the spectral composition of solar radiation $S_0(\lambda)$ and integral solar constant S_0. Recent extremely complicated, long-term satellite measurements have shown the following special features.

1. integral solar radiation shows variations on different time

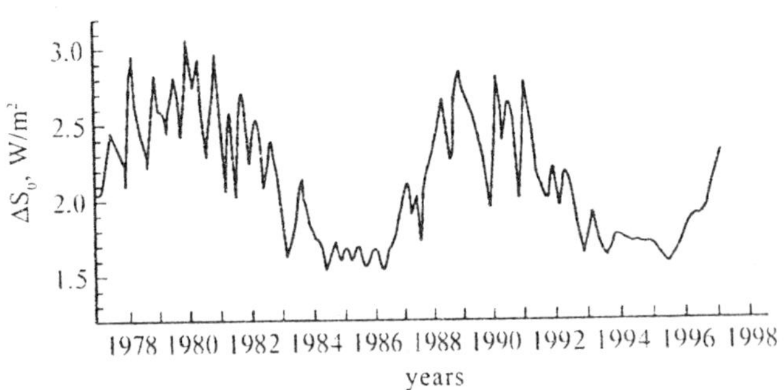

Fig.1.6. Variations (in W·m^{-2}) of the solar constant S_0 in the period 1978–1997 according to the results of measurements on five satellite devices [90].

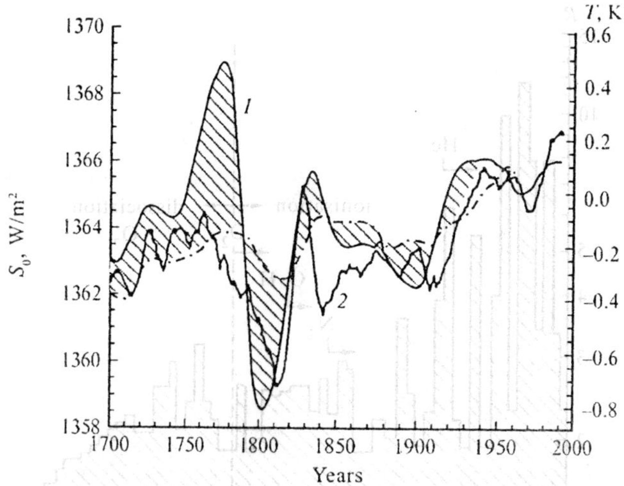

Fig.1.7. Results of reconstruction of solar constant S_0 from the year 1700 (1) and the variability of the mean temperature of dry land in the northern hemisphere (2) [113].

scales. Figure 1.6. shows the variation of S_0 (in W·m^{-2} in relation to the value S_0=1363 W·m^{-2}) in the period 1978–1997, according to the results of measurements of 5 satellite instruments. The graph also·gives the curve of the empirical model of variability of S_0. As indicated by the graph, in the examined period, the variation of S_0 was approximately 1.5 W·m^{-2}.

2. Analysis of the results of long term measurements of the solar constant enables the values of S_0 to be reconstructed to the year 1700. Figure 1.7 shows the results of such a reconstruction. The cross hatched areas show the amount of indeterminacy in the reconstructed values of S_0 which is especially large in the 18th century and the beginning of the 19th century. The figure shows that in the examined period the variation of the solar constant may reach 10 W·m^{-2}. Figure 1.7. also shows the curve characterising the variability of the mean temperature of the dry land in the northern hemisphere. As indicated by the data, there is a clear relationship between the variation of S_0 and temperature.

3. The solar radiation spectrum greatly changes at the edges of the electromagnetic spectrum at short wave lengths (x-ray and ultra violet radiation) and in the radio range. In these regions, the variations reach thousands of per cent. Regardless of the small relative contribution of these spectral ranges to the value of the integral solar constant, this variability has a strong effect on the physical processes in the upper atmosphere. For example, x-ray and ultraviolet radiation lead to dissociation and ionisation of the gas

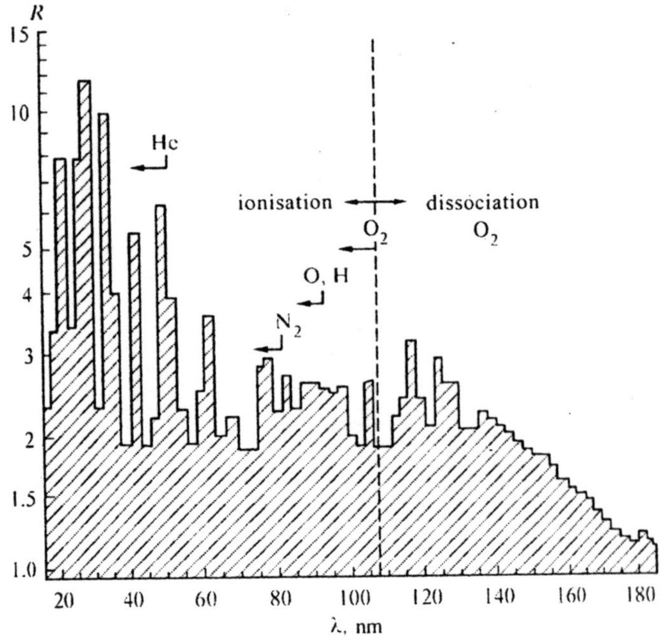

Fig.1.8. Ratio R of the energies of solar radiation in the period of maximum (Wolf number W = 178) and minimum (W = 0) solar activity for the spectral range 15–185 nm [91].

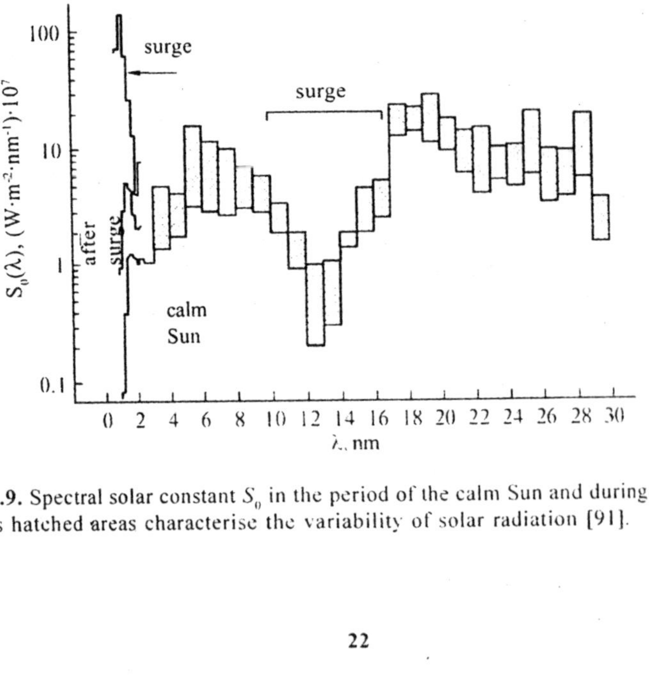

Fig.1.9. Spectral solar constant S_0 in the period of the calm Sun and during solar surges. Cross hatched areas characterise the variability of solar radiation [91].

components of the upper atmosphere, and form ions and radicals which take an active part in a large number of chemical and photochemical reactions. These aspects of the interaction of the Sun with the atmosphere will be examined in Chapter 8. Quantitative data of the variability of $S_0(\lambda)$ in the short wave range are shown in Fig.1.8. and 1.9.

Figure 1.8. shows the ratio *R* of the energy of solar radiation measured from a satellite in the maximum (22.01.1979, Wolf number W=178) and the minimum of solar activity (June 1976, W=0) for the spectral range 15–185 nm. The graph shows that the relative variation of solar radiation in the shortwave region of the spectrum exceeds 10. It should be mentioned that Fig.1.8 shows spectral intervals in which radiation is capable of dissociating and ionising different atmospheric components – He, N_2, O_2, O, H. Even larger variations of solar radiation are found in the spectral range 1–30 nm (Fig. 1.9). For example, during surges on the Sun the intensity of radiation with a wavelength of 1 nm may change up to three orders of magnitude.

Taking into account the previously described large changes of solar radiation in relation to solar activity, we can supplement the definition of the solar constant S_0: the value of the solar constant corresponds to the minimum solar activity.

In addition to the previously examined different variations of solar radiation in the disk and with time we mention another phenomenon referred to in the literature as the 'darkening' effect ('brightening') to the edge of the disk. In this phenomenon, moving from the centre of the Sun to the edge of the Sun disk, we detect changes of its brightness. At some wavelengths we may detect a decrease of brightness on the disk ('darkening') on others – increase of brightness ('brightening'). This phenomenon is caused by the presence of vertical temperature gradients in the photosphere and chromosphere and by the fact that when examining different points of this Sun disk we in fact observe outgoing radiation at different angles in relation to the local vertical on the Sun. These observation angles correspond to different altitude regions of the formation of outgoing solar radiation. Thus, scanning the Sun disk, we scan the photosphere and the atmosphere of the Sun in the direction of altitude. Depending on the sign of the vertical temperature gradient on the Sun in the regions of formation of outgoing radiation (different for different wavelengths) we can detect both a decrease of brightness – 'darkening' and increase of brightness – 'brightening' to the edge of the disk.

CHAPTER 2

ATMOSPHERE OF THE EARTH

2.1. Division of the atmosphere into layers

Structural parameters of the atmosphere

The quantitative characterisation of the state of the Earth atmosphere is carried out using meteorological quantities [2, 43]. They include a wide range of different parameters of the physical state of such a complex system as the atmosphere–Earth surface: temperature, pressure, density of air, content of different gases (for example, water vapour); the velocity and direction of wind; characteristics of clouds – amount, altitude and thickness of clouds; intensity of precipitation; meteorological range of visibility; water content of mist and clouds, etc. We shall be also interested in a number of characteristics of the underlying surface, the temperature of the land and oceans, the moisture of soils, etc.

The structural parameters of the atmosphere usually include pressure, temperature and air density. It should be mentioned that these quantities are linked by two relationships.

1. The equation of the state of the ideal gas available in thermodynamics:

$$pV = \frac{m}{\mu} RT, \qquad (2.1.1)$$

where p is pressure; T is temperature; m is mass; μ is the molecular mass of air (its exact definition is given in the Appendix); V is the volume of air; R is the universal gas constant. Dividing both parts of (2.1.1) by volume V, we obtain the first relationship of the link of the structural parameters

$$p = \frac{\rho}{\mu} RT, \qquad (2.1.2)$$

where ρ is the density of air.

2. Hydrostatics equation

$$dp = -\rho g dz, \tag{2.1.3}$$

where g is the free fall acceleration; z is altitude. Using the hydrostatics equation (2.1.3), the density of air is determined by the relationship:

$$\rho = -\frac{1}{g}\frac{dp}{dz}. \tag{2.1.4}$$

For an ideal gas, the equation of state may be written as:

$$p = nk_B T, \tag{2.1.5}$$

where n is the number of molecules in the unit volume; k_B is the Boltzmann constant (equation (2.1.5) is derived in the Appendix).

Parameters determining division of the atmosphere into layers

The division of the atmosphere into layers is carried out on the basis of different features [2, 43, 79]:

1. Distribution of temperature with altitude
2. Gas composition and the presence of charged particles
3. Nature of interaction with the Earth's surface
4. Effect of the atmosphere on flying systems
5. Effect of the magnetic field on the state of the atmosphere

Temperature stratification of the atmosphere

The difference in the atmospheric layers is most evidently reflected in the variation of the temperature of air with altitude. On the basis of this feature, the atmosphere is divided (stratified) into five main layers.

1. Troposphere – the layer of the atmosphere adjacent to the surface of the Earth. This layer is characteristic by an almost linear decrease of temperature with altitude. The mean rate of decrease of temperature with altitude in the troposphere is 6 deg/km. The troposphere is characterised by the concentration of approximately 80% of mass of the entire atmosphere. This is the most important layer of the atmosphere. It determines the formation of weather and many aspects of the life activity of mankind.

2. Stratosphere – the next layer after the troposphere in which the temperature increases with altitude.

3. Mesosphere – the layer following the stratosphere where temperature again decreases with altitude and reaches the lowest values in the Earth's atmosphere (to 130 K at high altitude in summer).

4. Thermosphere – the layer, above the mesosphere, where the temperature starts to increase again and reaches approximately 1000 K. It should be mentioned that this temperature is determined as the measure of the mean speed of movement of the molecules (kinetic temperature) but because of the considerable rarefaction of air in the thermosphere the typical processes of heat exchange are not detected there, and spacecrafts flying in this layer are not overheated even at air temperatures of 1000 K.

5. Another part of the thermosphere – exosphere. This is the outer sphere of the atmosphere where the mean free path of the molecules amounts to hundreds of kilometres, and their mean velocity (kinetic temperature) is sufficient for overcoming the gravitational field of the Earth and travelling into the cosmos forever – the phenomenon of dissipation of the atmosphere.

The boundaries (regions), separating the troposphere from the stratosphere, the stratosphere and the mesosphere, the mesosphere and the thermosphere, the thermosphere and the exosphere, are referred to respectively as tropopause, stratopause, mesopause, and thermopause. The mean altitudes of the boundaries of the layers and the transitions zones are given in Table 2.1. It should be mentioned that there may be large deviations from the given mean values, depending on the latitude, the time of year, the meteorological situation, etc. In particular, the altitude of the tropopause depends on the latitude: in the tropics it is approximately 16–17 km, in middle latitudes 9–12 km, in polar latitudes 8–9 km.

The typical mean vertical distribution of the temperature of the atmosphere and its layers are shown graphically in Fig. 2.1.

Table 2.1 Main and transition layers of the atmosphere

Layer	Mean altitude of the lower and upper boundaries, km	Transition layer
Troposphere	0–11	Tropopause
Stratosphere	11–50	Stratopause
Mesosphere	50–90	Mesopause
Thermosphere	Above 90	Exosphere
Exosphere	Above 450	

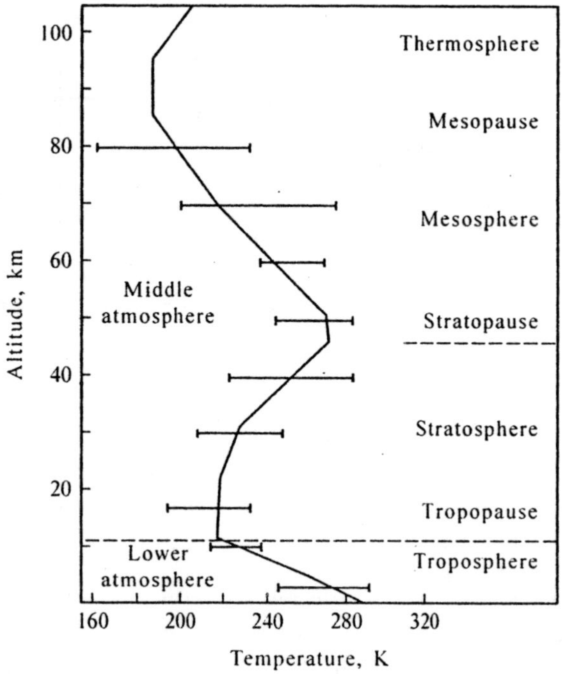

Fig.2.1. Vertical structure of the atmosphere and the mean temperature profile $T(z)$ [92].

Horizontal sections in the figure show the variations (possible deviations) of the actual values of the temperature from the mean values. It may be seen that these variations are large and reach tens of degrees. Table 2.2 gives the mean values of the main meteorological quantities for moderate latitudes in the atmosphere layer 0–600 km.

Division of the atmosphere into layers on the basis of other features

On the basis of the composition of air, the atmosphere is divided into the homosphere and the heterosphere [3, 8, 43, 81]. In the first of these layers (0–95 km) the relative content of the main atmospheric gases (nitrogen, oxygen, argon) and molecular mass of air (μ = 28.9645 g/mol) changes only slightly with altitude.

In the heterosphere, above 95 km, in addition to the molecules of N_2 and O_2 there are also large quantities of atomic oxygen O as a result of the processes of the dissociation of the molecules of O_2 by the shortwave radiation of the Sun. Therefore, the molecular

Table 2.2 Mean values of meteorological quantities [81].

z, km	p, gPa	T, K	ρ, kg m^{-3}	μ, g/mol
0	1.01×10^{3}	288	1.23×10^{0}	28.96
5	5.40×10^{2}	256	7.36×10^{-1}	28.96
10	2.65×10^{2}	233	4.14×10^{-1}	28.96
20	5.53×10^{1}	217	8.89×10^{-2}	28.96
40	2.87×10^{0}	250	4.00×10^{-3}	28.96
60	2.20×10^{-1}	247	3.10×10^{-4}	28.96
80	1.05×10^{-2}	199	1.85×10^{-5}	28.96
100	3.20×10^{-4}	195	5.60×10^{-7}	28.40
150	4.54×10^{-6}	634	2.08×10^{-9}	24.10
200	8.47×10^{-7}	855	2.59×10^{-10}	21.30
300	8.77×10^{-8}	976	1.92×10^{-11}	17.73
400	1.45×10^{-8}	996	2.80×10^{-12}	15.98
500	3.02×10^{-9}	999	5.22×10^{-13}	14.33
600	8.21×10^{-10}	1000	1.44×10^{-13}	11.51

mass of air in the heterosphere decreases with altitude (Table 2.2).

The gas composition of the atmosphere is also used to separate the ozonosphere (15–55 km) with the concentration of the main mass of ozone which is an important atmospheric gas.

Starting at 60 km, the content of charged particles (ions and electrons) in the atmosphere rapidly increases. Therefore, the atmospheric layers above 60 km are referred to as the ionosphere. The ionosphere itself is also subdivided into individual layers (layers D, E and F).

On the basis of the interaction of the atmosphere with Earth's surface, the atmosphere is divided into the boundary layer (friction layer) and the free atmosphere. In the boundary layer (height up to 1–1.5 km) the nature of motion is greatly affected by the Earth's surface. The forces of turbulent friction play a significant role in the same layer. The boundary layer is characterised by distinctive daily variations of the meteorological quantities. Inside the boundary layer there is also a layer of the atmosphere near the ground with a thickness of 50–100 m, within which the meteorological quantities (for example, temperature and the speed of wind) may rapidly change with altitude. In the free atmosphere, i.e. above 1.0–1.5 km the forces of turbulent friction may be ignored to a first approximation.

As a result of the construction of spacecrafts the atmosphere is divided into dense layers (atmosphere) and the terrestrial space close to the Earth whose lower boundary is at an altitude of approximately 150 km. Thus, the terrestrial space at the Earth's surface starts in the thermosphere and 'covers' naturally the exosphere (Table 2.1). The resistance (friction) of the dense layers of the atmosphere is so high that within the limits of these layers a spacecraft with the switched off engine cannot perform even a single orbit around the Earth. At the same time, at altitudes greater than 150 km, the duration of existence of artificial satellites of the Earth is longer than the time required for carrying out a single orbit.

From the viewpoint of the effect of the magnetic field of the Earth on the state of the atmosphere, there is another layer – the magnetosphere which includes the outer part of the thermosphere. In the magnetosphere, particles of the gases (ions) are sustained not only by the gravitational but also magnetic field of the Earth.

Finally, the division of the atmosphere into lower and upper should be mentioned. The former includes the troposphere, and the latter includes all layers situated above the tropopause. The term 'middle' atmosphere has been used recently in scientific literature and international scientific programs. The middle atmosphere is the region of the atmosphere at an altitude of 15–120 km.

Upper boundary of the atmosphere

As indicated by these considerations, it is not possible to provide an unambiguous answer to the question as to which altitude represents the end of the atmosphere of the Earth, i.e. its upper boundary. All depends on a specific purpose for which it is necessary to specify some upper limiting altitude.

For the problem of propagation of solar radiation in the visible part of the spectrum, the upper boundary of the atmosphere should be considered as very low, 50–60 km; the overlying layers have almost no effect on this process. In examination of the outgoing thermal radiation of the Earth and the atmosphere the upper boundary is lifted to 100–150 km. Finally, when examining the ionisation of the atmosphere and its interaction with the short-wave radiation of the Sun and the fluxes of cosmic particles, the upper boundary of the atmosphere is lifted to 500–1000 km.

2.2. Spatial and time variability of the structural parameters of the atmosphere

Variability of temperature

The main structural parameters of the atmosphere change in space (in the direction of altitude, along Earth's surface) and with time. Physically, these variations are determined by dynamic, radiation (the transfer of solar and atmospheric radiation), photochemical processes and phase transformations of water.

It is quite evident that temperature, like the other related structural parameters of the atmosphere, strongly depend on the geographic latitude (in the tropics, its hot, at the poles – cold). In fact, as explained in paragraph 1.3, the latitude controls the amount of solar energy received by the atmosphere and the surface in different periods of the year.* Therefore, the zonal temperature is used as the simplest characteristic of the spatial field of temperatures. The temperature averaged out in respect of longitude (like other quantities) is referred to as zonal. It is only a function of latitude and altitude. Figure 2.2 shows the vertical–meridional sections of the field of the mean zonal temperature for January and July.

Figure 2.2 shows, as also shown in Fig.2.1, that the temperature in the atmosphere decreases with increasing altitude, reaching the minimum values in the tropopause. The altitude of the tropopause and its temperature depend on the time of year and latitude. The highest (16–17 km) and coldest (–80°C and less) tropopause is found in the equatorial part of the Earth. In moderate (mid) latitudes, the tropopause spreads at a altitude of 9–10 km in the winter and 11–12 km in the summer. Its temperature at 50° northern latitude (N) is equal to approximately –55°C throughout the year. The lowest tropopause (8–9 km) is found in polar regions. At the surface of the Earth, temperature is maximum in the region of the equator, and minimum in the polar regions in winter. However, only experts know that the pattern for the temperature of the tropopause is reversed: it is maximum in the winter in the vicinity of pole and minimum on the equator. The variability of temperature in the upper layers of the atmosphere is very large.

To characterise the seasonal variability of the temperature of the atmosphere at different altitudes and latitudes in the northern hemisphere, Table 2.3 gives the temperature differences in the

---

*The equation for this dependence is derived in Chapter 9.

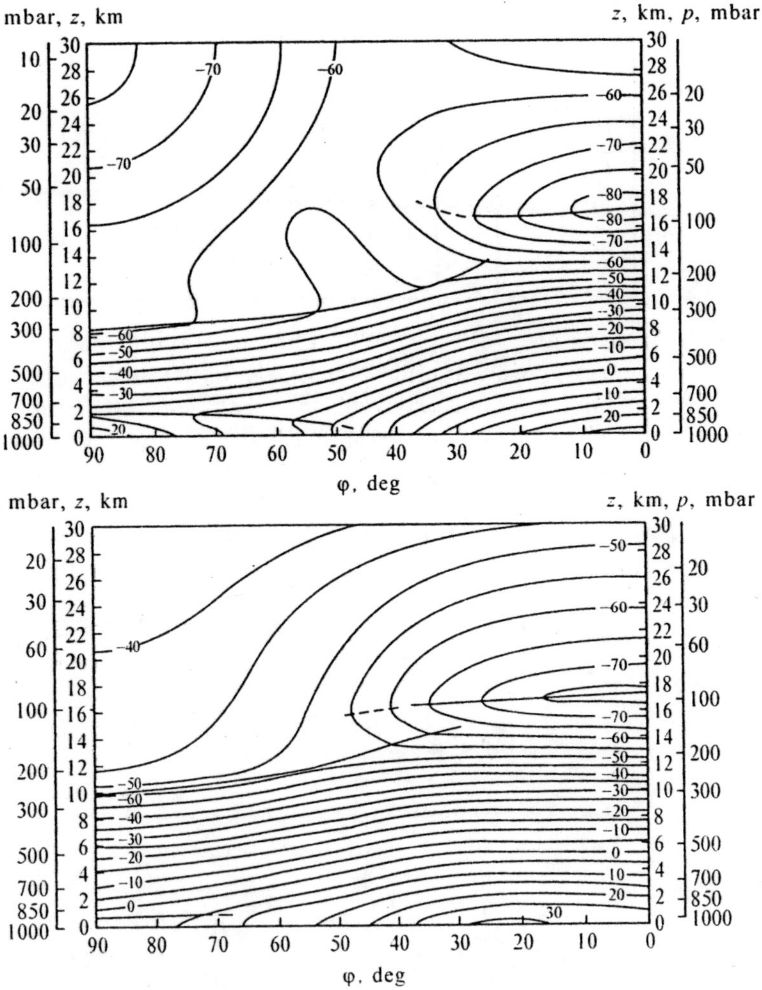

Fig.2.2. Vertical–meridional sections of the field of zonal temperatures [91] for January (*a*) and July (*b*) in the northern hemisphere. φ – latitude; *z* – altitude; *p* – pressure.

summer–winter period.

The data in Table 2.3 show that:

1. The seasonal variability of the temperature of the subtropic region of the Earth (30° N) is small at all altitudes from 10–100 km. This variability is maximum for polar latitudes (80° N).

2. At altitudes of 10–60 km the temperature in the summer is higher than in the winter, and at large altitudes the situation is reversed for the latitude range φ > 30°.

The temperature regime of the Earth has changed a number of

Table 2.3 Winter–summer temperature differences (K) [43]

Latitude, deg	Altitude, km									
	10	20	30	40	50	60	70	80	90	100
80	12	25	5	49	35	15	-24	-57	-63	-49
60	13	11	25	32	22	6	-22	-44	-51	-32
50	13	0	12	20	14	0	-10	-23	-28	-14
30	8	2	4	5	0	4	5	0	0	4

times (for example, Ice Age). The changes taking place in 19th and 20th centuries were studied especially in detail because of the presence of long-term and relatively accurate measurements of the temperature near the surface of the Earth. Figure 2.3 shows the variation of the mean hemispherical and global annual temperatures of surface air for the period 1856–1998 in comparison with the mean temperatures in 1961–1990. Two periods of considerable warming: 1920–1944 and 1978–1998 are clearly visible. In the first

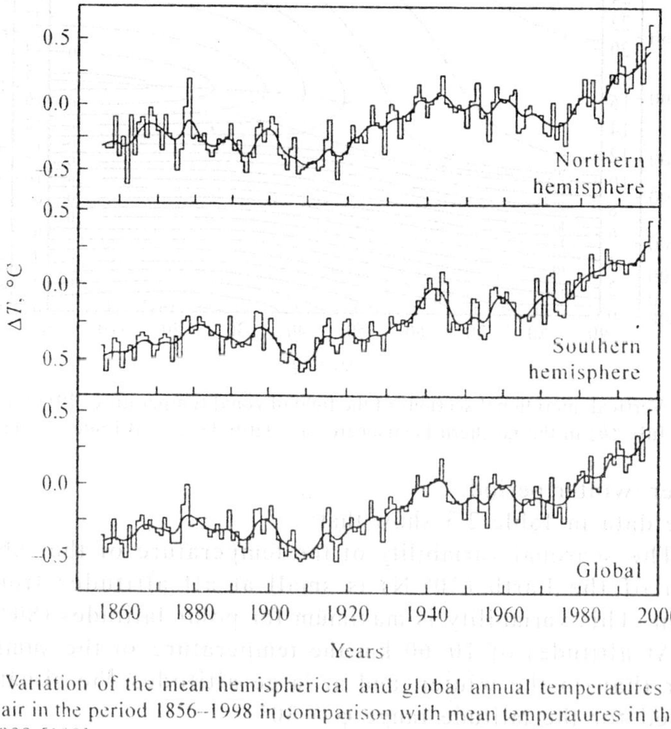

Fig.2.3 Variation of the mean hemispherical and global annual temperatures of near-surface air in the period 1856–1998 in comparison with mean temperatures in the period 1961–1990 [100].

period, T increased by 0.37°C, in the second period, the temperature increase almost reached 0.5°C.

The scientific literature discusses various reasons for 'warming of the Earth's climate' – astronomic factors, solar activity, changes in the gas composition of the atmosphere, and so on. In the latter case, it is assumed (as confirmed by the current climate models) that the increase of the mean temperature of our planet is caused by changes of the radiation (absorption) properties of the atmosphere as a result of the increase of the content in the atmosphere of radiation-active ('greenhouse') gases – CO_2, H_2O, CH_4, N_2O and so on (see later). Especially large long-term changes of temperature (temperature trends) have been detected in the last couple of decades in the upper layers of the atmosphere in a number of regions of the globe. For example, rocket measurements showed negative trends of temperature in the upper stratosphere and mesosphere reaching 0.3–1.7 deg/year [17].

Barometric equations

We shall return to the hydrostatics equation (2.1.3). Substituting into this equations the density from (2.1.2), gives

$$\frac{dp}{dz} = -\frac{\mu g p}{RT}. \tag{2.2.1}$$

The solution of the simplest differential equation gives

$$p(z) = p(0)\exp\left(-\frac{1}{R}\int_0^z \frac{\mu(z)g(z)}{T(z)}dz\right), \tag{2.2.2}$$

where $p(0)$ is pressure at $z = 0$. Thus, the vertical profiles of temperature and pressure in the atmosphere are not independent.* Knowing the temperature profile $T(z)$ and pressure $p(0)$ (or, in a general case, the pressure at the specific altitude $p(z_0)$ from which it is necessary to integrate in (2.2.2)), we can calculate the pressure profile. Conversely, knowing the pressure profile, equation (2.2.1) can be used to calculate the temperature profile.

Equation (2.2.2) considers the dependence of temperature, the molecular mass of air and free fall acceleration on altitude. Specifying the explicit form of these dependences from (2.2.2), we obtain different relationships between the profiles of pressure and

*In the hydrostatic approximation, i.e. neglecting the vertical motion of air.

temperature, referred to as barometric equations.

In the simplest case, ignoring in (2.2.2) the dependence μ, g and T on altitude, we obtain

$$p(z) = p(0)\exp\left(-\frac{\mu g}{RT}\right)z. \quad (2.2.3)$$

The constant $H = RT/\mu g$ in the exponent has the dimension of altitude and is referred to as the altitude of the homogeneous atmosphere or the altitude scale. For standard conditions μ = 28.96 g/mol, g = 9.81 ms^{-2} and T = 273.16 K we obtain H = 7.966 km, i.e. approximately 8 km. Through the altitude of the homogeneous atmosphere H, the barometric equation (2.2.3) is written in a form suitable for application in practice:

$$p(z) = p(0)\exp(-z/H) \quad (2.2.4)$$

Thus, the pressure in the atmosphere decreases almost exponentially with altitude.

In the troposphere, the temperature decreases almost linearly with altitude. Writing $T(z) = T(0) - \gamma z$, where γ is the vertical temperature gradient and again, not taking into account the dependence of μ and g on altitude, from (2.2.2) we obtain

$$p(z) = p(0)\left(\frac{T(0)-\gamma z}{T(0)}\right)^{\frac{\mu g}{R\gamma}} \quad (2.2.5)$$

The most general barometric equation (Laplace equation [2]) takes into account the arbitrary distribution of temperature in the direction of altitude, the dependence of the free fall acceleration on altitude in the atmosphere and geographic latitude, and also the presence of water vapour (humidity of air).

It should be mentioned that in computer calculations, the vertical profiles of temperature, μ and g are given in the form a table, and the pressure profile is calculated directly from (2.2.2) using a quadrature formula.

2.3. Gas composition of the atmosphere

All the gases, forming the Earth's atmosphere, are divided into three groups – main gas components, trace gases (TG) and free radicals [2, 30, 43, 79].

Main gas components

The main gas components of the atmosphere are nitrogen 78.1%, oxygen 20.9% and argon 0.9% by volume. They are found in the atmosphere in approximately a constant ratio to the altitude of the order of 95 km. It should be mentioned that these gases (with the exception of oxygen) are relatively inert in the chemical sense and absorb only slightly electromagnetic radiation. In the upper layers of the atmosphere, the content of the main gases slightly differs from the content in the lower atmosphere (Table 2.4). In this case, the sum of their concentrations (in percent) becomes less than 100% as a result of disintegration of part of the molecules of N_2 and O_2 to atomic nitrogen and oxygen.

Trace gases (TG)

These gas components are constantly present in the atmosphere but their content may vary in time and space [2, 25, 30, 37]. The TG include mainly H_2O (water vapour), CO_2 (carbon dioxide) and O_3 (ozone) – compounds strongly absorbing electromagnetic radiation and taking an active part in different reactions and chemical transformations. Because of these properties, they play a significant role in the formation of the climate of our planet. A detailed list of the TGs includes at present tens of compounds.

Sometimes, a special group is formed by gases (impurities) having powerful antropogenous sources. These are compounds which penetrate into the atmosphere in large quantities as a result of the activities of mankind, but in many cases these impurities may also contain significant natural sources.

Table 2.5 gives the data on the most important atmospheric impurities having significant antropogeneous sources.

The importance of the examined impurities is associated with the fact that many of them actively adsorb thermal radiation and take part in the formation (and changes) of the so-called greenhouse effect which will be discussed in detail later, and also in the

Table 2.4 Content (in volume %) of main gases in air at large altitudes [43]

Gases	Altitude, km				
	0	75	85	95	105
N_2	78.1	78.1	78.1	77.9	77.4
O_2	20.90.9	20.9	20.0	18.0	16.0
Ar_2	0.9	0.9	0.8	0.8	0.7

Table 2.5 Most important trace gases with antropogenous sources

Gas	Technical notation	Concentration at surface, ppmV	Trend of concentration per annum	Lifetime, years
CO_2	Carbon dioxide	350 (1989)	0.4	250
CH_4	Methane	1.7	1.0	10
CO	Carbon monoxide	0.12 N latitudes 0.06 S latitudes	1.0 N latitudes 0 S latitudes	0.3
N_2O	Nitrogen oxide	0.31	0.3	150
NO_x	Total nitrogen oxides (NO, NO_2)	$(1–2) \times 10^{-5}$	?	≤ 0.02
$CFCl_3$	Freon-11 (CFC-11)	2.6×10^{-4}	4	70
CF_2Cl_2	Freon-12 (CFC-12)	2.6×10^{-4}	4	120
$C_2Cl_3F_3$	Freon-113 (CFC-13)	3.2×10^{-5}	10	90
CH_3CCl_3	Methylchloroform	1.2×10^{-4}	4.5	6
CF_2ClBr	(H-1211)	1.0×10^{-6}	12	12–15
CF_3Br	(H-1301)	1.0×10^{-6}	12	12–15
SO_2	Sulphur dioxide	$1–20 \times 10^{-5}$	?	0.02
COS	Sulphur carboxide	5×10^{-4}	≤ 3.0	2–2.5

The name 'technical notation' which is in common use in the physics of the atmosphere does not always coincide with the chemical notations; for freons and their substitutes the 'notations' are technical marks; the trend is the general long-term tendency of the variation of the quantity from which the short-period (daily, seasonal) have been excluded; the lifetime in the atmosphere – the mean time of existence of the gas molecule in the atmosphere (for example, the time during which the number of molecules decreases *e* times).

photochemistry of ozone. Finally, it should be stressed that compounds such as CO_2, CH_4, SO_2 and CO have also powerful natural sources and sinks.

Free radicals

These are very active in the chemical sense, although they are short-life compounds. They include atomic oxygen O, hydroxyl OH, perhydroxyl HO_2, etc. It should also be mentioned that the literature contains other classifications of the gas impurities of the atmosphere, for example, on the basis of the values of the mixing ratio.

Vertical profiles of gas concentrations

Figure 2.4 shows the vertical profiles of the volume concentration of different atmospheric gases. Taking into account the observed variation of the content of gases, these profiles relate to the mean, background state of the atmosphere. Figure 2.4 demonstrates differences in the vertical distribution of the concentration for different gases. For example, oxygen and carbon dioxide, have a constant mixing ratio up to altitudes of the order of 80–100 km (they are uniformly mixed). The content of the water vapour rapidly decreases in the troposphere, is approximately constant in the stratosphere and again decreases in the mesosphere. Methane is uniformly mixed in the troposphere. The maximum concentration of the ozone is detected in the stratosphere. A similar altitude distribution (with the maximum content in the stratosphere) is also recorded for a number of other atmospheric gases (NO, HCl, and so on). The mixing ratio of CO in the lower layers of the atmosphere decreases with altitude, and in the stratosphere and mesosphere it increases. The nature of the vertical distribution of the content of the TG is determined by the complicated interaction of the processes of turbulent mixing, presence of different sources and sinks, etc.

The 'greenhouse effect' and 'greenhouse' gases

We shall discuss in greater detail the long-term trends in the content of a number of trace gases. This problem is currently important because of changes in Earth's climate referred to as the 'greenhouse effect', and also depletion of the ozone layer. The 'greenhouse effect' is the effect on the climate of a number of gases (CO_2, H_2O, CH_4, etc.), adsorbing thermal radiation of the

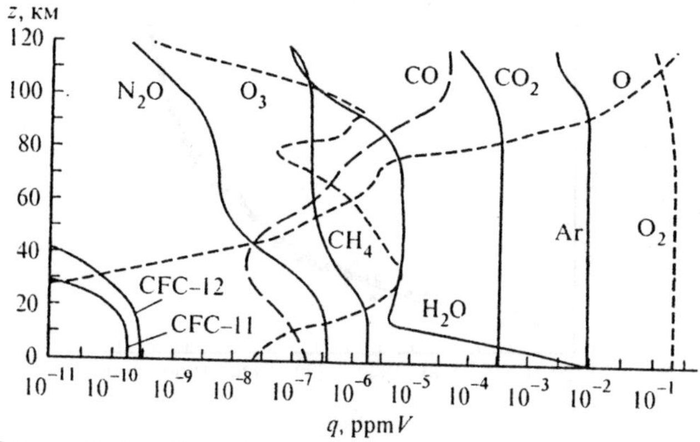

Fig.2.4 Mean vertical profiles of volume concentrations of different atmospheric gases [91]. z – altitude; q – concentration.

Earth and, consequently, preventing cooling of the Earth during the night when there is no heating by solar radiation. Correspondingly, these gases are sometimes referred to as 'greenhouse gases'.

As a result of rapid development of industry from the end of the 18th century, the content of carbon dioxide in the Earth's atmosphere has been increasing. If the volume concentration in the pre-industrial era was 280 ppmV, at present it has reached approximately 360 ppmV (Fig.2.5). Special investigations of the content of CO_2, carried out from the end of the 50th year of the 20th century in the Mauna Loa observatory (Hawaii) recorded the trend of CO_2, equalling 1.5 ppmV/year (Fig.2.6). It is interesting to note that this trend decreased in the 1990s, although the reasons for this are not completely known.

Another important greenhouse gas is methane which is responsible for approximately 15% of the greenhouse effect. There is a large number of natural sources of methane, and according to a number of estimates the antropogenous sources may reach 50–60% of its total emission. Investigations have shown that the pre-industrial (prior to the rapid development of industry) the mixing ratio of methane was 0.72–0.74 ppmV, i.e. less than half of the current value (Table 2.5).

The variation of the methane concentration in the Earth's atmosphere in the last 1000 years is shown in Fig.2.7. The graph shows that starting in approximately the 19th century, there has been a rapid increase of the methane content. In the 80s of the 20th century, the trend of methane in the northern hemisphere reached

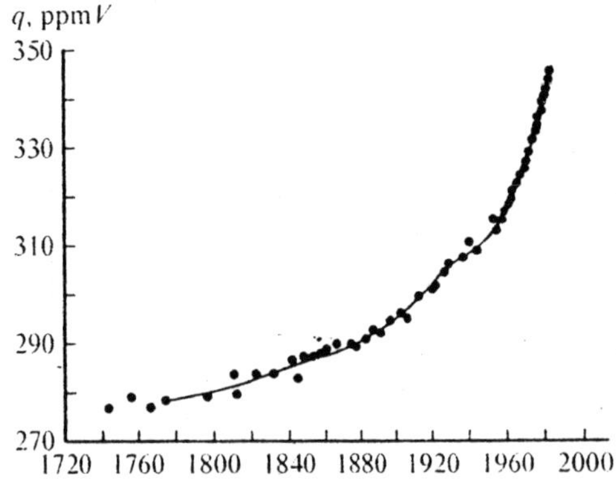

Fig.2.5 Variation of the volume concentration of CO_2 in 18th, 19th and 20th centuries [99]. q – concentration, t – time.

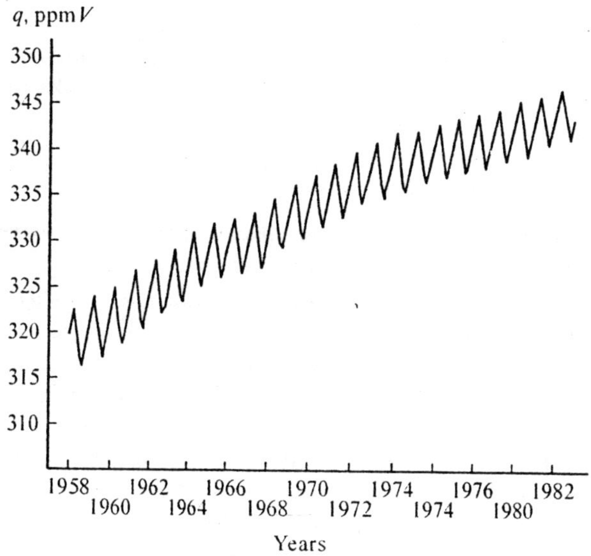

Fig.2.6 Variation of the volume concentration of CO_2 in the 2nd half of the 20th century according to observations at the Mauna Loa observatory (Hawaii) [91]. *q* – concentration, *t* – time.

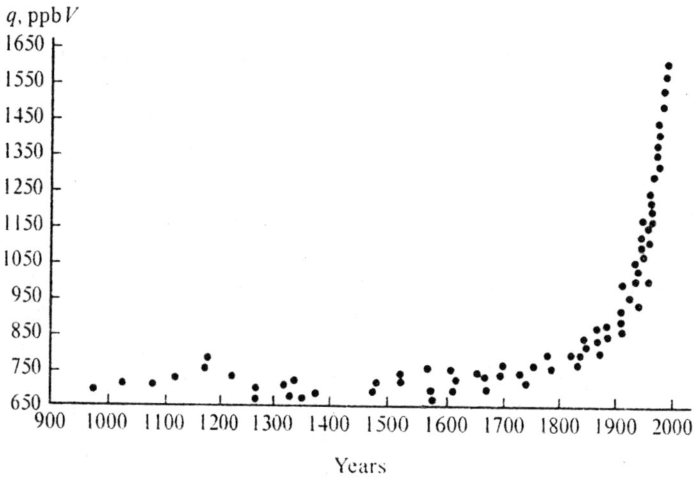

Fig.2.7 Variation of the volume concentration of CH_4 in the atmosphere in the period of 1000 years [99]. *q* – concentration, *t* – time.

1% per annum. This increase was associated with the burning of fossil fuels, gas extraction, rice harvesting, development of animal husbandry, burning of tropical forests. During the 90s, the methane trend decreased.

Water vapour in the atmosphere

The main source of water vapour in the atmosphere is the process of evaporation of water from the surface of seas and oceans. H_2O is concentrated mainly in the vicinity of the Earth's surface, and its concentration in the troposphere rapidly decreases with altitude. Water vapour is an important greenhouse gas, according to estimates its contribution to the greenhouse effect is greater than that of CO_2. The most important property of the water vapour is its capacity for condensation resulting in the formation of clouds and mist which were discussed in paragraph 2.5. The concentration of water vapour in the stratosphere is very low but in recent years it has been increasing which, if the trend continues, may result in significant changes of the optical properties of the stratosphere, for example, as a result of condensation of H_2O.

Ozone in the atmosphere

Special attention should be given to the changes in the content of atmospheric ozone – a compound which is extremely important both from the viewpoint of the radiation properties of the stratosphere and protecting the biosphere against the dangerous ultraviolet radiation of the Sun which is intensively adsorbed by ozone. The most dramatic changes in the ozone content in recent years took place over the Antarctica – the so-called ozóne holes appeared. Starting from the middle of the 80s, extremely low values of the total ozone content were detected in the autumn period above the Antarctica (Fig.2.8). In the 90s, the total ozone content of the ozone holes was less than 50% of the mean value previously recorded in the Antarctica. A decrease of the total content of O_3 was also recorded in other regions of the Earth, in particular in the northern hemisphere, above the territory of Russia. Estimates of the trends in the ozone content according to the results of Earth and satellite measurements show that they equal at present 0.1–1% per annum.

The variations of the ozone content differ at different altitude in the atmosphere. In the stratosphere in the periods of formation of ozone holes above the Antarctica measurements of the ozone content showed almost complete disappearance of ozone at altitudes

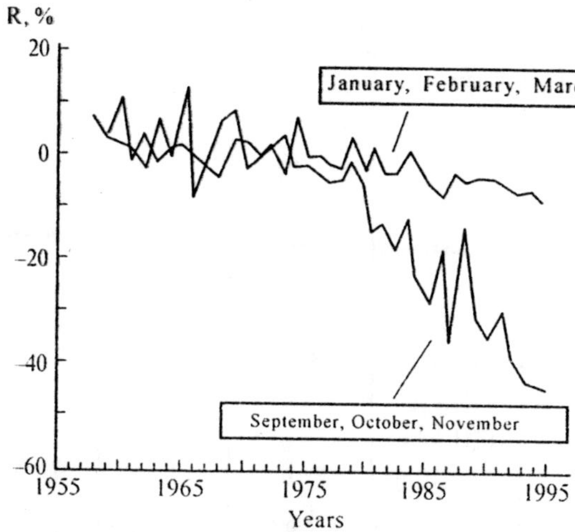

Fig.2.8 Relative variation (R) of the total ozone content in the atmosphere above the Antarctica in the 80s and 90s [88].

of 15–20 km. The statistical processing of the data of a large number of ground-based, ozonesonde and satellite measurements showed two altitude regions of the maximum trends of the ozone content: a altitude of 15–20 km and above 40 km. On the other hand, a number of regions of the Earth showed an increase of the ozone content of the troposphere, reaching 1–4%/annum.

Ozone and freons

In the 70s and 80s of the 20th century there was the fastest increase in the content in the atmosphere of purely antropogenous impurities – different freons and other compounds generated by refrigeration systems used in houses and industry. According to the current views, these compounds are particularly responsible for the depletion of the ozone layer. Table 2.6 gives the trends for these compounds on the basis of the estimates obtained in 1991 and 1994.

The measures aimed to reduce the emissions of a number of freons, taken on the basis of the decision of the International Society in recent years, have been successful to a certain degree. The increase in the content of freons -11, -12 and -13 greatly reduced and in the troposphere it was even interrupted. However, the content of their substitutes (HCFC-22, HCFC-142b, HCFC-141b), which are less 'dangerous' for ozone, greatly increased.

On the whole, in the 1960–1999 observation period, the content of individual gases in the stratosphere notably changed. For

Table 2.6 Trends of antropogenous impurities [99]

Compound	Chemical formulae	1991		1994	
		pptV/year	%/year	pptV/year	%/year
Freon-11 (CFC-11)	$CFCl_3$	9.3–10.1	3.7–3.8	2.5	0.9
Freon-12 (CFC-12)	CF_2Cl_2	16.9–18.2	3.7–4.0	13	2.6
Freon-113 (CFC-113)	$C_2F_3Cl_3$	5.4–6.2	9.1	2.5	3.1
Carbon tetrachloride	CCl_4	1.0–1.5	1.2	–1	–0.8
Methylchloroform	CH_3CCl_3	4.8–5.1	3.7	3.5	2.2
HCFC-22	CHF_2Cl	5–6	6.7	7.0	6.9
HCFC-142b	$C_2H_3F_2Cl$	–	–	1	30
HCFC-141b	$C_2H_2FCl_3$	–	–	0.75	200
H-1211	CF_2ClBr	0.2–0.4	15	0.075	3
H-1301	CF_3Br	0.4–0.7	20	0.16	8

example, the total level of the content of chloride compounds increased from ~1 ppbV to ~3.3 ppbV. There have also been long-term changes in the stratospheric content of H_2O, methane, NO_2, etc. It should be mentioned that the oberved changes in the gas composition of the atmosphere take place not only as a result of antropogenous but also natural factors – volcanic emissions, solar proton surges (SPS), etc. For example, in the period of powerful SPS the upper layers of the atmosphere are characterised by the formation of a large amount of compounds of hydrogen (H, OH, H_2O) and nitrogen (N, NO, NO_2, NO_3, etc.).

2.4. Atmospheric aerosol

Aerosols and their role in Earth's atmosphere

In physics, aerosols are mixtures of air and particles (solid, liquid) that are in equilibrium. In meteorology and atmospheric physics, the aerosol is represented by particles suspended in air [2, 33, 43, 79]. Thus, aerosol (aerosols) are solid and liquid fine particles suspended in air, of greatly differing composition, shape, dimensions and properties: water droplets and ice crystals of clouds, dust, rising from the surface of the Earth, and also ejected by volcanoes or formed from volcanic gases, meteorite dust, particles of salts of sea water, particles formed as a result of production processes, etc. The rate of their falling (settling) is low and the surface is large which in particular is the reason for their active participation in chemical and photochemical reactions with trace gases and radicals.

The aerosol particles play a significant role in the transfer of solar and thermal radiation, affecting the radiation regime of the atmosphere–Earth's surface system, and consequently, the weather and climate on the Earth. Aerosols play an especially important role in the absorption and scattering of solar radiation. The contribution to these processes is very large, and the optics and energetics of the atmosphere cannot be examined without considering the optical properties of atmospheric aerosols [2, 20, 24, 26, 33].

Aerosols also play a significant role in the processes of cloud and fog formation, etc., where they act as nuclei of condensation – nuclei for the condensation of saturated water vapours. Without aerosols this process will be difficult and, therefore, the presence of clouds and precipitation on the Earth are directly linked with the presence of aerosols in the atmosphere. It should be mentioned that, according to definition, particles of the clouds are also aerosols but they are usually separated from other (non-water) aerosols. We shall follow this tradition and examine the properties of 'non-water' aerosols and particles of clouds and precipitation in different sections.

Classification of atmospheric aerosols

Depending on its composition or sources, the atmospheric aerosol is subdivided into the following types [23, 26]:

1. the aerosol of natural origin:

a. products of evaporation of sea splashes – salt particles;

b. mineral dust lifted into the atmosphere by the wind;

c. volcanic aerosol, both directly ejected into the atmosphere (ash) and formed as a result of gas phase reactions (sulphuric acid particles);

d. particles of biogenous origin, both those directly ejected into the atmosphere and formed as a result of condensation of volatile organic compounds and chemical reactions between these compounds;

e. products of natural gas phase reactions (for example, sulphates, formed as a result of oxidation of SO_2).

2. aerosol of the antropogenous origin:

a. industrial emissions of particles (soot, smog, road dust, etc.);

b. products of agricultural activity (for example, dust formed during plowing);

c. products of gas phase reactions (formed in the same manner as natural products in reactions of antropogenous trace gases).

The atmospheric aerosol is also subdivided into tropospheric and stratospheric (on the basis of the features of location in the appropriate layer of the atmosphere) and also into primary aesorols – falling into the atmosphere directly, and secondary – formed in the atmosphere as a result of chemical reactions.

Dimensions of aerosol particles in the atmosphere

The actual atmospheric aerosol is polydispersed, i.e. consists of particles of different sizes. The shape of these particles may also greatly differ. One of the classifications of aerosols, based on their dimensions (dispersion) and also their role in different processes was proposed by Ch. Junge. The diagram of this classification is shown in Fig. 2.9 showing the sources of aerosol particles, lifetime and also information on their effect on different atmospheric processes.

The range of the sizes of aerosol particles is very wide – from less than 10^{-4} to more than 10^{2} μm. Their lower limit is determined by the dimensions of molecular complexes, the upper limit – by the rate of gravitational settling of the particles in the field of gravitational forces. Different meteorological phenomena and processes are associated with different ranges of the dimensions of the aerosol particles. For example, particles with radii of $r < 0.1$ μm (the so-called Aitken particles) strongly affect the electrical characteristics of the atmosphere. The particle with $0.1 \leq r \leq 1.0$ μm have a strong effect on the transfer of short-wave and infrared radiation, formation of clouds and precipitation, and the chemical composition of the atmosphere. The aerosol in the size range $0.1 \leq r \leq 1.0$ μm have the strongest effect on the transfer of shortwave radiation, on visibility in the atmosphere.

Concentrations of aerosol particles in the atmosphere

According to the classification, there is number (according to the number of particles) and mass (by mass) concentration of aerosol particles. Their accurate definitions and measurement units are given in the Appendix.

Table 2.7 gives the data on the relative number and mass concentration of aerosol particles, according to classification proposed by Ch. Junge, and also on the processes in which they take part.

Table 2.7 shows that the main type of particles according to the number concentration of the Aitken particles (90%), whereas the

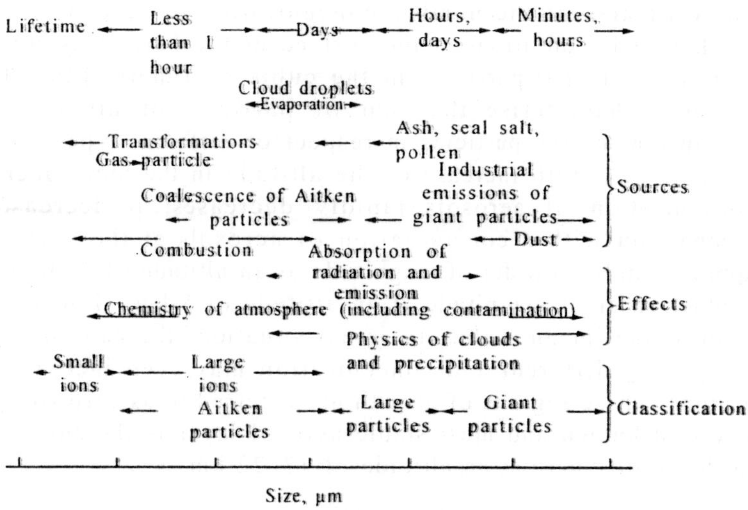

Fig.2.9. Atmospheric aerosol: dimensions, classification, effect on atmospheric processes, sources, lifetime (according to Ch. Junge) [120].

Table 2.7 Relative concentrations of aerosol particles of different sizes according to the classification proposed by Ch. Junge

Size range, μm	Type of particle	Physical phenomena and processes	Relative number, %	Relative mass, %
$r < 0.1$	Aitken particles	Atmospheric electricity	90.0	20
$0.1 \leq r \leq 1.0$	Large	Attenuation and scattering of optical radiation Cloud formation processes	9.9	31
$r > 1.0$	Giant	Cloud formation and precipitation processes Effect on chemical composition Optical phenomena	0.1	49

mass concentration is determined by the large and giant particles (80%).

As regards the absolute values of the number concentration of aerosol particles in the atmosphere, it is difficult to discuss them in detail because the number concentration of the aerosols varies greatly and depends on the altitude in the atmosphere, the proximity to the sources of aerosols (for example, in a town it is several orders of magnitude higher than away from it), and the time of day.

The estimates of the number concentration of aerosols for the 'background' conditions at the surface of the Earth may give the values of 10^3–10^5 particles in the cubic centimetre [26]. These numbers characterise the sum of particles of all sizes and 'expansion' of the particles in respect of different types is given in Table 2.7. With increase of the altitude in the atmosphere the concentration of aerosols rapidly decreases: it decreases in comparison with a concentration of aerosols at the surface by approximately an order of magnitude at an altitude of 2 km and by another order of magnitude at an altitude of 5 km. However, this again is the 'mean' value, in actual situations the pattern may be completely different and concentration may even increase with altitude. These regions of growth are referred to as aerosol layers. The best known and most stable aerosol layer is the Junge layer in the stratosphere at an altitude of 17–22 km.

Sources and sinks of aerosol particles. Variability of aerosols

The aerosols may also be classified from the viewpoint of processes of their formation to condensation and dispersion. Condensation aerosols form during vacuum condensation of supersaturated vapour of substances or cooling of gaseous products of combustion and subsequent aggregation of molecules. Dispersion aerosols form during dispersion (disruption) of solid and liquid phases of matter, for example, in disintegration of rock and minerals and their splashing, weathering of soil, agricultural operations, dispersion of sea water droplets.

An important source of aerosol particles in the atmosphere are the processes of chemical interaction of substances, found in the gas phase, in particular the interaction on the main atmospheric gases (nitrogen, oxygen), and also of water vapour with some gases, for example, sulphur dioxide, chlorine, ammonia, ozone, etc. Suitable examples are the formation of droplets of sulphuric acid in oxidation of SO_2 into SO_3 and subsequent interaction of the latter with water vapour, the formation of ammonium sulphate in interaction of sulphuric acid with ammonia, etc.

An important role in the formation of aerosols is played by volcanic emissions, forest fires, biological processes, production activities of mankind, etc. The main gases, causing the formation of aerosols, are SO_2, H_2S and NH_3. After powerful volcanic eruptions the number of aerosol particles in the stratosphere increases many times resulting in a change of the optical characteristics of the stratosphere. These changes in the stratosphere remain for 1–2 years after eruptions. The main sinks, i.e. reasons for the removal of aerosols from the atmospheres, are gravitational settling (simply speaking, precipitation of aerosols on the surface) and elusion by precipitation absorbing aerosol particles by rain droplets (snow particles).

Of course, the main quality of atmospheric aerosols which must be mentioned is their changeability. As mentioned previously, the sources of aerosols differ, and each source produces particles of different sizes, shape and chemical composition; the spatial distribution of the concentration of the particles has a distinctive local nature; sources may 'operate' discontinuously, and only for a specific period of time (for example, volcanoes); the power of sources and sinks may also greatly vary. All these factors determine the complicated nature of adequate description of physical systems such as atmospheric aerosol and, consequently, the complicated description of its optical properties.

Size distribution function of aerosol particles

Since the aerosol particles are characterised by different sizes (polydispersion), then in contrast to gases the total concentration is not sufficient for describing their amount in the atmosphere and it is necessary to mention concentrations of particles of different radii. The entire range of possible radii of aerosol particles (from 0 to ∞) will be divided into subintervals Δr_i and in each subinterval we determine the number of particles $N(r_i, r_i+\Delta r_i)$ which evidently depends on the boundary radius of the subinterval Δr_i and the width of the subinterval Δr_i (example of such a division is given in Table 2.7). In order to be able to compare the concentrations of different subintervals, we shall normalise them in respect of width, i.e. divide by Δr_i. We denote $n(r_i)=\frac{N(r_i\Delta r_i)}{\Delta r_i}$. The concentration of particles in the range of the radii from 0 to r_i will be:

$$N(r_i)=\sum_{i=1}^{M} n(r_i)\Delta r_i, \tag{2.4.1}$$

where M is the number of the subintervals with right boundary r_i. Tending Δr_i to 0, we obtain

$$N(r)=\int_0^r n(r')dr' \tag{2.4.2}$$

and

$$n(r) = \frac{dN(r)}{dr}. \tag{2.4.3}$$

Function $n(r)$ is the function of the size distribution of aerosol particles. It is defined by the relationship (2.4.3). These considerations show that $n(r)$ has the meaning of the relative number concentration of particles of different radii (as in Table 2.7), i.e. it shows the number of particles with specific radii.

The total concentration of all aerosol particles, according to (2.4.2) is

$$N = \int_0^\infty n(r)dr. \tag{2.4.4}$$

To ensure that the distribution function $n(r)$ does not depend on total concentration N, we normalise it, introducing $f(r) = n(r)/N$. The normalised distribution function is very useful because it makes it possible to define the dependence of the concentration on the radius only. In subsequent considerations, the size distribution function of aerosol particles will refer only to the normalised function $f(r)$. Dividing (2.4.4) by N, we directly obtain the normalisation condition for $f(r)$:

$$\int_0^\infty f(r)dr = 1. \tag{2.4.5}$$

A large number of different functions $f(r)$ describing aerosol particles has been proposed. We examine only two of them.

The logarithmico-normal distribution (briefly log-normal distribution). This is the distribution of the quantity whose logarithm is governed by the normal (Gaussian) law:

$$f(r) = \frac{1}{\sqrt{2\pi}\,\sigma r}\exp\left(-\frac{1}{2\sigma^2}\ln^2\frac{r}{r_0}\right), \tag{2.4.6}$$

where r_0 and σ are the distribution parameters. The log-normal distribution was determined theoretically by A.N. Kolmogorov for the sizes of particles in random disintegration of solid matter.

Junge distribution. In approximation of the results of experimental measurements, Ch. Junge established that, starting from a specific radius r_0, the dimensions of the aerosol particles decrease in accordance with a power law:

$$f(r) = Cr^{-b} \tag{2.4.7}$$

where C is a constant; b is the distribution parameter. In contrast to the theoretical log-normal distribution (2.4.6), the empirical Junge distribution (2.4.7) is mathematically incorrect because $f(r)\to\infty$ at $r\to 0$. To remove this incorrectness and calculate the normalisation factor C one can use different assumptions on the behaviour of the function at $r < r_0$, for example, it is assumed that $f(r)$ = const, and so on.

One of the main characteristics of the distribution function is the modal radius – the radius corresponding to the local maximum $f(r)$. If the modal radius has one value, the distribution is referred to as single modal, if there are several values (i.e. $f(r)$ has several local maxima), the distribution is multimodal (or polymodal). The distribution of atmospheric aerosols in nature is usually multimodal. This is associated with the fact that the natural aerosol is the sum of particles from several sources (Fig.2.9) each o. which is characterised by its own modal radius. In approximation of the distribution function by the relationships of the type (2.4.6) and (2.4.7) the multimodal distributions are presented in the form of the sum of individual single modal distributions – modes with appropriate weights. In this case, if we take into account differences in the chemical composition of particles from different sources, the modes are usually referred to as fractions.

2.5. Clouds and precipitation

Water in the atmosphere

In the Earth's atmosphere, water is present in all three phases: gaseous (amount of water vapour in air determines its humidity), liquid (in the form of droplets in clouds, mist, rain) and solid (cloud crystals, snowflakes, hail). The clouds are an important compound element of the circulation of water in the nature, affect the energy exchange in the Earth–atmosphere system, the radiation balance of the planet, the redistribution of heat on the globe and the general circulation of the atmosphere. The clouds are one of the most important weather- and climate-forming factors of our planet. They also affect the photochemical processes in the atmosphere.

Morphological classification of clouds

The classification of clouds is based on their morphology (external appearance) and their altitude above the Earth's surface (Table 2.8). The division of clouds according to altitude (upper, middle, and lower) is determined by the lower boundary of the clouds. In tropics, the lower range includes clouds below 3 km. It should be mentioned that there may also be clouds of several types and levels. The complete classification of clouds with their photographs is given in the Atlas of Clouds [1].

The special groups of cloudiness include polar stratospheric clouds which sometimes form in high latitudes at altitudes of 17–20 km and silvery (mesospheric) clouds, formed at altitudes of 85–90 km. In the group of the polar stratospheric clouds there are nacreous clouds, termed because of their characteristic rainbow colour. All these are very thin, usually invisible clouds which are detected by devices or become clearly visible at sunrise and sunset when they are illuminated by the Sun. According to current views,

Table 2.8. Classification of clouds [1, 51]

Latin name	Notation
Upper stratum clouds (> ~6 km)	
Cirrus	Ci
Cirrostratus	Cs
Cirrocumulus	Cc
Medium stratum clouds(~2–6 km)	
Altostratus	As
Altocumulus	Ac
Lower stratum clouds (<2 km)	
Stratus	St
Stratocumulus	Sc
Nimbostratus	Ns
Vertical development clouds	
Cumulus	Cu
Cumulonimbus	Cb

these clouds form during the appearance of layers in the atmosphere with a very low (relatively characteristic for given altitudes) temperature when the moisture content of air may reach 100% even at a very low content of the water vapour. The chemical composition of their crystals may differ.

The degree of covering of the sky with clouds is characterised by the amount of clouds – the ratio of the total area of the clouds to the area of the sky: 0 points – clear sky, 10 points – continuous cloudiness.

Microstructure of clouds

According to the phase state, the clouds are divided into liquid droplet (lower and partially middle level), crystalline (upper level) and mixed, containing both droplets and ice crystals. The phase composition of the clouds is determined mainly by their temperature. At positive temperatures, clouds consist of water droplets. Droplets are often present in them even at relatively low negative temperatures down to –40°C. At subzero temperatures, the clouds may be of the droplet, crystalline or mixed type.

Important microstructural characteristics of the clouds as varieties of the aerosols are the number and mass concentrations of the particles and the size distribution functions of particles. The number concentration in the clouds of different forms varies from 10 to 1000 cm^{-3}. The mass concentration of particles in the cloud is its water content. The size distribution functions of cloud particles are approximated using different empirical relationships, for example, the Khrgian–Mazin distribution [51]:

$$f(r) = \frac{b^3}{2} r^2 \exp(-br) \qquad (2.5.1)$$

with unique parameter b. For functions $f(r)$ it is often useful to use modified gamma distribution or log-normal distributions.

The form of the crystals in mixed and crystalline clouds is determined mainly by the temperature and moisture content of the clouds at which they form, and may greatly differ [23, 51]. The differences in the shapes of the crystals complicate modelling of their optical properties. The clouds are also found in atmospheres of other planets. Their physical nature greatly differs from the clouds on the Earth. For example, in the atmosphere of Venus, a thick cloud cover consists (at least partially) of droplets of concentrated solutions of sulphuric acid.

Precipitation

Clouds of specific types (mainly Ns and Cb, less frequently St, Sc and As) are accompanied by precipitation. In accordance with synoptic and thermodynamic conditions of precipitation, the latter are divided into drizzle, continuous rain and shower [2, 51]. They differ in the intensity and duration of precipitation. Drizzle has the intensity lower than 0.1 mm/min and duration expressed in hours and days, and shower 0.03–0.05 mm/min and the duration of minutes or hours. In addition to the intensity of precipitation, they are also characterised by the size distribution function of particles of precipitation (spectrum of precipitation), and in the case of precipitation it has been decided to use non-normalised functions $n(r)$ because they are linked with the intensity of precipitation. As an example, we present the empirical Marshall–Palmer equation:

$$n(r) = n_0 \exp(-br)$$

where $n_0 = 8 \times 10^3$ mm$^{-1} \cdot$ m^{-3}, and the parameter b is linked with the intensity of precipitation R by the following equation: $b = 8.2\ R^{-0.21}$ mm^{-1}.

The form of the liquid droplets in the clouds and precipitation is often approximated by spheres. However, this is not so for large droplets and their form depends on their dimensions. The droplets with a diameter smaller than 1 mm do not differ greatly from the spheres and have the form of oblate spheroids. With increase of the size of the droplets, their shape differs more and more from spherical. The shape of the droplets affects their optical properties.

CHAPTER 3

PROPAGATION OF RADIATION IN THE ATMOSPHERE

3.1. Electromagnetic waves

Modern optics, including atmospheric optics, is based on the considerations of radiation as electromagnetic waves and also the flux of photons in examination of quantum phenomena [39, 62]. The electromagnetic wave is transverse and is a system of mutually orthogonal vectors of the strength of electrical field **E** and magnetic field **H** propagating in vacuum with the velocity of light. In turn, these vectors are orthogonal to the direction of propagation of the wave **v**. We shall write the equation of a plane electromagnetic wave, i.e. a wave where the strength of the electrical field E oscillates in a single plane, in the following form (selecting the axis X in the direction of movement):

$$E(x,t) = E_0 \cos\left(2\pi\nu t - \frac{2\pi}{\lambda}x + \delta\right). \qquad (3.1.1.)$$

here x is the spatial co-ordinate; t is time; E_0 is amplitude; ν is frequency; λ is the wavelength; δ is some phase of the wave given by the initial condition $E(0,0) = E_0 \cos \delta$.

In (3.1.1) we fixed the co-ordinate x and obtained harmonic oscillations E with the time period $T = 1/\nu$. Fixing time t in (3.1.1), we obtain the distribution of E along the axis x with spatial period λ. The relationship between these periods is evident: during a single time period, the wave passes through one spatial period, i.e. $\lambda = vT$, where v is the velocity of propagation of the wave. Consequently, we obtain the relationship of wavelength with frequency:

$$\lambda = \frac{v}{\nu}. \qquad (3.1.2)$$

For vacuum $v = c$, where c is the velocity of light in vacuum.

In matter, the velocity of propagation of electromagnetic waves is lower than c and is defined by the relationship $v = c/n$, where n is the refractive index of matter. This results in the appropriate reduction of the wavelength of radiation in the matter. To avoid misunderstanding, in the optics it is accepted to operate with the wavelength in vacuum and, consequently, the term 'the wavelength' denotes 'the wavelength in vacuum' ($\lambda = c/\nu$), otherwise if we are discussing the wavelength in the matter, this must be stressed. This is in transition from vacuum to matter the wavelength changes, whereas the frequency of electromagnetic waves remains constant.

Using (3.2.2) we shall write the equation of the wave (3.1.1) in a less descriptive but more suitable form for further analysis:

$$E(x,t) = E_0 \cos\left(2\pi\nu t - \frac{2\pi n\nu}{c}x + \delta\right). \qquad (3.1.3)$$

The length of the electromagnetic waves is measured in the units of length, in atmospheric optics these are usually micrometers (μm), nanometers (nm) and angstroems (Å). The frequency of the wave of the optical range is very high and, consequently, is used only seldom. In many cases, the spectral characteristics are expressed in wave numbers, i.e. the quantities inverse to the wavelength.

Electromagnetic radiation with strictly one frequency (wavelength) is monochromatic. In fact, there are no monochromatic waves and, as shown in Chapter 4, the impossibility of their existence is a principal factor. However, in examining many processes in the atmosphere it is fully possible to get away with monochromatic approximation.

The scale of electromagnetic waves

Depending on the wavelength (frequency), electromagnetic waves are divided into a number of ranges (Table 3.1).

Modern atmospheric optics examines the propagation, transformation and also generation of electromagnetic waves from ultraviolet (UV) to radiowave ranges. In addition to the division of the spectrum, shown in Table 3.1, in atmospheric optics the entire spectrum of electromagnetic radiation in the atmospheres of planets is divided into the solar and thermal regions. The solar region includes UV, visible and near infrared (NIR) ranges. In these ranges in daytime the energy of solar radiation exceeds the energy of atmospheric (in particular, thermal) radiation of the atmosphere and the surface. The thermal region extends from the NIR range

Table 3.1. The ranges of the spectrum of electromagnetic radiation

Range	Characteristic wavelengths, μm	Characteristic frequency, Hz
Gamma	10^{-5}	3×10^{19}
X-ray	10^{-2}	3×10^{16}
Ultraviolet	3×10^{-1}	10^{15}
Visible	0.4 – 0.7	$(4.3 - 7.5) \times 10^{14}$
Infrared (IR):		
near (NIR)	1 – 4	$(0.8 - 3) \times 10^{14}$
mid	4 – 50	$10^{12} - 10^{14}$
long	50 – 1000	$3 \times 10^{10} - 10^{12}$
Microwave (MW)	10^{3}	3×10^{10}
Television	10^{7}	3×10^{7}
Radiowave	$10^{8} - 10^{9}$	$3 \times 10^{5} - 3 \times 10^{6}$

to radiowaves. Here, on the other hand, the energy of thermal radiation (in daytime and, of course, during the night) is greater than the solar component. The spectrum boundary (or more accurately the region) in which the energy of solar and thermal radiation in the Earth's atmosphere is approximately identical, is located in the range 3–4 μm. For other planets, this boundary is displaced to the short-wave or long-wave region depending on the distance from the Sun and the equilibrium effective temperature of the planet. It should be mentioned that both solar and thermal radiation is present in all ranges of electromagnetic radiation but their role, for example, in the formation of outgoing radiation, greatly differs.

It is important to mention in particular the visible range of the spectrum – the wavelength range perceived by the human eye. Its boundaries are defined as 0.4–0.7 μm. The concept 'light' conventionally refers to 'visible light', whereas for electromagnetic rays not visible to the eye we use the term 'radiation' (UV radiation, microwave radiation). The term 'radiation' as a synonym of 'emission' is often used in the scientific literature. This word is linked with stable word combinations 'the radiation regime of the atmosphere', 'radiation transfer in the atmosphere', 'atmospheric radiation', etc.

Geometric optics

The law of straight propagation of light has been known from old times. This term explains the presence of shadows from objects and narrow light beams produced in passage of light through holes. This leads to the concept of beam lights – straight lines, along which it propagates. The interaction of beams with different objects, taking into account the laws of reflection and refraction, can be described using geometrical constructions and, consequently, the optics, using the concept 'light beam' is referred to as geometrical optics.

The first foundation of geometrical objects within the framework of a wave theory was presented in the 17th century by Huygens who formulated this principle: every element of the front of the lightwave is a source of secondary waves. The interference of these secondary waves determines further propagation of the wave in space. The Huygens principle made it possible to explain the straight form of the light beams. The law of the straightness of the light beams is derived from Maxwell equations as a limiting case for the scales of the examined phenomena considerably greater than the wavelength of light [62, 84]. Therefore, the 'approximation of geometrical optics' is considered in this case.

The geometrical optics is applicable to phenomena whose scales are comparable with the wavelength of radiation. In addition to this, as shown in Chapter 5, it may produce inaccurate results even for cases in which the scale is considerably greater than the wavelength. It should be mentioned that in contrast to geometrical optics, the optics examining the wave properties of light is often referred to as wave optics.

Energy of electromagnetic waves

It is well known that electromagnetic waves transfer energy in the direction of their propagation. It's value – the Pointing vector – is proportional to the vector product of the strength of electrical field **E** and magnetic field **H**. The modulus of the Pointing vector taking into account the relationship between **H** and **E** from the Maxwell equations is proportional to the square of the strength of the electrical field E. The frequency of radiation in the optical range is so high (Table 3.1) that it is not possible to measure the instantaneous value of E. The effect on any measuring device is exerted only by the square of the strength of electrical field E averaged in respect of time [7, 26, 39, 62]. Consequently, the energy of the electromagnetic wave is proportional to the mean square of its electrical strength. Taking into account the periodicity of oscillations of E with time, the averaging interval is the period $T = 1/\nu$, and dividing the energy by this interval T, we transfer to power W:

$$W = \frac{1}{T}\int_0^T E^2(x,t)\,dt. \tag{3.1.4}$$

Substitution of (3.1.1) into (3.1.4) gives

$$W = \frac{1}{T}\int_0^T E_0^2 \cos^2\left(2\pi\nu t - \frac{2\pi}{\lambda}x + \delta\right)dt = \frac{1}{2\pi\nu t}E_0^2 \int_{-\frac{2\pi}{\lambda}x+\delta}^{2\pi-\frac{2\pi}{\lambda}x+\delta} \cos^2 y\,dy,$$

and

$$W = \frac{1}{2}E_0^2. \tag{3.1.5}$$

3.2 Intensity and radiation flux

Radiation intensity

We examined radiation as electromagnetiç waves. However, many problems of atmospheric optics çan be solved without setting the question of the nature of radiation, but examining it only as an energy flux. Let us assume that in space there is a radiation field, i.e. the electromagnetic wave (light beam) is given in a general case at every point of the space in every selected direction. A classic example of the radiation field is the atmosphere, illuminated with the Sun where at every point there is direct solar radiation, the

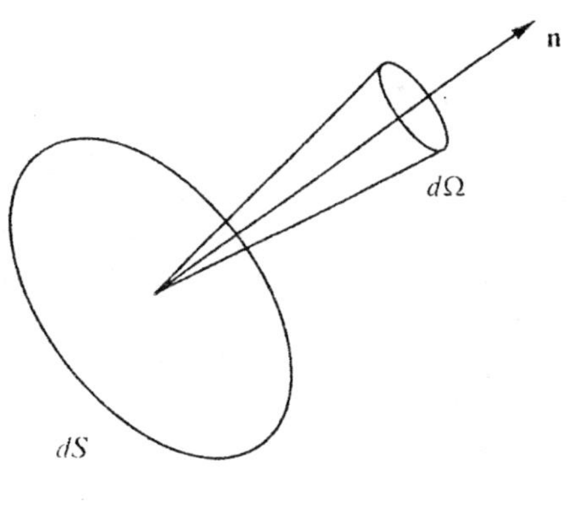

Fig. 3.1. Determination of radiation intensity.

scattered light of the sky and the radiation of the atmosphere and the surface.

The main characteristic of the radiation field is radiation intensity. We select in the space an elementary area with area dS, normal to **n** and the solid angle $d\Omega$, described around the normal (Fig.3.1). If the size of the area is dS, and radiation penetrates in the wavelength range from λ to $\lambda + d\lambda$, in the solid angle $d\Omega$, during time dt, the amount of energy dE_λ, falling on the area, will be proportional to dS, $d\lambda$, $d\Omega$, dt, i.e. equal to

$$dE_\lambda = I_\lambda \, dS d\lambda d\Omega dt. \tag{3.2.1}$$

The coefficient of proportionality I_λ, included in this equation, is the monochromatic intensity of radiation [47, 64]. This means that the monochromatic intensity of radiation I_λ, is the amount of electromagnetic energy dE_λ in the unit wavelength interval, falling on the unit area (or passing through it) in the direction normal to it from the single solid angle per unit time:

$$I_\lambda = \frac{dE_\lambda}{dS\, d\Omega\, dt\, d\lambda}. \tag{3.2.2}$$

The formal mathematical approach requires writing the intensity (3.2.2) as a partial derivative

$$I_\lambda = \frac{\partial^4 E_\lambda}{\partial S\, \partial\Omega\, \partial t\, \partial\lambda}. \tag{3.2.3}$$

The index λ at the intensity and energy denotes the dependence on the wavelength.*

In a general case, the intensity is the function of a point in the space (x, y, z), with the direction determined by the normal **n**, and time t: (I_λ (x, y, z, **n**, t)). If intensity does not depend on time, the radiation field is regarded as stationary. Usually, in atmospheric optics we are concerned with stationary fields (or more accurately with fields in which the dependence of intensity on time can be ignored). If the intensity is not dependent on direction **n**, the radiation field is isotropic. If intensity does not depend on any or on several spatial co-ordinates, we are concerned with the uniform radiation field. For example, in atmospheric optics, the radiation field

*Instead of wavelength in the definition of intensity we can use frequency of wave number).

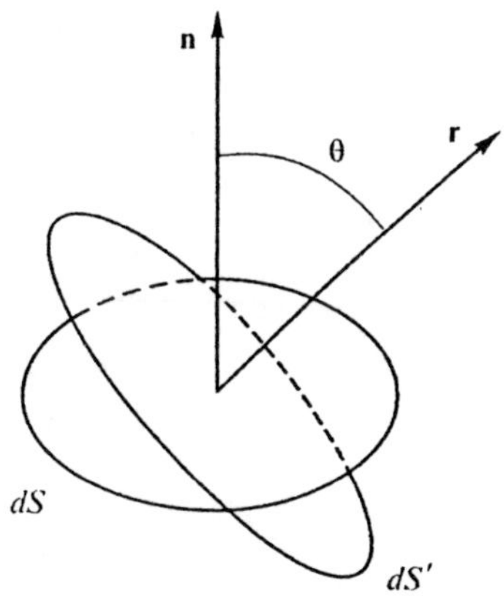

Fig 3.2. Determination of radiation flux.

is often assumed to be horizontally uniform, i.e. it does not change along the horizontal co-ordinates x and y.

Radiation flux

The second most important characteristic of the radiation field is the radiation flux. The radiation flux (monochromatic) F_λ is the amount of the electromagnetic energy dE_λ' in the unit wavelength interval falling on the unit area (or passing through it) per unit time from all directions [47, 64], i.e.

$$F_\lambda = \frac{dE'_\lambda}{dS\,dt\,d\lambda}. \qquad (3.2.4)$$

Evidently, the flux should be linked with radiation intensity. In fact, we examine direction **r** (Fig.3.2). Accord-ing to definition, the intensity of radiation from direction **r** is the energy passing through the area dS' normal to **r**. However, if θ is the angle between the direction **r** and the normal **n** to the area dS, then $dS' = dS\cos\theta$, and substituting dS' into (3.2.1), we obtain

$$dE_\lambda = I_\lambda\, dS\, cos\theta d\Omega dt d\lambda. \qquad (3.2.5)$$

Now, in order to calculate the energy dE_λ' required for determination of the flux (3.2.4), energy dE_λ (3.2.5) should be integrated in respect of all directions $d\Omega$ and, subsequently, substituting dE_λ' into (3.2.4) we obtain the required relationship:

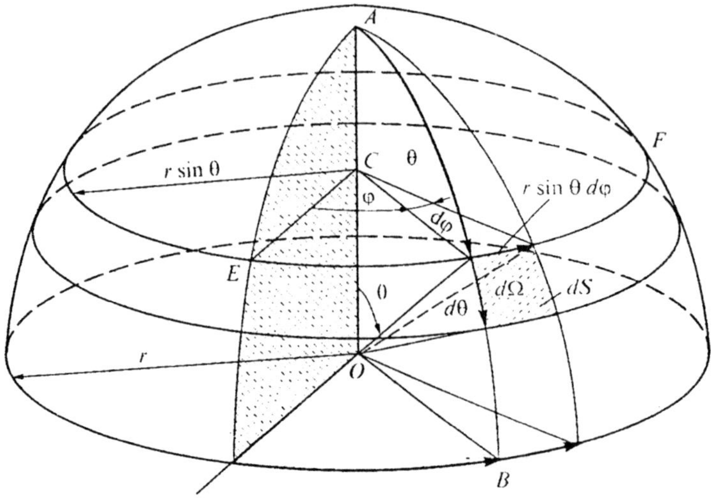

Fig. 3.3. Representation of the solid angle in spherical coordinates.

$$F_\lambda = \int I(\mathbf{r})\cos\theta(\mathbf{r})d\Omega. \quad (3.2.6)$$

The problem of the limits of integration in (3.2.6) will be left opened. Strictly speaking, according to the definition of the flux, it is necessary to carry out integration over the entire sphere (in respect of the total solid angle 4π). This value is referred to as the total flux. However, in atmospheric optics it is also accepted to examine integral (3.2.6) at half of the total solid angle. In fact, in the atmosphere there is a specific vertical direction. The unit vector **n** in the present case is the normal to the Earth's surface. Therefore, we examine the hemispherical descending flux – all directions of propagation of downward radiation are taken into account – and the hemispherical rising flux – all directions of upward radiation are considered. The descending flux is always negative ($\cos\theta<0$), and the rising flux is always positive ($\cos\theta>0$). The total flux is equal to the algebraic sum of these two fluxes.

In (3.2.6) it is useful to pass to integration in the spherical coordinates – angle θ and azimuth φ (Fig.3.3). The solid angle $d\Omega$ is determined as the ratio of the area dS of the spherical surface, visible from the centre of the sphere, to the square of the radius of the sphere r: $d\Omega = dS/r^2$. We measure the solid angle in steradians (sr), and for the sphere whose surface area is equal to $4\pi r^2$ the total solid angle is 4πsr. The surface area of the sphere dS, corresponding to the differential of the solid angle $d\Omega$, as indicated in Fig.3.3, is equal to the product of the length of the circle arc *OAB*, determined by the plane angle $d\theta$, and the length of the circle arc *CEF*, determined by the plane angle $d\varphi$. The length of the circular arc is equal to the product of its radius by the plane angle and, consequently, the first length is $r\ d\theta$, the second $r \sin\theta\ d\varphi$, since the radius of the circle arc *CEF* is $r \sin\theta$. Thus

$$dS = r\ d\theta \cdot r \sin\theta\ d\varphi,$$

which gives

$$d\Omega = \sin\theta\ d\theta\ d\varphi.$$

Thus, integration of any function $f(\Omega)$ in respect of the solid angle may be reduced to integration in the spherical co-ordinates

$$\int f(\Omega)d\Omega = \int\int f(\theta,\varphi)\sin(\theta)d\theta d\varphi. \tag{3.2.7}$$

Taking into account (3.2.7), from (3.2.6) we obtain for the total flux:

$$F_\lambda = \int_0^{2\pi} d\varphi \int_{\pi/2}^{\pi/2} I_\lambda(\theta,\varphi)\cos\theta\sin\theta d\theta, \tag{3.2.8}$$

and for the descending $F_\lambda^\downarrow$ and ascending $F_\lambda^\uparrow$ fluxes

$$F_\lambda^\downarrow = \int_0^{2\pi} d\varphi \int_{\pi/2}^{\pi} I_\lambda(\theta,\varphi)\cos\theta\sin\theta d\theta, \tag{3.2.9}$$

$$F_\lambda^\uparrow = \int_0^{2\pi} d\varphi \int_0^{\pi/2} I_\lambda(\theta,\varphi)\cos\theta\sin\theta d\theta. \tag{3.2.10}$$

Since the definition of intensity and radiation flux includes 'infinitely small spectral range' $d\lambda$, the intensity and flux are determined for the monochromatic radiation with the wavelength λ. In theoretical atmospheric optics it is often accepted to omit the spectral index at monochromatic quantities if we are not discussing the explicit dependence on wavelengths (frequency, wave number). Therefore, in subsequent considerations in the majority of cases we shall write notations for all quantities without spectral indices referring to their monochromatic values at the specific radiation wavelength, and also omit the attached 'monochromatic' term.

We examine a partial case in which radiation propagates only in one specific direction (θ_0, φ_0). This radiation falls on the elementary area dS under angle θ_0 to its normal. It is assumed that the intensity of this radiation is I. Consequently, in integration, the equation (3.2.8) retains only the contribution from one direction and the flux will be numerically equal to:

$$F = I \cos\theta_0. \tag{3.2.11}$$

We examine a case in which radiation passes from the limited solid angle $\delta\Omega$. For example, this radiation corresponds to the radiation of the Sun disc. It should be mentioned that its angular dimensions for Earth are equal to ~32'. At the upper boundary of the atmosphere, the solar radiation flux, falling on the unit area normal to the Earth–Sun direction is

$$F_0 = \int I_0(\mathbf{r})\cos\theta(\mathbf{r})\,d\Omega \approx \overline{I}_0 d\Omega_0, \tag{3.2.12}$$

where $\overline{I}_0$ is the mean intensity of solar radiation, $d\Omega_0$ is the solid angle under which the Sun disc can be seen from the Earth. Here, we have taken into account that for the selected perpendicular area $\cos\theta(\mathbf{r}) \approx 1$. As mentioned previously (Chapter 1), in atmospheric optics the quantity $\overline{I}_0 d\Omega_0$ has the special name – the solar constant.

The quantities presented previously and the relationships characterise monochromatic radiation. In practice, the quantities relating to the finite spectral intervals (for example, to individual absorption bands) or to the entire spectrum of electromagnetic radiation play a special role. In this case, they are referred to as integral. For example, the integral flux of radiation is determined by the equation

$$F = \int_0^{\infty} F_\lambda\, d\lambda. \tag{3.2.13}$$

We examine the layer (z_1, z_2) of a plane-parallel, horizontally homogeneous atmosphere. The difference of the total radiation fluxes $F(z_1) - F(z_2)$ at the boundaries of the layer characterises the energy of radiation absorbed (or emitted) by the layer. This energy is used for heating (or cooling) of the examined layer, i.e. for increasing (or reducing) its internal energy. The quantity

$$H(z_1, z_2) = F(z_1) - F(z_2) \tag{3.2.14}$$

is referred to as the radiant influx of energy to the layer (z_1, z_2). The radiant influx $H(z_1, z_2)$ may be positive (heating) or negative (cooling). The influx determines radiation changes of the temperature of the atmosphere with time:

$$\left(\frac{\partial T}{\partial t}\right)_r = \frac{1}{\rho c_p}\frac{dF(z)}{dz}. \tag{3.2.15}$$

Relationship of radiation intensity with the power of the electromagnetic wave

Equation (3.1.4) defines the relationship between the power of electromagnetic radiation and the mean square of its electrical strength. We examine again a partial case of radiation in a specific direction (θ_0, φ_0). Consequently, the energy flux to the unit area perpendicular to the given direction can be expressed by means of power W and radiation intensity (3.2.11). This results in the relationship of radiation intensity with the strength of the electrical field of the electromagnetic wave [62]:

$$I = W = \frac{1}{T}\int_0^T E^2(x,t)\,dt. \tag{3.2.16}$$

Some other characteristics of radiation energy

If the radiation intensity is given, we can also determine other quantities characterising the radiation field (for example, radiation flux). One of them is radiation density ρ, representing the amount of radiation energy in the unit volume. It is assumed that the radiation of intensity I_ν in the frequency range from ν to $\nu + d\nu$ falls on the area dS in the direction normal to it inside the small solid angle $d\Omega$ during time dt. Consequently, the amount of radiation energy falling on the area, is $I_\nu dS d\nu d\Omega dt$. During time dt, radiation passes the distance $c\,dt$, where c is the velocity of light. Therefore, the amount of radiation energy, arriving in the unit volume, is $I_\nu d\nu d\Omega/c$ ($c\,dS\,dt$ is the volume occupied by radiation). On the other hand, the same quantity according to definition is $\rho_\nu d\nu$. Consequently, the differential of radiation density is:

$$d\rho_\nu = \frac{I_\nu d\Omega}{c}. \tag{3.2.17}$$

In a general case when the investigated volume receives radiation from all sides, radiation density ρ_ν is determined by the expression:

$$\rho_\nu = \frac{1}{c}\int I_\nu d\Omega. \tag{3.2.18}$$

If the radiation density ρ_ν is divided by photon energy ($E = h\nu$, where h is the Planck's constant), radiation density gives the number of photons with frequency ν in the unit volume for the examined radiation field. This quantity determines many processes of interaction of radiation with the medium, for example, photochemical effect of radiation (Chapter 8).

The concepts 'intensity' and 'flux' are used only in relation to the radiation field. For identical characteristics of the energy, related to different objects or arising from such objects, we use the terms irradiance and brightness, respectively. For example, there is the brightness of the sky, i.e. the intensity of scattered solar radiation arriving from the sky, the irradiance of the surface, i.e. the flux of radiation falling on the surface, the irradiance of the input slit of a spectral device, etc. It is important to note that the concepts 'irradiance' and 'brightness' always refer to some elongated, not 'point', object.

3.3. Characteristics of interaction of radiation with a medium

In the propagation of electromagnetic radiation in vacuum the intensity of radiation does not change. In actual media (in particular, in the atmospheres of planets and on their surface) different complicated processes of interaction of radiation with the medium take place and change its intensity. We shall mention the main ones quite formally, leaving more detailed examination for subsequent chapters. The main mechanisms of interaction of radiation with the medium are:

- extinction
- scattering
- adsorption
- reflection
- refraction

The generation of radiation by the medium itself should be added to these processes.

Here, these interaction processes will be examined formally which is sufficient for obtaining one of the main equations of the theory of radiation transfer – the radiation transfer equation.

Extinction (scattering and adsorption) and radiation

The interaction of radiation with matter results in attenuation (extinction) of the radiation as a result of the absorption of radiation by the matter and its scattering to the sides away from the direction of propagation. From the physical viewpoint, the nature of absorption is related to the transition of radiation energy to the internal energy of the atoms and the molecules of atmospheric air and aerosol particles [7, 39, 40, 82]. The scattering of radiation is associated with the diffraction of electromagnetic waves on aerosol particles and on fluctuations of air density [39, 62]. Specific types of scattering of radiation are described on the basis of quantum mechanics by means of the interaction of photons with atoms and molecules. To these processes it is necessary to add the generation (emission) of radiation by the medium itself.

In a general case, the internal energy of the molecule E consists of kinetic (translational) E_k, electronic E_e, vibrational E_v, rotational E_r and the energies of the individual electrons and nuclei of the molecule (spin) E_s:

$$E = E_k + E_e + E_v + E_r + E_s.$$

Kinetic energy E_k is determined by the translational movement of molecules, E_e is linked with the position of electrons on orbital shells; E_v with the vibrations of atoms forming the molecule; E_r with the rotation of the molecule as an integral unit; E_s with internal energies of the electrons and nuclei forming the molecule. (In the literature there is another definition of the internal energy of the molecule in which the kinetic energy of the molecule is excluded from internal energy).

If the interaction of radiation with a molecule or particle is not accompanied by any change in the internal energy of the latter, this process is referred to as scattering (simple or elastic scattering). The intensity of interaction of radiation with the molecule depends on the type of internal energy of the molecule with which radiation interacts. If radiation interacts with matter whose internal energy is determined only by translational motion, the appropriate coefficients of interaction are very small under the conditions in the atmospheres of planets. For example, if a radiation quantum interacts with a free electron (Compton scattering), only 5×10^{-5} of quantum energy is transferred to the kinetic energy of the electrons [91]. This example illustrates the most efficient processes of interaction of radiation with the translational degrees of freedom of the particle in the atmosphere of the planets.

As mentioned previously, all components of the atmosphere (atoms, molecules, aerosols) have electronic, vibrational, rotational, etc., internal energies. The internal energy of the molecules and the atoms of the gases (with the exception of kinetic energy) is quantised, i.e. it may acquire only specific, discrete values.

The case of simple (elastic) scattering is realised in practice if a molecule (matter) is characterised by the narrow (quantised) states of internal energy, and the interacting photon has the frequency which greatly differs from the frequency of possible transitions between them. The interaction of the photon with the molecule may cause the transition of the molecule to a higher energy (excited) state. This excited state has a finite lifetime, and if the 'absorbed' photon is re-emitted with a negligibly small transfer of radiation energy to internal (electronic, vibration, etc.) energy of the molecule, this process also relates to the processes of elastic scattering.

If the transition of the molecule from the excited upper state to the lower initial state takes place in the form of a unique process (without intermediate transition between other levels of internal energy), the frequency of the re-emitted photon is identical with the frequency of the initially 'absorbed' photon. This process is the process of resonance or coherent scattering because in this case there are no changes in the frequency of the photon falling on the molecule. The process of re-emission of the 'absorbed' photon may take place in a more complicated manner when the molecule is transferred into the initial lower energy state in consecutively 'steps' passing through the intermediate energy states (see for more details Chapter 5). A photon forms in each step. In this case, the frequencies of re-emitted photons differ from the frequency of 'absorbed' photon, but the total energy 'absorption' and re-emission remains unchanged. This type of scattering is referred to as non-coherent scattering or scattering with frequency changes.

In nature there may be frequent cases in which prior to the excited molecule being capable of re-emitting a photon there may be a collision of the molecules with exchange of energy between them. In particular, these collisions may be accompanied by the deactivation of molecules, i.e. the transition of a molecule to a lower energy state without re-emission – emissionless transition. In this case, generally speaking, the type of internal energy to which radiation energy changes in this process is not important – to translational energy of the molecule or some other types of internal energy. It is important that the photon was 'taken' from the radiation field. These processes are referred to as absorption of radiation. The literature also uses the term true absorption which usually refers to the process of transition of radiation energy to kinetic energy of the molecules. The introduction of these definitions (absorption or true absorption) explains why we previously used quotation marks for the term 'absorbed' photon when discussing the processes of coherent and non-coherent scattering.

We shall discuss in greater detail the physical nature of these absorption processes [92]. To simplify considerations, we examine the interaction of a solar photon with a diatomic molecule AB. The solar photon with frequency ν has the energy $h\nu$. The interaction of the solar photon with a molecule can be described as follows:

$$AB + h\nu \rightarrow (AB)^*$$

where $(AB)^*$ denotes the molecule in the excited state, i.e. the molecule is not in the ground but in a high energy state. After a relatively large number of different processes of energy transformation (in collision with other molecules and re-emission) the molecule $(AB)^*$ returns to the initial, non-excited state. There are many possible mechanisms of deactivation of the excited molecule $(AB)^*$ which we started to discuss previously. Table 3.2 describes four possible initial stages of deactivation of a molecule. There are also other mechanisms (stages) but those presented in Table 3.2 are sufficient for explaining the main propositions.

Table 3.2 Initial stages of deactivation of the molecule

Process number	Process	Name of the process
1	$(AB)^* \rightarrow AB + h\nu$	Re-emission of photon (radiation decay)
2	$(AB)^* + M \rightarrow AB + M + E$	Emission-free transition (deactivation)
3	$(AB)^* \rightarrow A + B$	Dissociation
4	$(AB)^* + C \rightarrow A + BC$	Chemical reaction

Process 1 results in the re-emission of the absorbed solar photon and, as shown previously, this is the scattering process. This process is not accompanied by the exchange of energy between the radiation field and the molecule. The process 2 is the deactivation of the molecule in the process of collision with any other molecule M. In some cases, this process is also referred to as thermalisation. As a result of this process, the radiation energy transforms to the kinetic energy of movement of the molecule E, i.e. the atmosphere is heated. Both processes 1 and 2 return the molecule (AB)* to the initial state, but from the viewpoint of conversion of the energy of solar photon they greatly differ. In the first case, the excess energy of the excited molecule (AB)* again transforms to the energy of the radiation field. In the second case, this energy is transformed to the internal energy of the molecule. The processes 3 and 4 describe photochemical processes taking place in the atmospheres of the planets. For the process 3 to take place, the solar quantum must carry sufficient energy capable of, for example, dissociating molecule AB to the atoms A and B forming this molecule. The dissociation energy differs for different molecules. For example, for the dissociation of molecular oxygen it is necessary to provide photons with wave numbers higher than 41660 cm^{-1} (wavelength less than 242.2 nm), and for dissociation of an ozone molecule – photons with frequencies of 8656 cm^{-1} (wavelength less than 1155.3 nm). New components of the atmosphere, formed as a result of photochemical reactions, are usually highly chemically active and enter into new reactions with other components of the atmosphere. In addition to this, the products of the processes 3 and 4 may be in the excited state and cause subsequent re-emission of the photons. Here, we should only mention that as a result of processes 3 and 4 and also successive transformations, taking place during collisions of the molecules, the energy of the initial solar photon is transformed ultimately (at least partially) to the internal energy of the molecules and atoms forming the atmosphere. Consequently, from the viewpoint of radiation transfer all these processes are processes of radiation absorption. This absorption (generally speaking, not only solar but also atmospheric radiation) results in heating of the atmosphere, i.e. in changing its temperature.

Volume coefficients of extinction, scattering and absorption

Let us assume that radiation interacts with matter in the elementary volume $dV = dS\, dl$ (Fig.3.4). To characterise this interaction on the basis of analysis of the variation of the radiation energy, passing

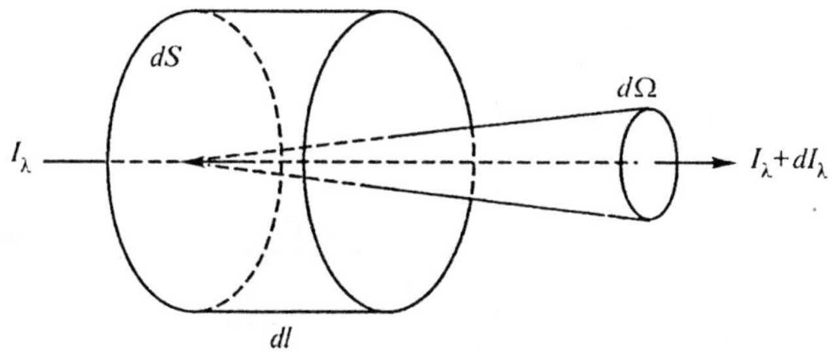

Fig.3.4. Determination of the volume of coefficient attenuation and derivation of the equation of transfer of radiation.

through the medium, we introduce the volume coefficient of extinction α_λ [47, 65].

Let us assume that the area δS, placed normal to the direction of propagation, receives, within the solid angle $d\Omega$, radiation with intensity I_λ in the wavelength range from λ to $\lambda + d\lambda$ during time dt (Fig.3.4). The amount of energy, falling on the area, will be equal to $I_\lambda\, dS\, d\Omega\, d\lambda\, dt$. If radiation is attenuated in propagation along the path dl (as a result of absorption and scattering), then along the path dl the attenuation of energy will be proportional to dl. The fraction of attenuated radiation-energy will be denoted by $\alpha_\lambda dl$. Thus, the amount of radiation energy, attenuated along the path of dl will be

$$dE_\lambda = \alpha_\lambda E_\lambda dl = \alpha_\lambda I_\lambda dl\, dS\, d\Omega\, d\lambda\, dt. \qquad (3.3.1)$$

Consequently, to determine the volume extinction coefficient, we have:

$$\alpha_\lambda = \frac{dE_\lambda}{E_\lambda dl} = \frac{dE_\lambda}{I_\lambda dl\, dS\, d\Omega\, d\lambda\, dt} = \frac{dI_\lambda}{I_\lambda dl}. \qquad (3.3.2)$$

Quantity α_λ has the dimension of the reversed length.*

As already mentioned, the radiation extinction is the sum of the scattering and absorption processes. Correspondingly, the attenuation of radiation energy dE consists of the scattering $dE_{\lambda s}$ and absorption $dE_{\lambda a}$. Thus,

$$dE_\lambda = dE_{\lambda s} + dE_{\lambda a}. \qquad (3.3.3)$$

*It appears that it is not logical to use for the quantity of such dimension the term 'volume' instead of 'linear'. However, we shall explain the meaning of this terminology.

Substituting (3.3.3) into (3.3.2) we obtain

$$\alpha_\lambda = \sigma_\lambda + k_\lambda, \tag{3.3.4}$$

where σ_λ and k_λ are the volume coefficients of scattering and absorption, respectively. These coefficients depend on the wavelength (frequency) and the examined point of the medium (for example, $\sigma_\lambda = \sigma_\lambda\ (x, y, z)$), but do not depend on the direction of radiation in the isotropic medium.

Extinction, scattering and absorption cross-sections

In examining the attenuation of radiation by the elementary volume dV, we have attributed the attenuating property to the volume as a whole. However, in the atmosphere of planets, radiation interacts with the molecules of air and aerosol particles present in the volume dV. Therefore, it is convenient to examine the processes of interaction of radiation with a single particle of matter (one molecule or aerosol particle) [6, 22].

Let us assume that the particle with the area of projection $\delta S'$ receives, on the plane normal to the direction of propagation of radiation received the directional monochromatic radiation with intensity I_0 in the unit solid angle per unit time. As a result of scattering and absorption, part of the incident radiation is weakened. The attenuated cross section C_e of the particle is the relationship:

$$C_e = \frac{\delta E_e}{I_0}, \tag{3.3.5}$$

where E_e is the energy of radiation attenuated by the particle. We shall use the definition of intensity of radiation (3.2.1). For the present case ($dt = 1$, $d\lambda = 1$, $d\Omega = 1$), the radiation energy falling on the area $\delta S'$ is equal to $\delta E = I_0 \delta S'$. Consequently, the relationship (3.3.5) may be written in the new form

$$C_e = \frac{\delta E_e}{\delta E} \delta S'. \tag{3.3.6}$$

Taking into account that the energy of radiation, attenuated by the particle, consists of the scattered and absorbed energies:

$$\delta E_e = \delta E_s + \delta E_a. \tag{3.3.7}$$

As in the case of the extinction cross-section, we introduce scattering C_s and absorption C_a cross-sections:

$$C_s = \frac{\delta E_s}{\delta E}\delta S', \ C_a = \frac{\delta E_a}{\delta E}\delta S'. \tag{3.3.8}$$

According to the law of conservation of energy

$$C_e = C_s + C_a. \tag{3.3.9}$$

The concepts of the cross-sections are very useful for describing the processes of interaction of radiation with matter. The main feature of their universal nature is that they are determined for a single arbitrary particle of matter irrespective of its nature (this may be the particle of aerosol or a gas molecule). The units of the cross-sections of interaction are represented by commonly used square centimetre (cm²).

Scattering phase function

We examine in greater detail the scattering of radiation on a single particle. In a general case, the energy of radiation, scattered in specific direction **r**, differs for different directions. This energy will be denoted by $\delta E_{sd}(\mathbf{r})$. The value of this energy is proportional to the solid angle $d\Omega$ (Fig.3.5). Dividing this energy by the intensity of radiation, falling on the area $\delta S'$ and normalising (dividing) by the value of the solid angle $d\Omega$, we obtain the cross-section of

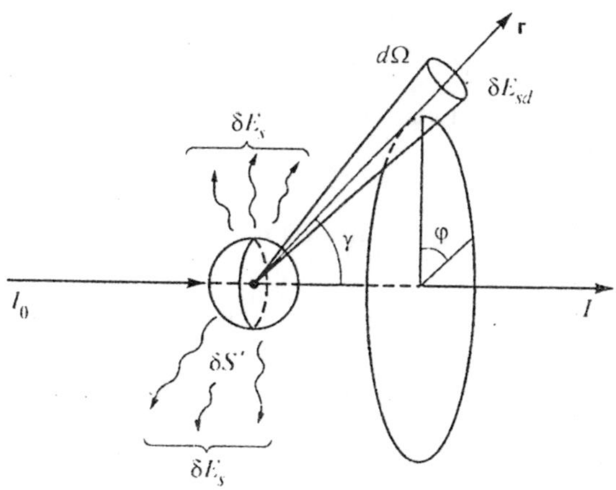

Fig.3.5. Determination of the extinction, scattering and absorption cross-sections and the scattering phase function.

directional scattering $C_{sd}(\mathbf{r})$:

$$C_{sd}(\mathbf{r}) = \frac{\delta E_{sd}}{I_0 d\Omega}. \tag{3.3.10}$$

Since the total energy of scattered radiation E_{sd} is equal to the sum of energies δE_{sd}, scattered in all directions, i.e. the integral in respect of the entire sphere (in respect of the solid angle 4π), from the law of conservation of energy we obtain the relationship between the scattering cross-section and the cross-section of directional scattering:

$$C_s = \int_{4\pi} C_{sd}(\mathrm{r}) d\Omega. \tag{3.3.11}$$

It should be mentioned that in some cases the scattering cross-section C_s is referred to as the 'integral or total scattering cross-section' and the cross-section of directional scattering $C_{sd}(\mathbf{r})$ as the 'differential scattering cross-sections' [24, 26].

The cross-section of directional scattering characterises the 'scattering force': it increases with increase of the degree of scattering of radiation in the given direction. For comparative characterisation of the 'scattering force' in different directions by different particles it is convenient introduce a dimensionless quantity and, for this purpose it is sufficient to normalise $C_{sd}(\mathbf{r})$ by it's integral over the entire sphere which according to (3.3.11) is C_s. This dimensionless characteristic is the scattering phase function $x(\mathbf{r})$ and is expressed by the relationship:

$$x(\mathbf{r}) = 4\pi C_{sd}(r)/C_s. \tag{3.3.12}$$

The definitions of (3.3.11) and (3.3.12) lead to the condition of normalisation of the scattering phase function:

$$\frac{1}{4\pi} \int_{4\pi} x(\mathbf{r}) d\Omega = 1. \tag{3.3.13}$$

If the scattering is isotropic, i.e. the same in all directions, from condition (3.3.12) we obtain $x(\mathbf{r}) \equiv 1$.

The scattering direction $\mathbf{r}$ is determined by the scattering angle and the scattering azimuth φ (Fig.3.5). Consequently, denoting $x(\mathbf{r}) = x(\gamma, \varphi)$, the condition (3.3.13) may be written in the new form:

$$\frac{1}{4\pi}\int_0^{2\pi} d\varphi \int_0^{\pi} x(\gamma,\varphi)\sin\gamma\, d\gamma = 1. \tag{3.3.14}$$

where we use the equation (3.2.7).*

In atmospheric optics we are usually concerned with processes for which the cross-section of directional scattering and, correspondingly, the phase function depends only on the scattering angle and not on the azimuth. Consequently, the integral in respect of the azimuth in (3.2.7) is equal to 2π and the normalisation condition is:

$$\frac{1}{2}\int_0^{\pi} x(\gamma)\sin\gamma\, d\gamma = 1. \tag{3.3.15}$$

The scattering phase function x (γ) maybe given the probability meaning. We determine the probability of scattering in the angle range from 0 to specific angle γ_0. According to the definition of probability, for this purpose it is necessary to count the number of events with the required outcome and divide it by the number of all events. The event with the required outcome is the scattering by the angle $0 < \gamma < \gamma_0$, all events – scattering by any angle (from 0 to π). The scattering energy in the angle range $[0, \gamma_0]$ is denoted by $\delta E_{sd}(0, \gamma_0)$, the total scattering energy is $\delta E(0, \pi) = \delta E_s$.

Each photon carries energy $h\nu$, and therefore the number of photons, scattered in the relevant range $[0, \gamma_0]$, is $\frac{\delta E_{sd}(0,\gamma_0)}{h\nu}$, and the number of all scattered photons is $\frac{\delta E_s}{h\nu}$. The required probability $P(\gamma_0)$ is determined as their ratio:

$$P(\gamma_0)\frac{\delta E_{sd}(0,\gamma_0)}{\delta E_s} = \frac{\int\limits_{\Omega=\{0\le\gamma\le\gamma_0,\ 0\le\varphi\le 2\pi\}} \delta E_{sd}(\gamma)\, d\Omega}{\delta E_s} =$$

*Strictly speaking, the cross-section of directional scattering and the phase function are not determined for the scattering angle $\gamma = 0$ because at the zero angle there is no scattered radiation, and radiation is transmitted (Fig.3.5). However, since we can formally make the scattering angle as close to zero as necessary, this indeterminacy is always ignored and the phase function is regarded as given for all angles.

$$= \int_0^{\gamma_0} \frac{C_{sd}(\gamma)\sin\gamma d\gamma}{C_s} = \frac{1}{2}\int_0^{\gamma_0} x(\gamma)\sin\gamma d\gamma$$

Thus, the probability of scattering in the angle range from 0 to γ_0 is equal to the integral of the scattering phase function in respect of the given angle range. Thus, the scattering phase function $x(\gamma)$ is the density of the probability of scattering in the angle γ. In this case, the presence of multipliers 1/2 and sin γ is determined by the integration in the spherical co-ordinate system (according to (3.2.7)). Evidently, applying the same considerations to the phase function of a general type, which depends both on the angle and the azimuth of scattering, we obtain that the scattering phase function is the density of the probability of scattering in the given direction.

Lidar backscattering cross-section

Lidar (Light Detection and Ranging) transmits a laser pulse to the atmosphere and records backscattered radiation. In the problems of lidar sounding of the atmosphere we are usually interested in the cross-section of directional scattering only for the angle of 180°. Therefore, in these problems it is convenient to introduce formally the lidar backscattering cross-section $C_{sd}(\pi)$ [6]. According to (3.3.11) and (3.3.12) we have

$$C_\pi = 4\pi C_{sd}(\pi) C_s x(\pi). \qquad (3.3.16).$$

Relationship between the volume extinction coefficient and scattering cross-section

It is evident that there should be some relationship between the volume coefficient of extinction and the previously introduced extinction cross-sections of the individual particles located in the volume. It is assumed that all the particles of the volume interact with radiation independently of each other. As the starting point, we examine the simplest case in which all particles in the elementary volume are the same. Consequently, the total extinction cross-section of the entire volume is equal to the sum of the cross-sections of all its particles. Let us assume that the volume $dV = dS\, dl$ contains dN particles. According to the definition of the extinction cross-section (3.3.5) the energy of monochromatic radiation, attenuated (per unit time, in unit solid angle) by the system of particles dN, is:

$$\delta E_e = C_e I_0 \, dN. \qquad (3.3.17)$$

On the other hand, in accordance with equation (3.3.2) the same energy may be written in the form:

$$\delta E_e = \alpha_e E dl = \alpha_e I_0 dl \, dS. \qquad (3.3.18)$$

Equating both expressions for the attenuated energy, we obtain

$$\alpha_e = \frac{dN}{dl\,dS} C_e = \frac{dN}{dV} C_e. \qquad (3.3.19)$$

According to the definition, dN/dV is the number concentration of attenuating particles, i.e. the number of particles in the unit volume. We denote it by n. Finally, we obtain

$$\alpha_e = nC_e. \qquad (3.3.20)$$

The volume extinction coefficient is equal to the product of the number concentration of particles by the extinction cross-section of a single particle, or, in other words, it is the total cross-section of the particles in the unit volume (this is why the coefficient is not 'linear' but 'volume').

Let us now assume that the elementary volume contains different particles, and we have particles of M type with cross-sections $C_{e,i}$ and concentrations n_i and, as previously, they all interact independently with radiation. Consequently, using the same procedure, finding the total extinction cross-section as the overall cross-section in respect of all particles, we obtain:

$$\alpha = \sum_{i=1}^{M} n_i C_{e,i}. \qquad (3.3.21)$$

Equation (3.3.21) is highly suitable for practical calculations because it makes it possible to calculate separately the volume extinction coefficients for particles of every type as $\alpha_i = n_i C_{e,i}$, and subsequently simply add them up. In particular, a standard procedure is the separate calculation of the volume coefficients of molecular extinction α_m and the volume coefficients of aerosol extinction α_a and the determination of the total extinction coefficient as their sum $\alpha = \alpha_m + \alpha_a$.

The relationships (3.3.4) and (3.3.9) give directly the identical equations, in comparison with (3.3.20) and (3.3.21), of the relationship of the volume scattering and absorption coefficients with the appropriate cross-sections:

$$\sigma = \sum_{i=1}^{M} n_i C_{s,i}, \quad k = \sum_{i=1}^{M} n_i C_{a,i}. \tag{3.3.22}$$

In particular, for the addition of the molecular and aerosol characteristics:

$$\sigma = \sigma_m + \sigma_a, \; k = k_m = k_a,$$

where σ_m is the volume coefficient of molecular scattering; σ_a is the volume coefficient of aerosol scattering; k_m is the volume coefficient of molecular absorption; k_a is the volume coefficient of aerosol absorption. The method of calculating these coefficients will be discussed in Chapters 4 and 5.

3.4. Radiation transfer equation

Volume radiation coefficient

We have examined the attenuation of radiation by the elementary volume of air. We now take into account that in addition to attenuation there may be also an increase of the radiation intensity as a result of atmospheric radiation inside the volume. Suitable examples are thermal radiation in the infrared range and also different glow in the atmosphere (this will be examined in Chapter 8). For characterisation of the atmospheric radiation of the medium we introduce the volume radiation coefficient ε_λ [47, 64]. If the medium is capable of emitting energy, the amount of energy, emitted by the volume $dV = dS\, dl$ in the solid angle $d\Omega$ during the time dt in the wavelength range $d\lambda$, will be proportional to $dV\, d\Omega\, d\lambda\, dt$. This amount of energy will be denoted by:

$$dE_\lambda = \varepsilon_\lambda\, dV\, d\Omega\, d\lambda\, dt \tag{3.4.1}$$

and ε_λ will be referred to as the radiation coefficient. Equation (3.4.1) gives that

$$\varepsilon_\lambda = \frac{dE_\lambda}{dV\, d\Omega\, d\lambda\, dt}. \tag{3.4.2}$$

Consequently, the volume radiation coefficient at the wavelength λ is the amount of energy emitted by the unit volume into the unit solid angle per unit time. In a general case, the coefficient depends on the wavelength, on the co-ordinate of the point and generally speaking, on the direction of radiation ε_λ (**r**).

Regardless of the identical notation, the volume coefficients of extinction and emission greatly differ: the volume extinction coefficient α_λ is determined as the ratio of the energies (see the previous paragraph), and the volume coefficient of radiation ε_λ as energy. They also have different dimensions. It should be mentioned that the determination of the radiation intensity (3.2.1) gives a simple relationship with the volume coefficient of radiation: $dI_\lambda = \varepsilon_\lambda dl$.

Derivation of the transfer equation

Let us now assume that after the passage of radiation through the elementary volume, the intensity of radiation is $I_\lambda + dI_\lambda$ (Fig.3.4). Consequently, according to the definition of intensity (3.2.1), the energy, incident on the left phase of the elementary volume is $I_\lambda\, dS\, d\Omega\, dt\, d\lambda$. The energy emitted through the right face is $(I_\lambda + dI_\lambda)\, dS\, d\Omega\, dt\, d\lambda$. According to the law of conservation of energy, the variation of the energy inside the volume is equal to the reduction of energy as a result of extinction and the increase as a result of radiation. According to the definitions of intensity (3.2.1) and the volume extinction coefficient (3.3.1), the reduction of energy is equal to

$$dE_e = \alpha_\lambda I_\lambda dl\, dS\, d\Omega\, dt\, d_\lambda,$$

and the increase of energy as a result of radiation is given by equation (3.4.1). Consequently

$$(I_\lambda + dI_\lambda) dS\, d\Omega\, dt\, d\lambda - I_\lambda\, dS\, d\Omega\, dt\, d\lambda =$$

$$= -I_\lambda \alpha_\lambda\, dl\, d\Omega\, dt\, d\lambda + \varepsilon_\lambda\, dl\, dS\, d\Omega\, dt\, d\lambda.$$

This gives the differential equation of transfer of energy

$$\frac{dI}{dl} = -\alpha I + \varepsilon, \tag{3.4.3}$$

where the spectral indices for monochromatic quantities are omitted.

Bouguer law, optical thickness

We examine the equation of transfer of radiation (3.4.3) with special reference to atmospheric optics problems. We start with the simplest case of attenuation in the absence of radiation in the medium, i.e. when $\varepsilon \doteq 0$. The differential equation of transfer has the form:

$$\frac{dI}{dl} = -\alpha I.$$

Its general solution is

$$I(l) = I_0 \exp(-\alpha l), \tag{3.4.5}$$

where I_0 is the initial value of intensity (at $l = 0$). Thus, the intensity in the attenuating medium decreases in accordance with an exponential law. This statement is referred to as the Bouguer law (or, more accurately, Bouguer–Beer–Lambert law – according to the names of scientists who derived this equation independently of each other).

In the transfer equation, intensity I, the volume coefficients of extinction α and radiation ε depend on the co-ordinate of the point l. In particular, in the atmosphere of planets all these quantities change with altitude. The explicit consideration of the circumstance, for example, in the Bouguer law, leads to the expression:

$$I(l) = I_0 \exp\left(-\int_0^l \alpha(l')dl'\right). \tag{3.4.6}$$

Integration in (3.4.6) is carried out along the trajectory of propagation of radiation which in a general case may be curvilinear because of the refraction phenomenon (associated with the variation of the refractive index of air with altitude). However, in the majority of atmospheric optic problems refraction may be ignored and it may be assumed that radiation propagates in the atmosphere along straight lines. (Atmospheric refraction will be examined in Chapter 5).

We now examine the geometry of light rays in the atmosphere (Fig.3.6). The vertical co-ordinate is represented by the altitude, i.e. the vertical axis is normal to the surface of the Earth. The direction of the light beam is characterised by the zenith angle θ (Chapter 1). Because of the spherical form of the atmosphere, the zenith angle continuously changes along the beam (angles θ and θ' in Fig.3.6). However, since the radius of the Earth is considerably larger than the thickness of the atmosphere, we can use the

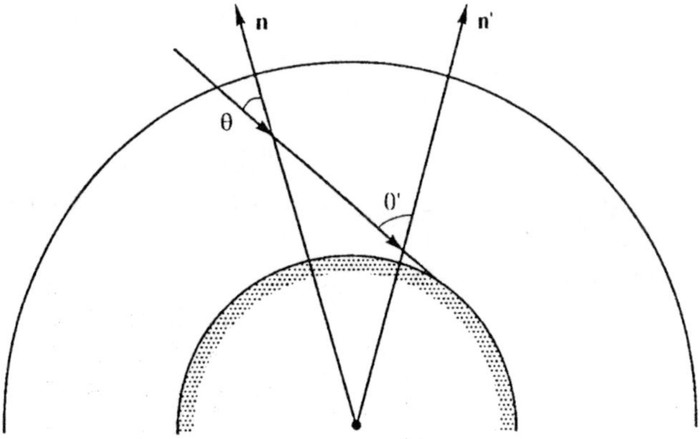

Fig.3.6. Determination of the zenith angle in the spherical atmosphere.

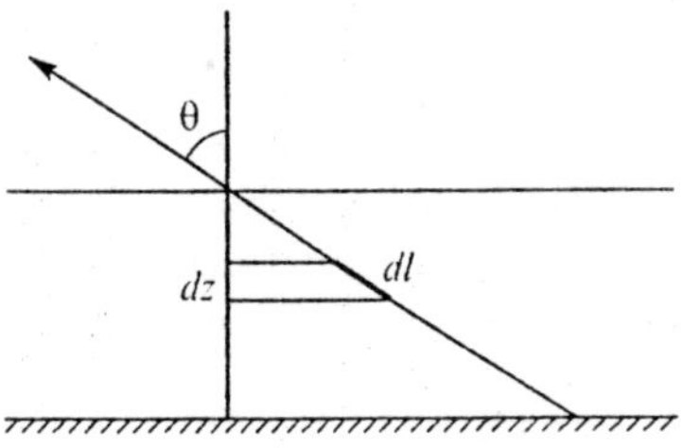

Fig.3.7. Plane-parallel atmosphere.

approximation of the plane-parallel atmosphere for a wide range of problems (Fig.3.7). In such an atmosphere, the zenith angle θ of any beam is constant and the element of the trajectory along the beam is $dl = dz \cos\theta$. In further explanation (if not stated otherwise) we shall always refer to the plane-parallel atmosphere. Consequently, the transfer equation for this atmosphere (3.4.3) may be written on the basis of the height z:

$$\cos\theta \frac{dI}{dz} = -\alpha(z)I + \varepsilon(z). \tag{3.4.7}$$

For the solution of equation (3.4.6) we have

$$I(z) = I_0 \exp\left(-\sec\theta \int_0^z \alpha(z')dz' \right). \tag{3.4.8}$$

The solution of (3.4.8) is used for the calculation of radiation when, as mentioned previously, we can ignore the atmospheric radiation of the medium and when, as shown later, we can also ignore the contribution of scattering to $I(z)$. The problems of this type include the calculation of attenuation of radiation from different sources in different spectrum ranges: the Sun, stars, lasers, etc.

The integral in the the exponent in (3.4.8) is a dimensionless quantity referred to as the optical thickness (depth) $\tau(z)$. In astrophysical literature, quantity $\tau(z)$ is often referred to as the optical distance between two points $(0, z)$. In atmospheric optics, we count the optical thickness from the upper boundary of the atmosphere

$$\tau(z)=\int_{z}^{\infty}\alpha(z')dz'.$$

An important characteristic is the optical thickness (depth) of the entire atmosphere $\tau_0 = \tau(0)$:

$$\tau=\int_{0}^{\infty}\alpha(z)dz,$$

we may also consider the optical thickness of individual atmospheric layers. After introducing the optical thickness, the solution of the transfer equation (3.4.8) is written in the simple form which does not depend explicitly on $\alpha(z)$:

$$I(z) = I_0 \exp(-\sec\theta\,\tau(z)).$$

Since the volume extinction coefficient in a general case is determined by different attenuation mechanisms and different attenuation atmospheric components, the optical thickness is the sum of optical thicknesses

$$\tau(z)=\sum_{i=1}^{n}\tau_i(z). \qquad (3.4.9)$$

In particular, τ_i refers to the optical thicknesses of the layer $(0, z)$ as a result of scattering and absorption, molecular and aerosol scattering and absorption.

Quantity $P(0, z) = I(z)/I_0$, characterising the fraction of radiation intensity, passed through the atmosphere (or layer), and equal to

$$P(0,z)=\exp\left(-\sec\theta\int_{0}^{z}\alpha(z')dz'\right) \qquad (3.4.10)$$

is the transmission function of the layer (0, z). Correspondingly, the fraction of absorbed radiation ($1 - P$) is the absorption function. These concepts are used widely when describing the processes of radiation transfer and for characterisation of the absorption and radiation spectra.

Linearity of processes of attenuation and radiation

The main attenuation law, i.e. the Bouguer law, formulated in the previous paragraph

$$dI = -\alpha I \, dl \qquad (3.4.11)$$

contains an important assumption according to which the processes of attenuation are linear and do not depend on the intensity of incident radiation and the amount of the attenuating substance. This is especially clear if we transfer to the extinction cross-section:

$$dI = -n \, C_e \, I \, dl. \qquad (3.4.12)$$

Equation (3.4.12) shows that the attenuation of radiation intensity along the path dl is linear in relation to the radiation intensity and the amount of the attenuating substance along this path. Such linear dependence is found in cases in which the attenuation cross-section itself does not depend on the intensity of incident radiation nor on the amount of the attenuating substance.

The first assumption is fulfilled with high accuracy for the majority of atmospheric optics problems, in particular, the problems of transfer of solar and atmospheric radiation. Large deviations from the Bouguer law are detected in the propagation of high power laser radiation. In this case, one can expect the dependence of the extinction cross-section on the radiation intensity. At a radiation power of a laser source of the order of 10^7 W/cm^2 we observe the spectroscopic effect of saturation for molecular absorption in atmospheric gases, characterised by a decrease of absorption in comparison with the absorption following from the Bouguer law. In the saturation conditions, the Bouguer law is not fulfilled and the attenuation of intensity takes place in accordance with a linear (not exponential) law. There are also other non-linear mechanisms of the interaction of high power laser radiation with the gas medium. These non-linear effects will not be discussed in this book.

The second aspect of satisfiability of the Bouguer law is associated with the absence of the dependence of the extinction cross-section on the concentration of the attenuating substance. The independence of the extinction cross-section of the concentration of the substance indicates that every molecule or particle attenuates the radiation independently of other molecules or particles. As shown by a large number of investigations, this assumption is fulfilled with high accuracy at low concentrations of the attenuating material.

In fact (this will be examined in greater detail later), the optical properties of individual molecules depend on the presence in their vicinity of other molecules and on the condition of the medium (temperature and pressure). Thus, the collisions of molecules lead to the broadening of the contour of the absorption line. Nevertheless, on the assumption that the physical state of the medium is given and constant, the processes of interaction of the molecules in the gas medium are also fixed. In these conditions, the attenuation of radiation is governed by the linear dependence on the concentration of the attenuating substance. If we consider a medium with changing parameters, we find a dependence of, for example, the cross-section of molecular absorption on the partial pressure of the absorbing gas, i.e. on the concentration of the attenuating substance.

As with the assumption of the linear nature of the attenuation processes we can introduce an assumption on the linearity of the radiation processes. As a formal statement we may write

$$dI = \alpha\tilde{B} = Cn\tilde{B}dl, \tag{3.4.13}$$

where $\tilde{B}$ is some function of sources (source) of radiation (equation (3.4.13) is its definition). Using the introduced function of the sources, the equation of radiation transfer (equation (3.4.3)) may be written in the following form:

$$\frac{dI}{dl} = -\alpha I + \alpha\tilde{B}. \tag{3.4.14}$$

Comparison of the equations (3.4.3) and (3.4.14) shows a relationship between the radiation coefficient and the function of sources:

$$\varepsilon = \alpha\tilde{B} \text{ and } \tilde{B} = \frac{\varepsilon}{\alpha}. \tag{3.4.15}$$

Solution of the transfer equation with atmospheric radiation taken into account

We examine a situation in which there are both attenuation and radiation processes in the medium and there is no scattering of radiation. This is a standard case in describing the propagation of atmospheric radiation in the atmosphere in infrared and microwave ranges, when scattering is ignored because it is very small. In this variant, the differential equation (3.4.3) has the form

$$\frac{dy(x)}{dx} = -a(x)y(x) + b(x), \tag{3.4.16}$$

i.e. it is a non-uniform linear differential equation with variable coefficients. The general solution of this equation is available:

$$\begin{aligned} y(x) = y(x_0)\exp\left(-\int_{x_0}^{x} a(x')dx'\right) + \\ + \int_{x_0}^{x} b(x')\exp\left(-\int_{x'}^{x} a(x'')dx''\right)dx'. \end{aligned} \tag{3.4.17}$$

In the present case $x = z$, $y(x) = I(z)$, $a(x) = k(z)/\cos\theta$, where α was replaced by k because of the equality, according to (3.3.4) of the volume coefficients of extinction and absorption in the absence of scattering, and $b(x) = \varepsilon(z)/\cos\theta$. After simple substitution of these values into (3.4.17), we obtain the general solution of the transfer equation with attenuation and atmospheric radiation taken into account:

$$\begin{aligned} I(z) = I_0\exp\left(-\sec\theta\int_0^z k(z')dz'\right) + \\ +\sec\theta\int_0^z \varepsilon(z')\exp\left(-\sec\theta\int_{z'}^{z} k(z'')dz''\right)dz'. \end{aligned} \tag{3.4.18}$$

The first term in (3.4.18) describes the absorption of initial radiation with intensity I_0, the second term describes the generation of atmospheric radiation which is also absorbed along the path from the emission point z' to the final height z. The functions

$\exp\left(-\sec\theta\int_0^z k(z')dz'\right)$ and $\exp\left(-\sec\theta\int_{z'}^z k(z'')dz''\right)$ in (3.4.18) are the transmittace functions of different layers of the atmosphere.

Different types of atmospheric radiation

The atmospheres of planets and their surfaces generate radiation referred to as atmospheric radiation. From the viewpoint of classic electrodynamics, radiation forms at any accelerated motion of charged particles. Thus, the electrons in atoms and molecules move with acceleration in circular orbits inducing the radiation.

Quantum mechanics interprets the formation of radiation as a process of the transition of a molecule or atom from the excited internal quantised state to a lower energy state. In transferring the molecule from state E_2 to state E_1 (where $E_2 > E_1$) the energy quantum $E_2-E_1 = h\nu_{21}$ is emitted. In this interpretation, radiation does not occur if all molecules are in the ground (not excited) state. However, if a medium is not at the absolute zero temperature, there is always a specific fraction of molecules in the excited state and, consequently, radiation will be generated because there is always non-zero probability of transfer of the molecules from state E_2 to state E_1 (Chapter 4). In addition to this, the presence of molecules or atoms in the excited state may be caused by an external effect. Therefore, the atmospheric radiation of a medium may be classified on the basis of examination of processes leading to the formation of molecules in the excited state. Landsberg [39] proposed the following classification of atmospheric radiation.

Chemiluminescence is the radiation formed as a result of excitation of molecules during chemical reactions. In this case, the radiation process is accompanied by changes of the chemical composition of matter and by a decrease of its internal energy. We shall present examples of these processes in Chapter 8 when examining the glow of the atmosphere.

If the process of radiation of the medium is determined by the incident external electromagnetic radiation, this process is referred to as photoluminescence. In this case, to sustain this radiation, it is necessary to provide continuously energy to the matter in the form of radiation.

If the excitation of a medium takes place as a result of an electrical effect, this type of radiation is referred to as electroluminescence. This is, for example, the flow of gases under the effect of electric current pulsing through them – glow discharge, electric arc, spark. The excitation of molecules and atoms may also take place under the effect of different particles with high energies. This may also be a reason for different atmospheric glows.

A special type of radiation, playing a fundamental role in the optics of the atmosphere, in particular in the formation of radiation in the infrared range of the spectrum is equilibrium or thermal radiation. This radiation forms if the condition of thermodynamic (thermal) equilibrium is fulfilled. In this case, radiation remains unchanged if the emission of radiation energy is compensated by the inflow of corresponding equal amount of energy in the form of heat. In thermal radiation, the medium is in a state in which the distribution of energy between the medium and the radiation field does not change with time.

From this viewpoint, the above types of atmospheric radiation of the medium – chemiluminescence, photoluminescence, etc., are not equilibrium. For example, chemiluminescence is accompanied by chemical changes in the medium. The process of this continuous radiation carries on until the chemical reaction takes place, and the medium moves further and further away from the initial state.

Case of equilibrium (thermal) radiation

The solution (3.4.18) should be regarded as formal because we have not yet explained the physical meaning of coefficient ε. The atmospheric radiation can be determined by different physical reasons (it should be mentioned that it is assumed, as previously, that there are no scattering processes taking place in the medium). Atmospheric radiation is subdivided into two types: thermal (equilibrium) and non-equilibrium. The non-equilibrium radiation of the atmosphere may be of a different nature, and we shall examine it later in greater detail (Chapter 7). Prior to examining thermal radiation, we shall mention several important definitions.

The thermodynamic equilibrium is the state of the medium in which the temperature of matter is constant everywhere, the mass of the medium does not move, and is mixed in such a manner that diffusion or any other movement of matter cannot form in it [82]. Strictly speaking, the thermodynamic equilibrium forms in a closed cavity whose walls are heated to some constant temperature T. The walls of the cavity emit and absorb electromagnetic radiation. The condition of thermodynamic equilibrium is characterised by the fact that every process is balanced by an opposite process. In particular, from this it follows that the radiation intensity in such a cavity (in the presence of thermodynamic equilibrium) is independent of the location and direction. If this was not the case, the energy would be transferred from one place to another in some directions. In addition to this, thermodynamic considerations show that the density of radiation in such a cavity depends only on frequency (wavelength) and temperature, but is independent of the nature of emitters – the walls of the cavity and the matter present in it [39, 62, 80, 82]. This radiation is referred to as equilibrium or thermal. If it is assumed that there is a small hole in the investigated cavity, the hole can be regarded as an hole in an absolutely black body since the external radiation, falling on the hole, is almost completely absorbed in this solid and does not exit from it. The probability of incident external radiation leaving the body is negligibly small. Thus, equilibrium radiation should be regarded as the radiation of an absolutely black body [39, 80, 82].

Analysis of special features of equilibrium radiation has played an exceptionally important role in the development of quantum theory. The point is that all the attempts to develop a non-contradicting theory of equilibrium radiation on the basis of classic representations, which would be in complete agreement with experiments, have not been successful. Only after introducing the concept of the Planck quantum at the beginning of the twentieth century, it has become possible to develop a successive theory of equilibrium (absolutely black) radiation. According to quantum theory, energy is emitted only in discrete portions – quanta, and the intensity of absolutely black radiation (Planck equations) is determined by the equation:

$$I_{eq}(\nu,T) = B(\nu,T) = \frac{2h\nu^3}{c^2}\frac{1}{\exp\frac{h\nu}{k_B T}-1}. \tag{3.4.19}$$

The intensity of equilibrium radiation does not depend on direction, i.e. equilibrium radiation is isotropic. Therefore, as indicated by equation (3.2.18), the density of equilibrium radiation is:

$$\rho_{eq}(\nu,T) = \frac{4\pi}{c}B(\nu,T) = \frac{8\pi h\nu^3}{c^3}\frac{1}{\exp\frac{h\nu}{k_B T}-1}. \tag{3.4.20}$$

The total flux of equilibrium radiation is evidently equal to zero.

The equation for the intensity of absolutely black radiation may also be written in the following form (if we use the wavelength):

$$B(\lambda,T)=\frac{\nu^2}{c}B(\nu,T)=\frac{c_1}{\pi\lambda^5\left(\exp\left(c_2/\lambda T\right)-1\right)}. \qquad (3.4.21)$$

Here

$$c_1 = 2\pi hc^2,$$

$$c_2 = hc/k_B$$

are the so-called first and second radiation constants. Both types of the Planck function are presented in Fig. 3.8. It should be noted that Fig.3.8 shows two scales of the abscissa for different radiation temperatures: 6000 K – the radiation of the Sun, 250 K – the radiation of the Earth atmosphere. It is also important to note a special feature. At the temperature of the absolute black radiation T = 6000 K (the radiation of the Sun), only 0.4% of the total radiation energy belongs to the wavelengths greater than 5 μm. At the radiation temperature of 250 K (irradiation of the Earth atmosphere), only 0.4% of the total energy radiation belongs to the wavelengths smaller than 5 μm. Thus, from the of practical viewpoint, the fields of the solar and atmospheric radiation may be examined independently. The Planck function has a single maximum at the values:

$$\frac{Tc}{\nu}=0.5099\,\text{cm}\cdot\text{K}$$

for expression (3.4.19) and

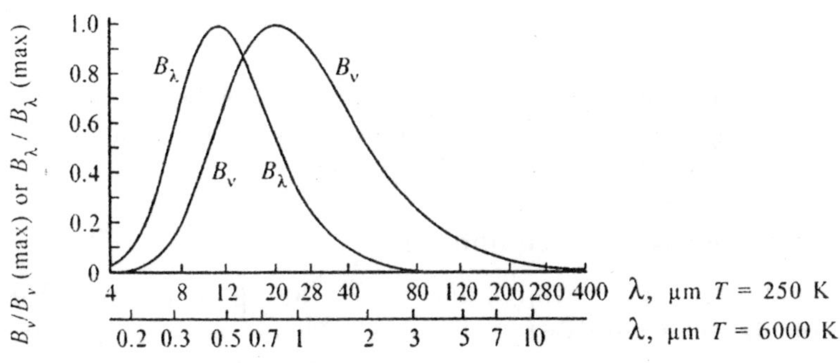

Fig.3.8. Planck function [91].

$$T\lambda = 0.28978 \text{ cm} \cdot \text{K}$$

for expression (3.4.21).

Integration of the equation (3.4.19) in respect of all frequencies gives the integral (total) intensity of equilibrium (absolutely black) radiation:

$$\int_0^\infty B(\nu,T)d\nu = \frac{\sigma_B T^4}{\pi}. \tag{3.4.22}$$

The equation (3.4.22) is the Stefan–Boltzmann law showing that the integral equilibrium radiation is proportional to the fourth degree of temperature (σ_B is the Stefan–Boltzmann constant). The behaviour of the Planck function away from its maximum values greatly differs. Therefore, two approximations are used often in atmospheric optics. At $\lambda \to \infty$ or $\nu \to 0$:

$$B(\nu,T) \to \frac{2k_B T \nu^2}{c^2}, \tag{3.4.23}$$

$$B(\lambda,T) \to \frac{2k_B Tc}{\lambda^4}. \tag{3.4.24}$$

The expressions (3.4.23) and (3.4.22) are known as the Rayleigh-Jeans law.

At $\lambda \to 0$ or $\nu \to \infty$:

$$B(\nu,T) \to \frac{2h\nu^3}{c^2}\exp\left(-\frac{h\nu}{k_B T}\right), \tag{3.4.25}$$

$$B(\lambda,T) \to \frac{2hc^2}{\lambda^5}\exp\left(-\frac{hc}{k_B \lambda T}\right). \tag{3.4.26}$$

These expressions are referred to as Wien's law. The areas of applicability of these approximations greatly depend on the temperature of the absolutely black body.

Local thermodynamic equilibrium

Undoubtedly, the atmospheres of the planets are not in thermodynamic equilibrium because of many reasons. Firstly, the temperature of the atmosphere changes from point to point and, secondly, the atmosphere is in continuous motion, etc. Consequently, the fields of radiation in the atmospheres of the planets greatly differ from the field of radiation in thermodynamic equilibrium. The presence of the upper external 'boundary' in the atmosphere of the planets and the outgoing radiation of the planets indicate that the atmospheres differ from the idealised closed cavity used for deriving the laws of equilibrium radiation. Regardless of this, the concept of thermodynamic equilibrium in the local sense is applicable to the atmosphere of the planets. It can be used efficiently for the limited volumes of the atmosphere in the lower, relatively dense layers of the atmosphere of the planets. This important assumption has been justified by astrophysicists with special reference to the photospheres of stars. We shall give the consideration [64] for the atmosphere of the planets.

Generally speaking, the conditions in the elementary volume of the atmosphere also greatly differ from the conditions of thermodynamic equilibrium. This is caused by the non-isotropic nature of incident radiation, for example, solar radiation. However, the radiation absorbed by the elementary volume is greatly 'processed' by the volume. As indicated by thermodynamics, this processing takes place in the direction of establishment of thermodynamic equilibrium. Therefore, it may be assumed that in every volume of the atmosphere, the radiation coefficient is linked with the absorption coefficient by the same relationship as in the case of thermodynamic equilibrium with some temperature T, characteristic for the given areas. This assumption is referred to as the assumption of local thermodynamic equilibrium (LTE) in the atmosphere of the planets, and it makes it possible to simplify greatly the examination of the problem of the transfer of atmospheric radiation. Naturally, an important task is the determination of the boundaries of applicability of LTE. In chapter 7, this assumption will be studied in greater detail. Here, it should only be mentioned that violations of LTE are detected at relatively high altitudes in the atmosphere of the planets, more accurately at low pressures. For example, for the Earth, the levels of violation of the LTE are usually situated above 30–40 km, and they differ for different atmospheric components and different absorption bands.

Kirchhoff law

We examine again the radiation under thermodynamic equilibrium (in an isolated cavity). We use the transfer equation (3.4.3) for this case. Since in this case $dI/dl = 0$, then $\varepsilon_\lambda = \alpha_\lambda I_\lambda$. Taking into account that in the investigated case $\alpha_\lambda = k_\lambda$, and $I_\lambda = B(\lambda, T)$, we obtain:

$$\varepsilon_\lambda = k_\lambda B(\lambda, T). \tag{3.4.27}$$

Equation (3.4.27) is the Kirchhoff law: in thermodynamic equilibrium, the ratio of the radiation coefficient to the absorption coefficient is equal to the intensity of radiation which is a universal function of frequency and temperature (Planck function). If the Kirchhoff law is satisfied, the solution of (3.4.18), i.e. the intensity of thermal radiation, maybe presented in the following form:

$$I_\lambda(z) = I_{\lambda,o} \exp\left(-\sec\theta \int_0^z k_\lambda(z')dz'\right) + \\ + \int_0^z k_\lambda(z') B_\lambda(T(z')) \exp\left(-\sec\theta \int_{z'}^{z} k_\lambda(z'')dz''\right) dz' \sec\theta. \tag{3.4.28}$$

Thus, the intensity of thermal radiation may be calculated if the temperature and the absorption coefficient are given as a function of the altitude in the atmosphere.

Taking into account scattering in the radiative transfer equation

We examine the case of the scattering of radiation in the atmospheres of the planets ignoring the atmospheric radiation. These problems form in the calculation of the field of solar radiation in the ultraviolet, visible and near infrared ranges of the spectrum. It would appear that in the absence of atmospheric radiation, the radiation coefficient should be equal to zero. However, this is not so, because now the medium contains the field of scattered solar radiation. Therefore, the element of the path dl is characterised by an increase of radiation intensity as a result of the additional scattering of radiation arriving in the volume of the medium from different directions and scattered in the direction of initial incident radiation I_0 (Fig. 3.9).

We determine an equation for the radiation coefficient, determined by scattering. It is assumed that radiation with the intensity I_0 falls on the elementary volume of the medium $dV = dl \cdot dS$ (Fig. 3.9).

To determine the amount of radiation energy in the unit interval

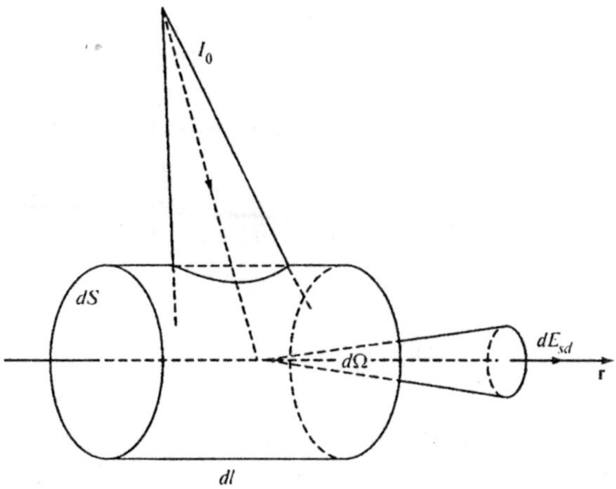

Fig.3.9. Derivation of the transfer equation for scattered radiation.

of the wavelength, scattered in the arbitrary direction r in the body angle $d\Omega$, in unit time, we use the definition of the cross-section of directional scattering. The energy of radiation, scattered by the elementary volume dV, containing dN scattered particles, in the direction r is equal to:

$$dE_{sd}(\mathbf{r}) = C_{sd}(\mathbf{r})I_0\, dN\, d\Omega. \qquad (3.4.29)$$

Taking into account that $dN = n \cdot dV$, where n is the concentration of scattered particles, we obtain:

$$dE_{sd}(\mathbf{r}) = C_{sd}(\mathbf{r})I_0\, ndV\, d\Omega. \qquad (3.4.30)$$

We also used the definition of the radiation coefficient (3.4.1). According to the definition, the energy of radiation in the direction r in the body angle Ω is

$$dE(\mathbf{r}) = \varepsilon(\mathbf{r})dV\, d\Omega. \qquad (3.4.31)$$

If the radiation 'forms' in the medium only as a result of scattering (there is no atmospheric radiation), the equations (3.4.30) and (3.4.31) can be equated to each other. Consequently, for the coefficient of radiation, determined by the scattering of radiation, we may write the following equation:

$$\varepsilon(\mathbf{r}) = C_{sd}(\mathbf{r})\, nI_0. \qquad (3.4.32)$$

The equation (3.4.32) shows that the radiation coefficient, associated with scattering, depends on the direction and intensity of incident radiation. This is the difference between this coefficient and the coefficient of equilibrium radiation which is expressed by means of the Planck function of absolutely black radiation. Using the relationship between the coefficient of directional scattering and the scattering phase function (3.3.12), the equation (3.4.32) may be presented in the following form:

$$\varepsilon(\mathbf{r}) = \frac{1}{4\pi} x(\mathbf{r}) \sigma I, \tag{3.4.33}$$

where σ is the volume scattering coefficient. The 'addition rules' and the normalisation of the phase function (3.3.13) show that to determine the general phase function of scattering of the particles of types M, it is sufficient to normalise its total value $\sum_{i=1}^{M} \sigma_i x_i(\mathbf{r})$. Consequently, we obtain the 'addition rule' for the phase function:

$$x(\mathbf{r}) = \frac{\sum_{i=1}^{M} \sigma_i x_i(\mathbf{r})}{\sum_{i=1}^{M} \sigma_i}. \tag{3.4.34}$$

In particular, for the 'sum' of molecular and aerosol indicatrices we obtain:

$$x(\mathbf{r}) = \frac{\sigma_m x_m(\mathbf{r}) + \sigma_a x_a(\mathbf{r})}{\sigma_m + \sigma_a}.$$

To determine the value of ε, included in the transfer equation, it is necessary to take into account the intensity of scattered radiation arriving from all directions, i.e. integrate $\varepsilon(\mathbf{r})$ in respect of the entire sphere. Consequently, we obtain the transfer equation for scattered radiation:

$$\cos\theta \frac{dI}{dz} = -\alpha(z) I + \frac{1}{4\pi} \sigma(z) \int_{4\pi} x(z, \mathbf{r}) I(\mathbf{r}) d\Omega. \tag{3.4.35}$$

For the complete determination of equation (3.4.35) it is necessary to introduce a specific geometry into the equation. However, this will be carried out in chapter 8.

In a general case, equation (3.4.35) does not have a solution in the form of the explicit analytical expression. Examination of the tsansfer equation of scattered radiation, determination of its particular analytical solutions, and the development of numerical methods of the calculation of intensity and fluxes of scattered radiation are the subject of the theory of radiative transfer. Some of the main assumption of the theory will be examined in chapter 8.

Finally, it should be mentioned that in the most general case in the presence of both scattering and atmospheric radiation, equation (3.4.35) conserves its form, but another member, responsible for atmospheric radiation, is added to its right-hand part. In particular, for thermal radiation we obtain:

$$\cos\theta\frac{dI_\lambda}{dz} = -\alpha_\lambda(z)I_v +$$

$$+\frac{1}{4\pi}\sigma_\lambda(z)\int_{4\pi} x(z,\mathbf{r})I_\lambda(\mathbf{r})dr + k_\lambda(z)B_\lambda(T(z)).$$

Similar equations form, for example, in the problems of the transfer of radiation in the spectral range 3–5 μm, where in daytime the components of solar thermal radiation for the Earth atmosphere are of the same order of magnitude, and also in the problems of transfer of infrared and microwave radiation in clouds and precipitation.

3.5. Complex refraction index. Polarisation of radiation. Stokes parameters

Why polarisation should be taken into account

Previously, we determined the radiation characteristics from the energy viewpoint (intensity, flux) and also the characteristics of interaction of radiation with the medium in which radiation propagates (attenuation, scattering, absorption coefficients, radiation coefficient, etc). In many problems of atmospheric physics it is sufficient, but not in all problems. In the scattering of radiation in the atmosphere and the reflection of the radiation from the surface of the planet it is important to take into account additional properties determined by the electromagnetic nature of radiation – its polarisation. This property of radiation is determined by the fact that the electromagnetic wave is transverse, i.e. the oscillations of the mutually perpendicular directions of the electrical and magnetic fields are in turn perpendicular to the direction of propagation of the electromagnetic wave.

Complex form of writing the equation for electromagnetic waves

For the mathematical description of polarisation, we should use in the initial stage the relationships linked with the complex form of expressing electromagnetic waves. We start with the equation (3.1.1) of a flat electromagnetic wave:

$$E(x,t) = E_0 \cos\left(2\pi\nu t - \frac{2\pi}{\lambda}x + \delta\right). \tag{3.5.1}$$

It should be mentioned that the 'cosine' function in (3.5.1) was selected conventionally. We could have used the sine, changing the initial phase to $\pi/2 + \delta$ or the linear combination of the sine and cosine, since $\sin(x + \delta) = a \sin x + b \cos x$, where $a = \cos \delta$, $b = \sin \delta$. Usually, the specific value of the initial phase is of no interest to us. It is therefore desirable to write the equation for the electromagnetic wave (3.5.1) as a single equation without explicit application of δ. For this purpose, we shall use the Euler equation for the complex exponent:

$$e^{ix} = \cos x + i \sin x. \tag{3.5.2}$$

Using the Euler equation (3.5.2), equation (3.5.1) may be presented in the complex form:

$$E(x,t) = E_0 \exp\left(i(2\pi\nu t - \frac{2\pi}{\lambda}x\right), \tag{3.5.3}$$

where $E(x, t)$ is the complex strength of the electrical field. For the transition from this strength to the real strength having a physical meaning it is sufficient to use the linear combination of the real and imaginary parts (3.5.3) determining δ. Usually, to simplify considerations, one real part is used, i.e. the cosine.

Introducing the complex form (3.5.3), we immediately find another advantage in comparison with the real form (3.5.1). Now we can easily separate the dependence of the strength of the electrical field from the spatial coordinate and time:

$$E(x,t) = E'(x) \exp(i2\pi\nu t), \tag{3.5.4}$$

where

$$E'(x) = E_0 \exp\left(-i\frac{2\pi}{\lambda}x\right) \tag{3.5.5}$$

is the complex amplitude of the strength of the electrical field of the electromagnetic wave. The form of presentation of (3.5.5) is of considerable importance because in optics we usually examine the stationary waves whose amplitude does not change with time. In this case, to analyse the waves, it is sufficient to use only the complex amplitudes $E'(x)$.

We find in the complex form of writing the expression for the intensity of the electromagnetic wave. As agreed, the real part of (3.5.4) has a physical meaning and, consequently, the intensity is proportional to the square of the real part $E(x, t)$. Expanding the part explicitly in respect of (3.1.4), we obtain

$$I = \frac{I}{T}\int_0^T \left[\mathrm{Re}\left(E(x,t)\right)\right]^2 dt = \frac{1}{T}\int_0^T E_0^2\left[\cos\frac{2\pi}{\lambda}x\cos\frac{2\pi}{T}t + \right.$$

$$\left. + \sin\frac{2\pi}{\lambda}x\sin\frac{2\pi}{T}t\right]^2 = E_0\left[\cos^2\frac{2\pi}{\lambda} + \sin^2\frac{2\pi}{\lambda}x\right].$$

As expected, we again obtain the relationship $I = \frac{1}{2}E_0^2$. However, in this case, we obtain a convenient equation for the intensity of radiation through a complex amplitude which will be used widely in future considerations. In fact,

$$E_0\left[\cos^2\frac{2\pi}{\lambda}x + \sin^2\frac{2\pi}{\lambda}x\right] = \left|E'(x)\right|^2 = E'(x)E'^*(x),$$

where * indicates the complex conjugation. Finally:

$$I = E'(x)E'^*(x). \tag{3.5.6}$$

Complex refractive index

The following advantage of application of complex numbers is evident when examining attenuating electromagnetic waves. For these waves, amplitude E_0 in (3.5.1) decreases in space in accordance with the exponential law:

$$E_0(x) = E_0 \exp(-\beta x), \qquad (3.5.7)$$

where β is the attenuation factor.

It will be shown that the relationship (3.5.7) is in complete agreement with the previously introduced Bouguer law. Actually, the intensity of radiation is $I(x)=\frac{1}{2}E_0^2(x)$. Consequently,

$$I(x)=\frac{1}{2}E_0^2\exp(-2\beta x)=I_0\exp(-2\beta x) \qquad (3.5.8)$$

Here $2\beta = \alpha$, where α is the previously introduced volume coefficient of attenuation, and $I=\frac{1}{2}E_0^2$.

We introduce the extinction of the electromagnetic wave into the expression (3.5.5)

$$E'(x)=E_0\exp\left[-i\frac{2\pi}{\lambda}x-\beta x\right] \qquad (3.5.9)$$

or, taking into account the form of (3.1.3)

$$E'(x)=E_0\exp\left[-i\frac{2\pi n\nu}{c}x-\beta x\right], \qquad (3.5.10)$$

where n is the refractive index of the medium.

We transform the expression for the exponent of equation (3.5.10) in the following manner, taking into account $i^2 = -1$:

$$-i\frac{2\pi n\nu}{c}x-\beta x=-i\frac{2\pi n\nu}{c}x+i^2\beta x=$$

$$=-i\left(\frac{2\pi n\nu}{c}-i\beta x\right)x=-i\frac{2\pi\nu}{c}\left(n-i\frac{\beta c}{2\pi\nu}\right)x. \qquad (3.5.11)$$

The expression in the brackets $n-i\frac{\beta c}{2\pi\nu}$ is the complex refractive index (CRI) of the matter of the medium and is denoted by $m = n-i\kappa$. Consequently, the complex amplitude of the attenuating wave is presented in the following form:

$$E'(x) = E_0 \exp\left[-i\frac{2\pi\nu m}{c}x\right],$$

i.e. it is completely identical with (3.1.3), but the refractive index of the medium should now be used in the complex form. Thus, the meaning of introducing the complex refractive index is the uniformity of the forms of writing the equations of electromagnetic waves in the media with and without absorption. The real part of the CRI is the 'usual' refractive index. Its imaginary part characterises the absorption of radiation in the medium. The imaginary part of the CRI $\kappa = \frac{\beta c}{2\pi\nu}$ is associated with the volume coefficient of molecular absorption of the medium k. Actually, from the equations presented previously (3.5.8)–(3.5.11) we obtain:

$$\kappa = k\frac{c}{4\pi\nu}. \tag{3.5.12}$$

The concept of radiation polarisation

Usually, the phenomenon of radiation polarisation is described by the vector form of the electromagnetic wave. Using as the model of the wave the displacement of the entire curve $y = A \sin (kx)$ along X with a constant velocity v (to simplify considerations, it should be assumed that the sinusoid is produced from a wire). In the two-dimensional case, the characteristic of the wave is represented by its amplitude A. However, in a three-dimensional space, the sinusoid with the same amplitude can be displaced along the axis X in different manners. For example, the sinusoid can be placed in different planes inclined in relation to the plane XY (Fig. 3.10a). The movement of the sinusoid along the axis X can be combined with rotation around the axis (Fig. 3.10b), and there are two directions of rotation – to the right and to the left. Finally, in rotation around the axis the sinusoid can also be compressed in one of the planes (Fig. 3.10c). Thus, it is clear that the waves with the same amplitude in the three-dimensional space may greatly differ. Consequently, one amplitude is not sufficient for describing the wave in space and it is necessary to introduce additional characteristics. The combination of these additional characteristics also determines the polarisation of the wave.

The case in which the sinusoid is situated in one plane (Fig. 3.10a) corresponds to linear polarisation, and the plane in which the sinusoid is located is the polarisation plane. The case of rotation of

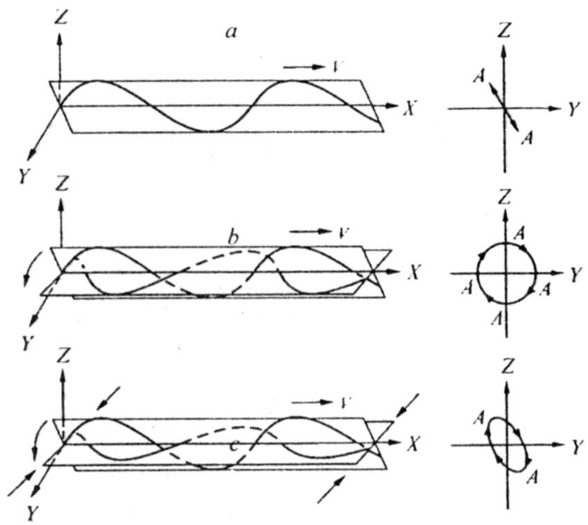

Fig.3.10. Model of a wave in space: a) movement in a single plane; b) movement with rotation; c) movement with rotation and compression.

the sinusoid without compression (Fig. 3.10b) is circular polarisation, and, depending on the direction of rotation, there is right and left polarisation. The case of rotation of the sinusoid with compression (Fig. 3.10c) corresponds to elliptical polarisation. This is a general case because circular polarisation is a partial case of elliptical polarisation, and the linear polarisation is a degenerate case (when the ellipse degenerates into a segment in the case of maximum compression).

Polarisation is significant when anisotropy is detected in the medium in which the wave propagates, i.e. the properties of the medium differ depending on the direction in space. Continuing the example with the movement of the sinusoid, we examine a conventional case in which the space contains a 'wall' in the *XZ* plane. Consequently, the only possible position of the sinusoid is the movement parallel to the 'wall', i.e. the wave is linearly polarised in the *XZ* plane.

Stokes parameters

Let us pass from the flat electromagnetic wave to a three-dimensional one. Equation (3.5.4) retains its form but the strength of the electrical field is now a vector (complex!) in the *YZ* plane:

$$\mathbf{E}'(x, t) = \mathbf{E}'(x) \exp(i2\pi\nu t).$$

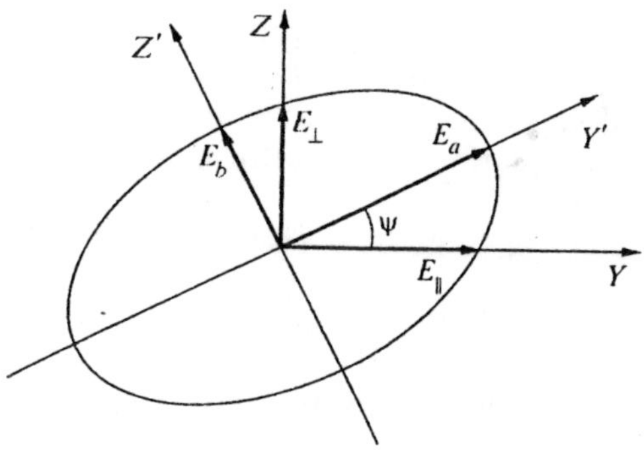

Fig.3.11. Polarisation ellipse.

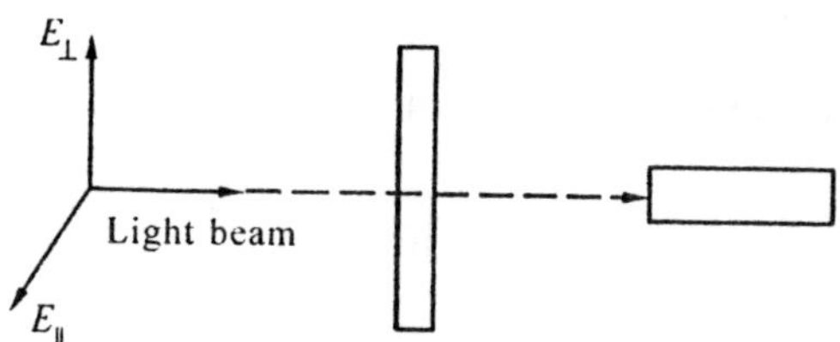

Fig.3.12. Determination of the Stokes parameters.

In further considerations, the prime and the vector at the complex amplitude will not be mentioned.

Vector E in the plane YZ describes, as mentioned previously when examining polarisation, the ellipse (Fig. 3.11). Its position is completely determined by the major and minor half-axes E_a and E_b and the angle ψ counted between the axis Y and the major half-axis of the ellipse. However, similar parameters of the description of the three-dimensional wave are not convenient for both experimental measurements and theoretical analysis because they are not homogeneous (they have different dimensions). In the experiments, the polarisation characteristics are usually measured during the passage of light through polarisers, i.e., special devices (crystal plates), transmitting the light polarised only in a specific manner. Therefore, we introduce the vector of the Stokes parameters – the vector of four real components (I, Q, U, V) which are determined in four (hypothetical) experiments in accordance with the scheme shown in Fig. 3.12 [6].

Experiment 1: there are no polarisers. Consequently, the first Stokes parameter is radiation intensity I. Its definition for the vector wave requires refinement. Actually, in this case, the square of the electrical vector should be averaged out not only in respect of time but also the spatial coordinates YZ. We examine the system of the coordinates $Y'Z'$ linked with the main axis of the polarisation ellipse (Fig. 3.11). The equation of the ellipse in the system is:

$$y' = E_a \cos\gamma,\; z' = E_b \sin\gamma,$$

where γ is the parameter that changes from 0 to 2π. We are interested in the square of the distance from the origin of the coordinates to the point of the ellipse in rotation through the angle γ:

$$y'^2 + z'^2 = E_a^2 \cos^2\gamma + E_b^2 \sin^2\gamma,$$

and averaging this parameter in respect of all angles, we obtain

$$\frac{1}{2\pi}\int_0^{2\pi}\left(E_a^2\cos^2\gamma + E_b^2\sin^2\gamma\right)d\gamma = E_a^2 + E_b^2.$$

Using this expression instead of E_a in derivation of (3.5.6), we obtain the same equation (3.5.6) but in this case the modulus refers to the modulus of the complex vector, and the square of the modulus, as indicated by linear algebra, is equal to the scalar product of the vector by the complex conjugate vector. Consequently

$$I = (E_a E_a^* + E_b E_b^*). \tag{3.5.13}$$

However, it is convenient to use the expression for intensity not through the half-axis of the ellipse E_a and E_b, but through the projection of the vector of the electrical field in any rectangular coordinate system YZ, rotated in relation to the main axes through the angle ψ (Fig. 3.11). These projections are denoted by $E_{\|}$ and $E_{\perp}$, and the coordinates of the projections in $Y'Z'$ are expressed by means of the half-axes of the ellipse:

$$\mathrm{E}_{\|} = (E_a\cos\psi,\ E_b\sin\psi),\ E_{\perp} = (E_a\sin\psi,\ E_b\cos\psi). \tag{3.5.14}$$

Equation (3.5.14) shows that:

$$E_a E_a^* + E_b E_b^* = E_{\|} E_{\|}^* + E_{\perp} E_{\perp}^*,$$

from which we finally obtain the expression for the intensity of radiation

$$I = (E_{\|}\mathrm{E}_{\|}^* + E_{\perp} E_{\perp}^*). \tag{3.5.15}$$

Since the angle ψ is selected arbitrarily, the equation (3.5.15) is a strict proof of the physically evident claim that the intensity is independent of the selection of the system of the coordinates in the plane normal to the direction of propagation of radiation.

Experiment 2: the horizontal and vertical polarisers. Initially, we measure the intensity passed through the horizontal polariser and, subsequently, through the vertical polariser. The second Stokes parameter Q is the difference of these intensities. Since only $E_{\|}$ passes through the horizontal polariser, and $E_{\perp}$ through the vertical polariser, from equation (3.5.6) we directly obtain:

$$Q = (E_{\|}E^{*}_{\|} - E_{\perp}E^{*}_{\perp}), \tag{3.5.16}$$

Experiment 3: polarisers, rotated through +45 and –45° in relation to the horizontal. Initially, we measure the intensity passed through the first polariser and subsequently through the second polariser. The third Stokes parameter U is the difference of these intensities. Each of these polarisers transmits only the component of the vector of the electrical field in the direction of its rotation. To determine the projection of the electrical vector on the direction +45°, it is taken into account that the projection of the vector to the direction is equal to its scalar product by the unit vector in the given direction or, which is the same, to the scalar product by any vector of the given direction divided by its length. The direction +45° has the coordinates (1, 1) and, consequently, the projection to be determined is $\frac{E_{\|}+E_{\perp}}{\sqrt{1+1}}=\frac{1}{\sqrt{2}}(E_{\|}+E_{\perp})$. Similarly, the coordinates of the direction –45° are (1,–1), and the projection $\frac{1}{\sqrt{2}}(E_{\|}+E_{\perp})$. For the difference of the intensities – parameter U, we obtain

$$U=\frac{1}{\sqrt{2}}(E_{\|}+E_{\perp})\frac{1}{\sqrt{2}}(E_{\|}+E_{\perp})^{*}-\frac{1}{\sqrt{2}}(E_{\|}-E_{\perp})\frac{1}{\sqrt{2}}(E_{\|}-E_{\perp})^{*}$$

and, finally

$$U = (E_{\|}E^{*}_{\|} + E_{\perp}E^{*}_{\perp}). \tag{3.5.17}$$

Experiment 4: the right and left circular polarisers. Initially, we measure the intensity passed through the right polariser and, subsequently, through the left polariser. The fourth Stokes parameter

V is the difference of these intensities. The work of the circular polarisers is not so distinctive as that of the linear ones and, consequently, the formal-mathematical approach may be useful in this case. The circular polarisers 'produce' from the initial wave a wave with the respectively right and left circular polarisation. In the set of the complex numbers, the circle may be regarded as a vector with the coordinates $(1,i)$. Actually, any complex number is geometrically interpreted as the point of the circle whose angle of rotation is determined by the real (projection on the vector (1, 0)) and imaginary (projection on the vector $(0, i)$) parts (remember the concept of the modules and the argument of the complex number). The right rotation corresponds to the vector $(1, i)$, the left rotation to the vector $(1, -i)$. Further steps are easy: it is necessary to find the projections of the electrical vector on the given circles remembering that the scalar product of the vectors a and b in the set of the complex numbers is defined as ab^*, and the length of the vector as $\sqrt{aa^*}$. The right polariser transmits

$$\frac{E_{\parallel}\cdot 1 + E_{\perp}\cdot(-i)}{\sqrt{1\cdot 1 + i(-i)}} = \frac{1}{\sqrt{2}}(E_{\parallel} - iE_{\perp}),$$

and the left polariser

$$\frac{E_{\parallel}\cdot 1 + E_{\perp}\cdot(i)}{\sqrt{1\cdot 1 + i(-i)}} = \frac{1}{\sqrt{2}}(E_{\parallel} + iE_{\perp}).$$

For the parameter V, according to (3.5.6), we obtain

$$V = -\frac{1}{\sqrt{2}}(E_{\parallel} - iE_{\perp})\frac{1}{\sqrt{2}}(E_{\parallel} - iE_{\perp})^* -$$

$$-\frac{1}{\sqrt{2}}(E_{\parallel} + iE_{\perp})\frac{1}{\sqrt{2}}(E_{\parallel} + iE_{\perp})^*.$$

Consequently

$$V = i(E_{\parallel}E_{\perp}^* - E_{\parallel}^*E_{\perp}). \tag{3.5.18}$$

It should be noted that, regardless of the complex form of writing (3.5.15)–(3.5.18), all the Stokes parameters are real numbers. They fully determine polarised radiation. In the group of the Stokes parameters, only three parameters are independent, because from (3.5.15)–(3.5.18) we obtain directly the identity:

$$I^2 = Q^2 + U^2 + V^2, \tag{3.5.19}$$

according to which the electromagnetic wave is unambiguously determined by three elliptical parameters.

Natural and polarised radiation. Degree of polarisation

We have examined the characteristics of polarisation for the idealised case of a monochromatic electromagnetic wave from a point radiation source. The actual emitter (for example, the Sun) may be treated as a set of point sources. In the process of emission, the radiation of the electromagnetic wave from each point of the source is independent (non-coherent) and, consequently, the polarisation ellipses are oriented randomly. This radiation is regarded as non-polarised or natural. Carrying out conceptual experiments with the measurement of the Stokes parameters for non-polarised radiation, and taking into account the averaging of random orientation in respect of time, we evidently obtain zeros for all differences of intensity. Thus, for non-polarised radiation, the Stokes vector is $(I, 0, 0, 0)$, where I is the intensity of non-polarised radiation. (Strictly speaking, a new notation should be used for this intensity, which differs from I in equation (3.5.19)).

In opposite to natural radiation, the radiation for which all the polarisation ellipses are oriented in the same manner, is referred to as completely polarised. A general case is partially polarised radiation which may be regarded as a mixture of natural and completely polarised radiations. For completely polarised radiation the equation (3.5.19) is fulfilled and it can be described using three Stokes parameters Q, U and V. Single intensity I is sufficient for describing non-polarised radiation. Thus, to describe any radiation (partially polarised), it is necessary to use all four Stokes parameters which are independent in a general case.

For any radiation, the ratio of the parts of the equation (3.5.19) is referred to as the degree of polarisation:

$$P = \frac{\sqrt{Q^2 + U^2 + V^2}}{I}. \tag{3.5.20}$$

(Again, it should be mentioned that the quantities in (3.5.20) relate to different radiation in comparison with equation (3.5.19) and, therefore, strictly speaking, different notations should be used for them). For the completely polarised radiation $P = 1$, for natural radiation $P = 0$, for partially polarised radiation $0 < P < 1$, and as

P comes closer to 1, the intensity of polarisation of radiation increases. The degree of polarisation is usually expressed in percent. The relationship (3.5.20) is not always suitable in practice because it is necessary to measure all Stokes parameters. A simpler procedure is used in many cases: we measure the dependence of the intensity of radiation during its passage through a linear polariser in relation to the angle of rotation of the polariser, making a full rotation, and calculate:

$$P_l = \frac{I_{max} - I_{min}}{I_{max} + I_{min}}, \qquad (3.5.21)$$

where I_{max} and I_{min} is the maximum and minimum value of the intensity. Value P_l is evidently determined by the difference between the major and minor half-axes of the ellipse. It is also frequently referred to as the degree of polarisation but to avoid confusion, it is better to use the more accurate term 'the degree of linear polarisation' (or the degree of linearity of polarisation). Actually, for completely linearly polarised radiation $P_l = 1$, but for elliptically polarised radiation $P_l < 1$, and for circular polarisation generally $P_l = 0$. Therefore, for any radiation (3.5.21) coincides with (3.5.20) only in the case of its linear polarisation.

3.6. Radiative transfer equation taking polarisation into account

Interaction matrix

Since the Stokes parameters are defined as the linear combinations of intensities, any variation of the state in elementary interaction with an optical element (reflector, scatterer, etc.) is also a linear combination. This gives an extremely useful body of mathematics:

$$\begin{pmatrix} I \\ Q \\ U \\ V \end{pmatrix} = \begin{pmatrix} D_{11} & D_{12} & D_{13} & D_{14} \\ D_{21} & D_{22} & D_{23} & D_{24} \\ D_{31} & D_{32} & D_{33} & D_{34} \\ D_{41} & D_{42} & D_{43} & D_{44} \end{pmatrix} \begin{pmatrix} I_0 \\ Q_0 \\ U_0 \\ V_0 \end{pmatrix}, \qquad (3.6.1)$$

where (I_0, Q_0, U_0, V_0) is the Stokes vector prior to interaction; (I, Q, U, V) is the Stokes vector after interaction; $\mathbf{D}_{ij}$ is the interaction matrix for which the term 'the Mueller matrix' is sometimes used [6].

Let the complex projections of the components of the electrical vectors after the interaction $E_{\|}$ and $E_{\perp}$ be linked with the components prior to interaction $E_{0,\|}$ and $E_{0,\perp}$ by the relationships:

$$E_{\|} = S_1 E_{0,\|} + S_2 E_{0,\perp},\ E_{\perp} = S_3 E_{0,\|} + S_4 E_{0,\perp}, \tag{3.6.2}$$

where S_1, S_2, S_3, S_4 are some coefficients (complex in a general case). Consequently, substituting (3.6.2) into the definitions of the components of the Stokes vector (3.5.15)–(3.5.18), dividing the products into 'halves', and grouping the members in the right-hand part, we easily obtain the expressions for the elements of the interaction matrix (3.6.1) through the coefficients S_1–S_4 of the transformation of the complex vectors [6]. However, we are interested only in the partial case (3.6.2) in which the parallel and perpendicular components of the vectors are transformed independently, i.e.

$$E_{\|} = S_1 E_{0,\|},\ E_{\perp} = S_4 E_{0,\perp}, \tag{3.6.3}$$

Carrying out the previously described procedure for (3.6.3), we obtain that the interaction matrix in this case has the following form:

$$\begin{pmatrix} D_{11} & D_{12} & 0 & 0 \\ D_{12} & D_{11} & 0 & 0 \\ 0 & 0 & D_{33} & D_{34} \\ 0 & 0 & -D_{34} & D_{33} \end{pmatrix}, \tag{3.6.4}$$

where

$$D_{11} = \frac{1}{2}\left(S_1 S_1^* + S_4 S_4^*\right),\ D_{12} = \frac{1}{2}\left(S_1 S_1^* - S_4 S_4^*\right),$$
$$D_{33} = \frac{1}{2}\left(S_1^* S_4 + S_1 S_4^*\right), D_{34} = \frac{i}{2}\left(S_1^* S_4 - S_1 S_4^*\right). \tag{3.6.5}$$

Thus, in the examined partial case, the interaction matrix is determined by only four parameters. Attention should be given to the fact that the expressions for the parameters (3.6.5) are mathematically equivalent to the definitions of the Stokes parameters (3.5.15)–(3.5.18). In particular, the following identity is valid

$$D_{11}^2 = D_{12}^2 + D_{33}^2 + D_{34}^2, \tag{3.6.6}$$

and, consequently, only three parameters are independent.

The scattering matrix

One of the most important cases of interaction of radiation with matter for the atmospheric optics is the scattering of radiation. We examine this phenomenon in greater detail. In this case, the matrix (3.6.1) is referred to as the scattering matrix. Its elements depend on the direction of scattering **r**. Examples of the scattering matrices (molecular and aerosol) will be presented in chapter 5 and here we examine only the associated general concepts.

We examine the scattering of the initial non-polarised radiation. The Stokes vector for the radiation is (I_0, 0, 0, 0), where I_0 is the intensity prior to scattering. Substituting this relationship into (3.6.1), we obtain $I(\mathbf{r}) = D_{11}(\mathbf{r})I_0$. However, this means that $D_{11}(\mathbf{r})$ has the meaning of the cross-section of directional scattering, if we are discussing scattering on a single particle. Usually (as explained in paragraph 3.3), it is convenient to separate the scattering cross-section and the scattering phase function, i.e.

$$D_{11}(\mathbf{r}) = C_{sd}(\mathbf{r}) = \frac{1}{4\pi} x(\mathbf{r}) C_s.$$

However, the same procedure can also be used in the general case, writing

$$\begin{pmatrix} D_{11}(\mathbf{r}) & D_{12}(\mathbf{r}) & D_{13}(\mathbf{r}) & D_{14}(\mathbf{r}) \\ D_{21}(\mathbf{r}) & D_{22}(\mathbf{r}) & D_{23}(\mathbf{r}) & D_{24}(\mathbf{r}) \\ D_{31}(\mathbf{r}) & D_{32}(\mathbf{r}) & D_{33}(\mathbf{r}) & D_{34}(\mathbf{r}) \\ D_{41}(\mathbf{r}) & D_{42}(\mathbf{r}) & D_{43}(\mathbf{r}) & D_{44}(\mathbf{r}) \end{pmatrix} =$$

$$= \frac{1}{4\pi} C_s \begin{pmatrix} d_{11}(\mathbf{r}) & d_{12}(\mathbf{r}) & d_{13}(\mathbf{r}) & d_{14}(\mathbf{r}) \\ d_{21}(\mathbf{r}) & d_{22}(\mathbf{r}) & d_{23}(\mathbf{r}) & d_{24}(\mathbf{r}) \\ d_{31}(\mathbf{r}) & d_{32}(\mathbf{r}) & d_{33}(\mathbf{r}) & d_{34}(\mathbf{r}) \\ d_{41}(\mathbf{r}) & d_{42}(\mathbf{r}) & d_{43}(\mathbf{r}) & d_{44}(\mathbf{r}) \end{pmatrix}, \tag{3.6.7}$$

where the matrix $\mathbf{d}_{ij}$ is the normalised scattering matrix, determined from $\mathbf{D}_{ij}$ by dividing all elements by the constant $\frac{1}{4\pi}\int_{4\pi} D_{11}(\mathbf{r})d\Omega$ for fulfilling the condition of normalisation of the phase function (3.3.13). The identical relationship with the replacement of the cross-section C_s by the volume scattering coefficient σ is also written for the scattering by the elementary volume. Actually, the scattering matrices are also calculated in the normalised form in practice. The element d_{11} of the normalised scattering matrix is the scattering phase function for non-polarised radiation. Of course, all these definitions remain in force for the partial type of the scattering matrix (3.6.4).

Derivation of the transfer equation taking polarisation into account

As in paragraphs 3.3 and 3.4, we examine the interaction of radiation with an elementary volume. It is assumed that in the initial stage only the processes of weakening of radiation take place in the volume and there are no processes of atmospheric radiation. **L** denotes the Stokes vector of radiation falling on the elementary volume. The variation $d\mathbf{L}$ after passage of the element dl of the path is linked in a linear manner, according to (3.6.1), with **L** and can be determined in the form of the product of the initial vector **L** by some interaction matrix $\mathbf{D}(dl)$:

$$d\mathbf{L} = \mathbf{D}(dl)\mathbf{L}, \tag{3.6.8}$$

where the form of $\mathbf{D}(dl)$ denotes the dependence of the interaction matrix **D** on the path dl.

The relationship (3.6.8) should be fulfilled for any radiation, including non-polarised radiation. Substituting the Stokes vector $(I, 0, 0, 0)$ into (3.6.8), we obtain $dI = D_{11}(dl)I$, and from the transfer equation without taking polarisation into account (3.4.3), we obtain $D_{11}(dl) = -\alpha dl$, where α is the volume coefficient of extinction. The matrix **D** in homogeneous, i.e., the dimensions of all the elements are the same. Consequently, like D_{11}, all these elements should be the product of some quantity with the dimensions of inverse length by dl. removing dl from the matrix, we can write $\mathbf{D}(dl) = -\mathbf{A}dl$, and

$$d\mathbf{L} = -\mathbf{A}\mathbf{L}dl, \tag{3.6.9}$$

where **A** is some matrix which is referred to as the attenuation matrix (since $A_{11} = \alpha$).

Let us pass to the general case in which atmospheric radiation is present in the elementary volume. The required transfer equation in the case of non-polarised radiation should change to (3.4.3). This means that in this case ε should be added to $d\mathbf{L}$ in (3.6.9). Because of the homogeneity of the Stokes parameters, generalisation of this can be only the addition of some vector **E** to the right-hand part of (3.6.8); the sector is referred to as the radiation vector (since $E_1 = \varepsilon$).

Finally, we obtain the transfer equation with polarisation taken into account:

$$\frac{d\mathbf{L}}{dl} = -\mathbf{AL} + \mathbf{E}. \tag{3.6.10}$$

It should be mentioned that (3.6.10) is equivalent to the equation with polarisation not taken into account (3.4.3), if all the scalars are replaced by the appropriate vectors and the matrices. Equation (3.4.3) is often referred to as the scalar transfer equation, and the more general equation (3.6.10) as the vector transfer equation.

The structure of the attenuation matrix and radiation vector

As explained in paragraph 3.3, the attenuation process includes absorption and scattering. Therefore, the attenuation matrix will be divided into the absorption matrix and the scattering matrix. Absorption by the atmospheric gases may be regarded as isotropic, i.e. identical for all orientations of the polarisation ellipse.* Examining the passage of radiation through the elementary volume in which only absorption takes place, we obtain the identical decrease of all intensities, included in the definition of the Stokes vector. However, this means that the absorption matrix is diagonal, and all the elements located on the diagonal are equal. Since the first element of the matrix is equal to the volume coefficient of absorption k, the absorption matrix is the product of the volume coefficient of absorption by the unit matrix. Similarly, the scattering matrix is the product of the volume coefficient of scattering σ by the normalised scattering matrix **d** (3.6.7).

Thus,

$$\mathbf{A} = \alpha \,\|1\| + \sigma\mathbf{d}, \tag{3.6.11}$$

where $\|1\|$ is the unit matrix.

For thermal radiation, assuming that it is isotropic, all the

*Absorption anisotropy effects are found in the upper atmosphere in the microwave range as a result of the Zeeman phenomenon – splitting of the levels of the molecule energies in the Earth's magnetic field.

intensities included in the definition of the Stokes vector show the same increase after passage through the elementary volume. However, this means that none of the differences of the intensities change as a result of radiation. Consequently, in this case, the first element of the radiation vector is the volume coefficient of radiation, and the remaining elements are equal to zero.

However, for additional radiation, caused by the scattering of radiation, the situation is already different. Here, we have the integral identical to the scalar case (3.4.35), but the integral already contains the Stokes vector and the scattering matrix, and also additional rotational matrices re-calculating the Stokes vector for every scattering azimuth (we will not give the explicit form of this equation which is presented, for example, in [47]).

As mentioned previously, the scattering of radiation is a highly anisotropic process in the transfer of radiation in the atmosphere. For this process the matrix is not diagonal. This will be proven in chapter 5. In chapter 6 we show the anisotropy of reflection of radiation from the surface.

From the possibility of writing the attenuation matrix in the form (3.6.11), the coincidence of the element A_{11} with the appropriate scalar characteristics, and the homogeneity of the matrix, we obtain the same consequences as for the scalar quantities. Namely: the possibility of the addition of the attenuation matrices of the particles of different types, in particular, calculated separately for the molecular and aerosol characteristics – the matrix analog of the relationships (3.3.21)–(3.3.22), and the possibility of addition of the normalised scattering matrices with the weights equal to the ratios of the volume coefficients of scattering – the matrix analog of the relationship (3.4.34).

The problem of taking into account polarisation in atmospheric optics problems

For the scalar transfer equation we obtained, in some partial cases, relatively simple analytical solutions. For the vector equation, if its parameters change with altitude in the atmosphere and the matrix **A** is not diagonal, we cannot obtain the general analytical solution even for the simplest case of calculating direct polarised radiation, i.e., the analog of the Bouguer law for polarised radiation is not available. Of course, there are no analytical solutions for the equations of transfer of polarised radiation with scattering taken into account. Investigations of the processes of transfer of polarised radiation in the scattering and absorbing atmosphere are a very complicated and cumbersome task. However, the importance of solving this problem has recently increased greatly because of the natural desire to use the measurements of the characteristics of polarisation for obtaining additional information on different parameters of the atmosphere and the underlying surface.

It should also be mentioned that the examination of the processes of multiple scattering of radiation without polarisation taken into account is approximate, even if we are interested only in the intensity of radiation. Actually, according to (3.6.1), natural radiation after the first scattering becomes partially polarised, and in subsequent scattering its intensity depends no longer only on I but also on other components of the Stokes vector. Consequently, the second and subsequent scattering of radiation is not determined only by the phase function but by the entire scattering matrix. Therefore, the classic theory of the transfer of non-polarised radiation taking multiple scattering into account (the fundamentals of the theory are examined in chapter 8) is only approximate description of the scattering phenomenon, regardless of its complicated nature.

Simplified description of polarisation

As mentioned previously, the role of polarisation is important in the processes of scattering and reflection of radiation. In many cases, the matrices of scattering and reflection have a simple structure (3.6.4). As shown in chapters 5 and 6, such matrices are the matrix of molecular scattering, the matrix of scattering of a spherical aerosol particles and the reflection matrix. For the matrices of the structure at the constant system of the coordinates, the components (I, Q) and (U, V) of the Stokes vector will be transformed independently. However, this means that if the initial radiation is natural ($\mathbf{L}$ = (I, 0, 0, 0)) then in the multiplication by similar matrices, only the components (I,Q) of the Stokes vector will be non-zero. Consequently, we can restrict ourselves to examination of the Stokes vector consisting of two components (I, Q), the attenuation matrix, the scattering matrix and the reflection matrix with the dimension 2 × 2 and the two-component radiation vector [22].

The vector transfer equation (3.6.10) is a system of linear differential equations. In the examined simplified case there are two such equations, for I and Q. The standard procedure of solving the system of linear differential equations is the reduction of the matrix of the system to the diagonal form. Consequently, the system breaks down into independent equations. To carry out this in the present case, it is important to mention some assumptions of linear algebra.

The transition from one system of coordinates to another is a linear transformation, i.e.

$$\mathbf{X}' = \mathbf{TX},$$

where **X** are the coordinates of the vector in the initial system; **X'** are the coordinates in the new system; **T** is the transition matrix. If the vectors **X** and **Y** in the initial coordinates are linked by the relationship

$$\mathbf{Y} = \mathbf{BX},$$

then in the new coordinates

$$\mathbf{Y}' = \mathbf{B}'\mathbf{X}', \text{ where } \mathbf{B}' = \mathbf{TBT}^{-1}.$$

There is a basis in which the matrix **B'** is diagonal and it is referred to as the eigenbasis. The columns of the matrix **T** of reduction to the eigenbasis are the eigenvectors of the matrix **B**, and eigen numbers **B** are located on the diagonal **B'**. The eigen numbers λ of the matrix **B** are determined from the equation for the determinant:

$$\left|\mathbf{B} - \lambda \|\mathbf{1}\|\right| = 0,$$

the matrix with the dimension n has n eigen numbers. The eigenvectors **t** are determined from the equations

$$\mathbf{Bt} = \lambda\mathbf{t}.$$

The matrix consisting of the eigenvectors is orthogonal, i.e. $\mathbf{T}^{-1} = \mathbf{T}^{+}$, where $\mathbf{T}^{+}$ is the transposed matrix.

We now find the transformation of the Stokes vector, reducing the attenuation matrix **A** for two-component polarisation to the diagonal form. The matrix **A** has the form (3.6.4)

$$\begin{pmatrix} a_1 & a_2 \\ a_2 & a_1 \end{pmatrix}.$$

The equation for its eigen numbers

$$(a_1 - \lambda)\,(a_1 - \lambda) - a^2_2 = 0$$

gives $\lambda_1 = a_1 + a_2$, $\lambda_2 = a_1 - a_2$. The eigenvector for λ_1 is determined from the condition:

$$a_1 t_{11} + a_2 t_{21} = (a_1 + a_2) t_{11},$$

$$a_2 t_{11} + a_1 t_{21} = (a_1 + a_2) t_{21},$$

which gives $t_{11} = t_{21}$, i.e. the elements of the first column of the matrix **T** are identical. Similarly, for the second eigenvector $t_{12} = -t_{22}$. Thus, the required transformation matrix has the form

$$\mathbf{T} = \begin{pmatrix} t_1 & t_2 \\ t_1 & -t_2 \end{pmatrix}.$$

The matrix is determined with the accuracy to constant multipliers t_1 and t_2. In linear algebra, the eigenvectors should have the unit length from which $t_1 = t_2 = 1/\sqrt{2}$. However, in this case, it is convenient to violate the formal requirements and select the normalisation of the eigenvectors on the basis of the physical meaning. We set $t_1 = t_2 = t$. Prior to transformation we have the natural light with the Stokes vector $(I, 0)$. After transformation, we obtain $(t\,(I + Q),\ t\,(I - Q))$. According to the definition of the Stokes parameters (3.5.15), (3.5.16)

$$t(I + Q) = 2tE_{\parallel}E^{*}_{\parallel},\quad t(I - Q) = 2tE_{\perp}E^{*}_{\perp}.$$

However, $E_{\parallel}E^{*}_{\parallel}$ is the intensity transmitted by the horizontal polariser. It will be denoted by $I_{\parallel}$. Similarly, $E_{\perp}E^{*}_{\perp}$ is the intensity transmitted by the vertical polariser. This means that, taking into account the physical meaning, it is convenient to set $t = 1/2$ [22]. It is now required to find the diagonal of the attenuation matrix in this normalisation:

$$\begin{pmatrix} \frac{1}{2} & \frac{1}{2} \\ \frac{1}{2} & -\frac{1}{2} \end{pmatrix} \begin{pmatrix} a_1 & a_2 \\ a_2 & a_1 \end{pmatrix} \begin{pmatrix} \frac{1}{2} & \frac{1}{2} \\ \frac{1}{2} & -\frac{1}{2} \end{pmatrix} =$$

$$= \begin{pmatrix} \frac{1}{2}(a_1 + a_2) & \frac{1}{2}(a_2 + a_1) \\ \frac{1}{2}(a_1 - a_2) & -\frac{1}{2}(a_2 - a_1) \end{pmatrix} \begin{pmatrix} \frac{1}{2} & \frac{1}{2} \\ \frac{1}{2} & -\frac{1}{2} \end{pmatrix} =$$

$$= \begin{pmatrix} \frac{1}{2}(a_1 + a_2) & 0 \\ 0 & \frac{1}{2}(a_1 - a_2) \end{pmatrix}.$$

Finally: the attenuation matrix becomes diagonal and has the following form:

$$A = \begin{pmatrix} \frac{1}{2}(a_1 + a_2) & 0 \\ 0 & \frac{1}{2}(a_1 - a_2) \end{pmatrix} \qquad (3.6.12)$$

in transition from the Stokes parameters (I,Q) to the new parameters $(I_{\parallel}, I_{\perp})$ using the equations:

$$I_{\parallel} = \frac{1}{2}(I + Q), \; I_{\perp} = \frac{1}{2}(I - Q). \qquad (3.6.13)$$

The new parameters $(I_{\parallel}, I_{\perp})$ have the meaning of the intensities in mutually perpendicular directions, and $I = I_{\parallel} + I_{\perp}$.

After this transformation, the vector transfer equations breaks down into a pair of scalar equations and for these equations we can obtain explicit analytical solutions for the case of attenuating of the light, the transfer of thermal radiation in the atmosphere taking into account the reflection from the surface and a number of other phenomena in which it is not necessary to calculate scattered radiation.

CHAPTER 4

MOLECULAR ABSORPTION IN THE ATMOSPHERE

4.1. The general characteristic of molecular absorption in the atmosphere of the Earth

In the previous chapters, we introduced, mostly formally, different characteristics of the interaction of radiation with a medium: the coefficients (and cross-sections) of weakening and scattering, the scattering indicatrix, emission coefficient. In this chapter and chapter 5 we examine specific physical mechanisms, determining these processes, so that it will be possible to determine the dependence of different characteristics of the interaction on the parameters of the physical state of the atmosphere – temperature, pressure, concentration of different gas components, the number of aerosol particles, etc. These relationships are important for calculating different characteristics of the radiation field in the atmosphere of planets (chapters 7–9), and also for solving the inverse problems of atmospheric physics (chapter 10).

It has been established that the extinction of emission is caused by two processes – absorption and scattering, and shown the principal differences between these two mechanisms of extinction from the viewpoint of transformations of radiation energy. In the absorption of emission by atoms, molecules or particles, the emission energy transfers to the internal energy of the medium, i.e. the photon 'dies'. In scattering, the photon changes the direction of its movement but there are no significant transitions of energy from the radiation field to the energy of the atoms, molecules or particles. And the exception is the case of Raman scattering (chapter 5).

Of fundamental importance in many problems of atmospheric physics is one of the mechanisms of attenuation of radiation– absorption of radiation by atoms and molecules, included in the composition of the atmosphere of the planets. To simplify further considerations, this mechanism will be referred to as molecular absorption, although absorption of different atoms plays also a certain role in the atmospheres of the planets [20, 24, 27, 33, 37, 50, 102].

Spectrum of molecular absorption of the Earth's atmosphere

Above all, we shall characterise the absorption spectrum of the atmosphere of the Earth referring to the spectral behaviour of the transmittance function $P(\lambda)$ (3.4.10) or the absorption function $A(\lambda) = 1 - P(\lambda)$ of different atmospheric layers. More detailed information on the quantitative characteristics of molecular absorption is presented in paragraph 4.6.

Figure 4.1 shows the spectral behaviour of the absorption function in a wide range of wavelength from 0.1 to 100 μm [91]. The upper part of the figure (Fig. 4.1a) shows the curves of absolutely black radiation for two temperatures: T = 6000 K and T = 250 K. The first temperature corresponds to the radiation of the Sun, the second temperature to the radiation of the Earth's atmosphere. These curves (they are normalised for the maximum values of quantity $B(\nu, T)$) clearly demonstrate the spectral ranges of transfer of solar and atmospheric radiation. As mentioned previously (chapter 1), the energy of solar radiation in the Earth's atmosphere is concentrated mainly in the ultraviolet, visible and near infrared ranges of the spectrum, and the energy of atmospheric radiation in the middle infrared range. It should be mentioned that in this case we are not concerned with the intensity of radiation, but with the radiation flux. (The intensity of solar radiation is many times greater than the intensity of radiation of the atmosphere or surface in any spectral range).

The middle part of the figure (Fig. 4.1b) shows the absorption functions (from 0 to the absence of absorption in the atmosphere, 100%-complete absorption of radiation) of the entire thickness of the atmosphere for the zenith angle of 40°, and Fig. 4.1c shows this for the layer of the atmosphere from its upper boundary to an altitude of 11 km. It should be stressed that the given absorption functions are related to the case of the atmosphere without any clouds. The graph shows that, like the entire atmosphere, part of the atmosphere above the tropopause also completely absorbs solar ultraviolet radiation. This is caused by the absorption mostly by molecules of oxygen and ozone playing the role of a shield for the biosphere against the radiation of high-energy photons, and also an important role in the formation of the profile of temperature in the stratosphere and mesosphere. More detailed information on the

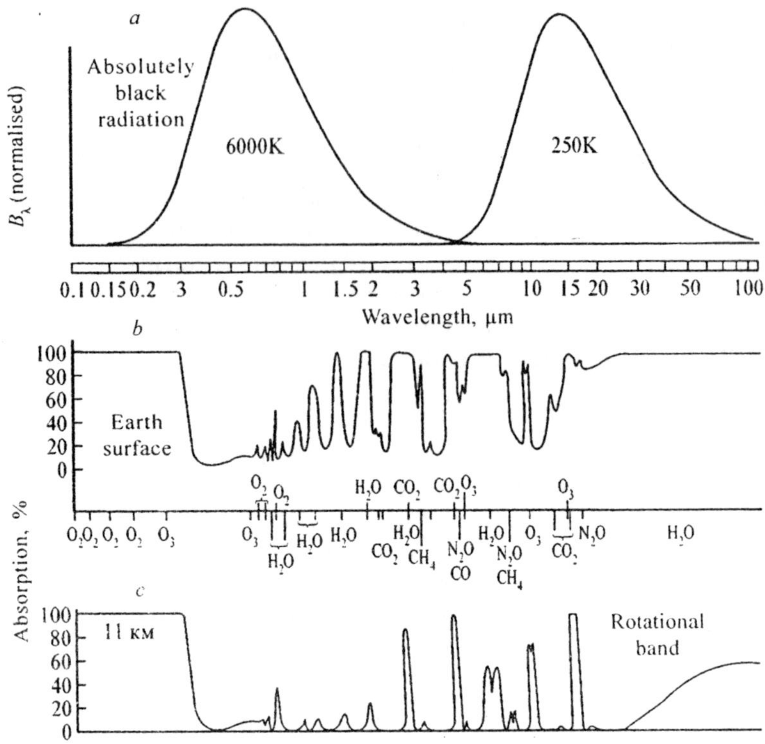

Fig. 4.1. Spectral behaviour of absorption functions for the case of the cloudless atmosphere [91]. a) curves of absolutely black radiation for two temperatures; b) absorption functions of the whole thicknesses of the atmosphere at the zenith angle of 40°; c) absorption function of the layer of the atmosphere from its upper boundary to 11 km.

penetration of solar ultraviolet radiation into the Earth's atmosphere is given in Fig. 4.2 which shows the altitude in the atmosphere for which the optical thickness of the atmosphere (from the upper boundary) is equal to unity. The figure shows that a large part of solar radiation for the wavelength in the range 1–150 nm is absorbed in the layers of the atmosphere above 100 km. For the wavelength of 200–300 nm, absorption takes place mainly in the layer of the atmosphere with the lower boundary at 30–40 km. Figure 4.2 shows the main gas components, responsible for the absorption of solar radiation in the ultraviolet part of the spectrum (O_2, O, N_2, N, NO, O_3).

Solar radiation starts to reach the Earth surface starting at a wavelength greater than 300 nm (Fig. 4.1a and 4.2). Approximately from this wavelength the atmospheric absorption rapidly decreases,

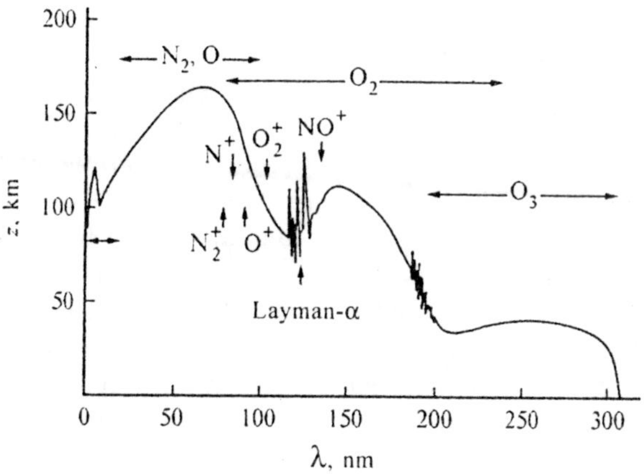

Fig. 4.2. Altitudes in the atmosphere at which the optical thickness (from the upper boundary) is equal to 1 [91].

and it may be seen that for the wavelengths of 0.3–0.9 μm (300–900 nm) a large part of solar radiation reaches the surface of the Earth (and, even more so, the altitude of 11 km). These spectral ranges in which the atmospheric absorption is not large are regarded as the transparency windows of the atmosphere. This spectral range corresponds to the transparency window in the visible part of the spectrum.

Starting from approximately 1 μm, atmospheric absorption increases in the mean, reaching 100% in the individual bands of absorption of different atmospheric gases (they are shown in Fig. 4.1). These absorption bands are specially clearly visible in Fig. 4.1c. Comparison of Fig. 4.1b and 4.1c shows that in the infrared range of the spectrum, atmospheric absorption is detected mainly in the troposphere (absorption is considerably smaller for a level of 11 km than for the surface of the Earth). It is associated with the fact that the troposphere in particular contains a large part of various important molecules absorbing infrared radiation, such as H_2O and CO_2. The spectral ranges between the absorption bands form transparency windows of the atmosphere in the infrared range of the spectrum. From approximately 13–14 μm the absorption functions of the entire thickness of the atmosphere approach 100 %, i.e. total absorption is found. This is associated mainly with the absorption by molecules of CO_2 and H_2O.

To illustrate the spectrum of absorption of the individual atmospheric gases in the infrared range of the spectrum, we examine

Fig. 4.3 which gives the transmittance functions for gases such as CO, CH_4, N_2O, O_3, CO_2, HDO, H_2O and of the entire multi-component atmosphere. As mentioned in chapter 2, the content of different atmospheric gases undergoes significant spatial and time changes. This results in the corresponding variations of molecular absorption in the Earth's atmosphere. Especially large variations of the atmospheric transmittance functions are found in the absorption bands of water vapours; this is associated with considerable spatial-time variations of the content of water vapours in the atmosphere of the earth. Since the most intensive absorption bands of this gas are found in the middle infrared range, the largest variations of the transmittance functions are found at wavelengths greater than 3 μm.

4.2. Different types of molecular absorption

The molecular absorption, formed as a result of the processes 2, 3, 4 and subsequent energy transformations (section 3.3, Table 3.2)

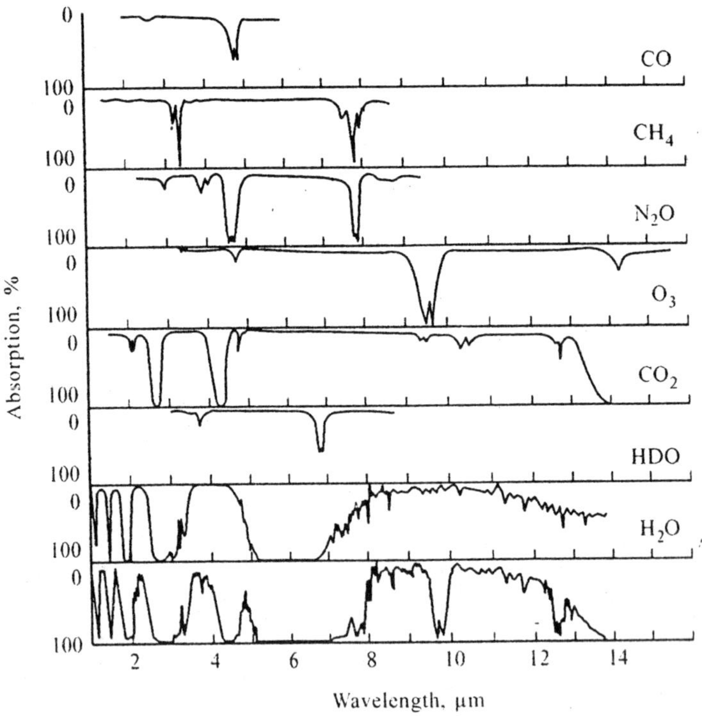

Fig. 4.3. Transmittance function for different gases and for the entire multicomponent atmosphere [114].

is sub-divided to different types to facilitate examination [33, 37, 91]. If the absorption of radiation leads to dissociation (or ionisation) of a molecule, the appropriate absorption spectra are referred to photodissociation spectra (or photoionisation). A characteristic feature of the spectra is the relatively weak spectral dependence (with certain exceptions) of the appropriate absorption coefficient. This is associated with the fact that, as mentioned previously, the types of energy of the molecule, such as electronic, vibrational, have discrete values, and the energy of translational motion is continuous. Consequently, the transition of the molecule, for example from the electronic (discrete) state, is possible to any level of the energy of translational movement. Therefore, the absorption in photodissociation (or photoionisation) may be detected at any frequency higher than its boundary value, determined by the minimum energy of the photons required for these processes to take place. The absorption of radiation, determined by transitions between the electronic, vibrational and rotational levels of energy of the molecule, has a distinctive spectral structure. Each such transition results, because of the discreteness of the energy levels, in the formation of an individual absorption line whose centre corresponds to the well-known relationship:

$$h\nu_{ij} = E_j - E_i, \tag{4.2.1}$$

where E_i and E_j are the energies of the lower and upper state of the molecule. The absorption spectra subdivided into electronic, vibrational and rotational, depending on the states of the internal energy taking part in the interaction of radiation with the molecule. In fact, since the electronic states of the molecule are 'split' into vibrational sub-levels, and vibrational sub-levels are split into rotational sub-levels, the variation of electronic state of the molecules is accompanied by changes in vibrational and rotational states. Therefore, it is more accurate to refer to the electronic absorption spectra as the electronic–vibrational–rotational spectra, but this long name is not used very often. Correspondingly, the vibrational absorption spectra should be referred to as the vibrational–rotational spectra.

Information on different types of the internal energy of the molecules and different interactions of radiation with molecules, and also on the spectral regions, corresponding to these interactions, is presented in Fig. 4.4. This figure indicates the following special features.

1. The most intensive exchange of energy between the

electromagnetic field and the molecule takes place in the case of changes in the configuration of nuclei, forming the atoms and molecules. This can be carried out only by photons with very high energy, corresponding to gamma radiation.

2. The transitions between different levels of the electronic energy may take place under the effect of x-ray, ultraviolet and visible radiation. The x-radiation, whose photons are characterised by very high energy, may result in changes in the position of the internal electrons. The absorption of ultraviolet and visible radiation results in transitions of the external electrons of the molecules and the atoms.

3. The vibrations of the atoms, forming the molecules, are responsible for the presence of vibrational levels of energy and absorption of infrared radiation.

4. The changes in the orientation of the molecule (rotation of the molecule) are associated with the absorption of long-range infrared irradiation or microwave radiation.

5. Finally, changes in the orientation of the spins of the nuclei and the electrons of the molecules, accompanied by very small changes of the internal energy of the molecule, are caused by the absorption of centimetre and metre radiowave radiation.

Discreteness of levels of internal energy of the molecule

The specific values of the internal energy of the molecule (electronic, vibrational, rotational, etc.) depend on its chemical nature (the atoms forming the molecule) and the structure. They determine possible transitions of the molecule from one state to another, accompanied by the absorption (or emission) of radiation, i.e. the radiation (absorption) spectrum of the molecule. Since every molecule consists of specific atoms and is characterised by a specific structure, it is absorption (and radiation) spectrum is strictly individual. This special feature is the physical basis of the application of absorption (and emission) spectra of the molecules for identifying their chemical nature and structure.

As an example, Fig. 4.5 shows the levels of the internal energy of a hypothetical diatomic molecule – electronic, vibrational, and rotational. The populations of the horizontal lines (A) and (B) in the graph represent two electronic states of the molecule. The indexes v and j characterise the vibrational and rotational energy, respectively. In a general case, absorption takes place in transition of the molecule from the state A, v'', j'' to the state B, v', j'. Each such transition results in the formation of an electronic–vibrational–

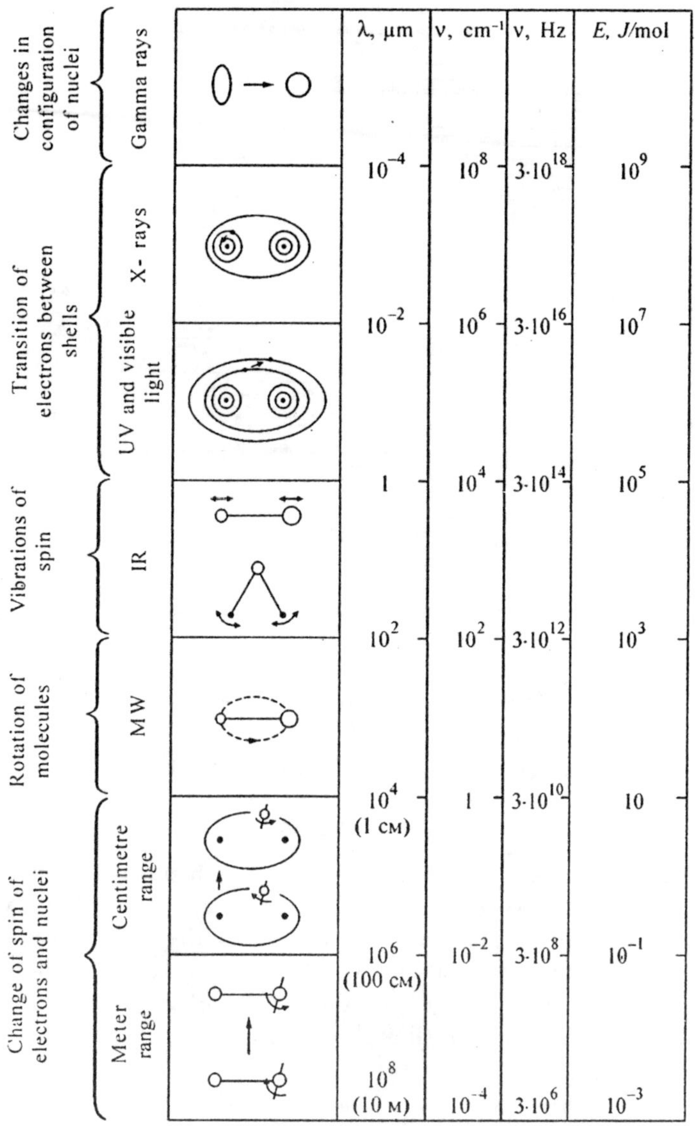

Fig. 4.4. Different types of the internal energy of molecules and interaction of radiation with molecules in different spectra ranges [114].

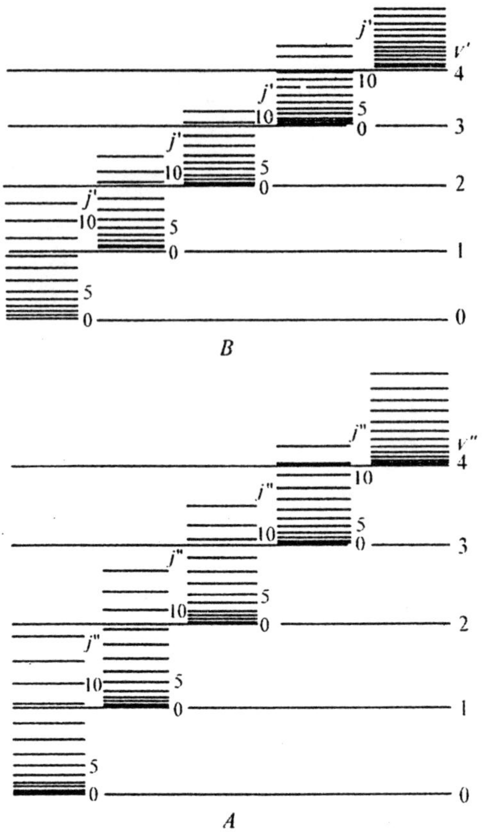

Fig. 4.5. Energy levels of an ideal diatomic molecule: (*A*) and (*B*) – electronic; *v'* and *v''* – vibrational; *j'* and *j''* – rotational [91].

rotational absorption line. The number of such transitions between different states may be very large (they are restricted by the special selection rules). The population of the absorption lines, formed in transitions between two specific electronic sub-levels, forms the electronic absorption band. If the electronic state of the molecule does not change (the photon energy is not sufficient for this to take place), we are talking about vibrational absorption bands. Taking into account the splitting of the vibrational levels of energy into rotational sub-levels of energy, the transitions between the vibrational states are also accompanied by changes of the rotational energy of the molecule. All these transitions form lines comprising the vibrational–rotational absorption band.

4.3. Absorption spectra of atmospheric gases

The absorption by atmospheric components may be characterised by means of the spectral dependence of the coefficients of molecular absorption. For atmospheric applications, it is more efficient to use characteristics of molecular absorption such as the transmittance function (or absorption function). Examples of these functions are presented in Fig. 4.1 and 4.3 for the case of considerable spectral averaging. In the given section, we examine the monochromatic functions, which characterise most convincingly the complicated spectral nature of molecular absorption.

The monochromatic absorption function is determined by the relationship (3.4.7):

$$P(\nu)=\exp\left(-\sec\theta\int_{z_1}^{z_2}\alpha(\nu,z)dz\right),$$

where $\alpha(\nu,z)$ is the volume extinction coefficient which depends on radiation frequency. In the examined case, in which this is concerned with the molecular absorption in the atmosphere, the extinction coefficient is equal to the volume coefficient of molecular absorption k (ν). Using the mass coefficient of absorption $\kappa_\nu(\nu,z)$, the transmittance function may be presented in the following form:

$$P(\nu)=\exp\left[-\sec\theta\int_{z_1}^{z_2}\kappa(\nu,z)\rho(z)dz\right], \tag{4.3.1}$$

where $\rho(z)$ is the density of the absorbing gas. (In subsequent sections, to denote the coefficient of molecular absorption, we shall use a single symbol k regardless of units used).

Figure 4.6 shows the example of calculated monochromatic transmittance functions (as a function of the wave number) for H_2O [91]. The minima of the values $P(\nu)$, corresponding to different absorption lines of the investigated gas, are clearly visible. The graph shows also clearly the very strong dependence of $P(\nu)$ on the wave number – the radiation is greatly absorbed in the vicinity of the centres of the lines and only slightly between them. This spectral nature of molecular absorption, determined by the marked variability of the coefficients of molecular absorption, is the reason for the use of the term 'selective absorption'. The spectral

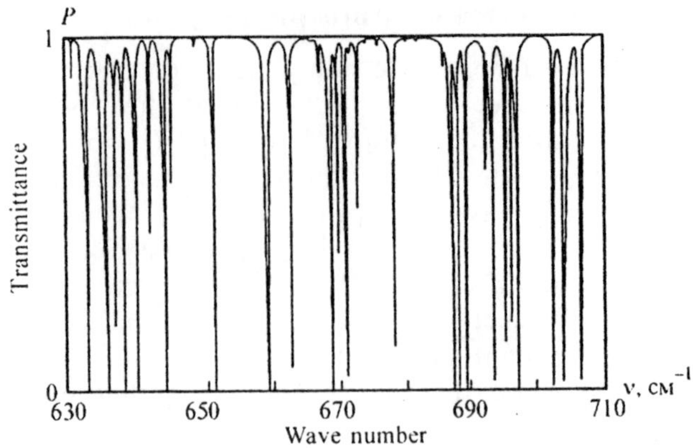

Fig. 4.6. Calculated spectrum $P(\nu)$ of the transmittance function of H_2O in the vicinity of 14.0 μm. Examination altitude z = 10 km, zenith observation angle θ = 30°.

behaviour of the absorption coefficients and of the transmittance functions of the atmosphere in photodissociation and photoionisation of the molecules greatly differ from the example shown in Fig. 4.6. They are characterised by the relatively weak dependence of the absorption coefficients on frequency and are referred to as non-selective (continuous) absorption. The same type of spectral dependence is also detected in the presence of the very large number of overlapping (superposed on each other) spectral absorption lines. This phenomenon is characteristic, for example, of electronic absorption bands of different atmospheric gases in ultraviolet and visible ranges of the spectrum characterised by the presence of a very large number of spectral absorption lines determined by the presence of vibrational–rotational sub-levels in the molecule. The same type of spectral dependence of the absorption coefficients is characteristic of the so-called molecular induced absorption of different gases, and also aerosol absorption.

4.4. Quantitative description of molecular absorption

Taking into account the presence of a large number of individual absorption lines in the spectra of molecular absorption formed in the transitions of the molecule from one energy state to another, the total coefficient of absorption for a specific molecule may be presented in the following form:

$$k_k(\nu) = \sum_i k_{ik}(\nu), \tag{4.4.1}$$

where $k_i(\nu)$ is the coefficient of absorption in the individual i-th spectral line. It should be stressed that the relationship (4.4.1), which appears to be evident, is in fact approximate. The relationship (4.4.1) is based on the assumption according to which the individual spectral absorption lines form independently of each other. In fact, as indicated by the general theory of molecular absorption in gases, this is not quite the case [91, 117]. The principle of additivity (4.4.1) is violated quite clearly if the corresponding absorption lines greatly overlap. This effect in molecular spectroscopy is referred to as the interference of spectral lines. In subsequent considerations in the majority of cases (to simplify considerations), we shall use the relationship (4.4.1) not forgetting, however, its approximate nature, and we shall use a stricter expression for the absorption in the set of closely spaced spectral lines in special cases (paragraph 4.5).

If it is taken into account that the atmospheres of the planets, for example, the Earth, contain a large number of different atmospheric gases (chapter 2), the coefficient of molecular absorption of air should be written in the following form:

$$k(\nu) = \sum_k k_{ik}(\nu) = \sum_k \sum_i k_{ik}, \qquad (4.4.2)$$

where the summation index k corresponds to different gases, included in the composition of air, and index i corresponds to the individual lines of each gas. Thus, the general coefficient of molecular absorption in the atmospheres of the planets is, in the first approximation, the sum of the coefficients of absorption in the individual lines of absorption of different atmospheric gases.

Intensity and shape of the spectral line

The coefficient of molecular absorption in the individual spectral line is described by the equation [20, 27, 37, 91, 102]:

$$k_{ij}(\nu) = S_{ij} f_{ij} (\nu - \nu_{ij}), \qquad (4.4.3)$$

where S_{ij} is the intensity of the spectral absorption line conditioned by the transition of the molecule from one state of internal energy (i) to another (j); f_{ij} is the shape of the absorption line, describing the spectral (frequency) distribution of the absorption coefficient. The same procedure can be used for introducing the shape of the radiation line. In this case, the shape of the line is described in such a manner as to fulfil the normalisation condition:

$$\int_{-\infty}^{\infty} f(\nu-\nu_{ij})d\nu = 1, \tag{4.4.4}$$

which leads to the relationship linking the intensity of the line and its absorption coefficient:

$$\int_{-\infty}^{\infty} k_{ij}(\nu)d\nu = S_{ij}. \tag{4.4.5}$$

Einstein coefficients (transition probabilities)

We examine the processes of molecular absorption and radiation from the viewpoint of the quantum-mechanics approach for the simplest model of a two-level molecule or atom (Fig. 4.7a). Einstein introduced the concept of the probabilities of transition of the molecule (atom) from one state to another. In the present case, these are transitions between the states 1 and 2, i.e. between the ground state with energy E_1 and the excited state with energy E_2. Einstein stipulated that all interactions of the molecule with the radiation may be described by means of the processes of spontaneous and forced emission and absorption. The number of spontaneous transitions from state 2 to state 1 in the unit volume during time dt is:

$$N_2\ A_{21}dt, \tag{4.4.6}$$

where N_2 is the number of molecules in the state 2 (population of state 2); A_{21} is the Einstein coefficient for spontaneous emission. Coefficient A_{21} expresses the probability of the molecule, transiting spontaneously from state 2 to state 1 per unit time.

If the examined volume of the medium is located in the radiation field with the radiation density of $\rho(\nu)$ in the vicinity of the frequency corresponding to the energy difference between the levels 1 and 2, i.e. $h\nu = E_2-E_1$. The number of photons absorbed during time dt (the number of transitions from state 1 to state 2) can be determined from the equation:

$$N_1 B_{12}\rho(\nu_{12})dt, \tag{4.4.7}$$

where B_{12} is the Einstein coefficient for absorption; N_1 is the population of state 1. The number of acts of induced emission (under the effect of radiation incident on the medium) may be determined from the equation:

$$N_2 B_{21} \rho(\nu_{21}) dt, \tag{4.4.8}$$

where B_{21} is the Einstein coefficient for forced radiation.

Einstein postulated that the three introduced coefficients A_{21}, B_{12} and B_{21} reflect the specific properties of the atoms and molecules and are independent of external conditions (for example, temperature and pressure). In this case, the relationships between these coefficients are of the universal nature, i.e. they are valid for any conditions.

Relationships between Einstein coefficients

These relationships can be determined by examining the case of thermodynamic (thermal) equilibrium in a medium in which three conditions are fulfilled [81, 91]:

1. Detailed balance, i.e. it is assumed that the number of transitions $1 \rightarrow 2$ is accurately equal to the number of transitions $2 \leftarrow 1$. In this case, the following equation can be written:

$$N_1 B_{12} \rho(\nu_{12}) dt = N_2 B_{21} \rho(\nu_{21}) dt + N_2 A_{21} dt \tag{4.4.9}$$

or

$$\frac{N_2}{N_1} = \frac{B_{12}\rho(\nu_{12})}{B_{21}\rho(\nu_{21}) + A_{21}}. \tag{4.4.10}$$

2. In thermal (thermodynamic) equilibrium, every element of the medium emits as a black body whose temperature is also the temperature of the medium. In this case, emission is isotropic and described by the Planck law. The following equation can be written for the density of radiation of a black body (paragraph 3.4),

$$\rho(\nu_{21}) = \frac{8\pi h \nu_{21}^3}{c^3} \frac{1}{\exp\left[\dfrac{h\nu_{21}}{k_B T}\right] - 1}. \tag{4.4.11}$$

3. The relative population of state 1 and 2 is described by the Boltzmann equation:

$$\frac{N_2}{N_1} = \frac{g_2}{g_1} \exp\left[-\frac{E_{21}}{k_B T}\right], \tag{4.4.12}$$

where $E_{21} = E_2 - E_1$, and g_1 and g_2 are the statistical weights of the states characterising the degree of their degeneration (i.e. the presence of several internal states of the molecule with the same energy).

The relationships (4.4.10) and (4.4.12) give the equation for the density of radiation:

$$\rho(\nu_{21}) = \frac{A_{21}}{\frac{g_1}{g_2}\exp\left[\frac{E_{21}}{k_B T}\right]B_{12} - B_{21}}. \qquad (4.4.13)$$

This equation should coincide (in the thermal equilibrium conditions) with the density of radiation of the absolutely black body (4.4.11). In order to ensure that they coincide, it is necessary to fulfil:

$$A_{21} = \frac{8\pi h\nu_{21}^3}{c^3} B_{21}, \qquad (4.4.14)$$

$$g_1 B_{12} = g_2 B_{21}. \qquad (4.4.15)$$

Thus, knowing only one Einstein coefficient, we can obtain the other two coefficients.

Quantum mechanics shows that the Einstein coefficient for spontaneous radiation is expressed by the matrix element of the dipole moment of the molecule R_{21} [20, 27, 91]:

$$A_{21} = \frac{64\pi^4\nu^3}{3hc^3}[R_{21}]^2. \qquad (4.4.16)$$

Thus, the data on the matrix elements of the dipole moment of the molecule make it possible to determine the Einstein coefficients and, as shown later, the intensity of spectral lines.

Equivalence of the shapes of spectral lines of absorption and radiation

The previously described optical processes of radiation and absorption of radiation take place in finite ranges of frequency as a result of different mechanisms of broadening of spectral lines (for more details see the following paragraph). It is therefore possible to introduce spectral analogues of the Einstein coefficients, taking this circumstance into account, using the equations

$$A_{21}(\nu) = A_{21}f_s(\nu),$$
$$B_{12}(\nu) = B_{12}f_i(\nu), \qquad (4.4.17)$$
$$B_{21}(\nu) = B_{21}f_a(\nu),$$

where f_s, f_i and f_a are the shapes of the lines of spontaneous and forced radiation and absorption. The introduced spectral Einstein coefficients have the same physical meaning of probability of appropriate transitions, but they are related to the single frequency interval.

By analogy with the relationships (4.4.14) and (4.4.15) it is possible to write spectral analogues of the relationships between the Einstein coefficients:

$$A_{21}(\nu) = \frac{8\pi h\nu_{21}^3}{c^3}B_{21}(\nu), \qquad (4.4.14a)$$

$$g_1B_{12}(\nu) = g_2B_{21}(\nu). \qquad (4.4.15a)$$

The previously introduced Einstein coefficients A_{21}, B_{12} and B_{21} are coefficients for all absorption or radiation lines. It will be shown that for all three types of the previously examined optical processes the shape of the lines is the same. For this purpose, we can use the equations (4.4.17) and write the relationships between the Einstein coefficients in the form:

$$A_{21}f_s(\nu) = \frac{8\pi h\nu_{21}^3}{c^3}B_{21}f_a(\nu), \qquad (4.4.14b)$$

$$g_1B_{12}f_i(\nu) = g_2B_{21}f_a(\nu). \qquad (4.4.15b)$$

The relationship (4.4.15b) shows directly that

$$f_i(\nu) = f_a(\nu) = f(\nu). \qquad (4.4.18)$$

It will also be taken into account that in the large part of the examined spectrum of electromagnetic radiation (from ultraviolet to far infrared) the width of the spectral lines is small in comparison with the frequency ν. Therefore, the quantity $8\pi h\nu^3/c^3$ remains almost constant along the width of the spectral lines . Consequently, equation (4.4.14b) gives:

$$f_s(\nu) = f_a(\nu) = f(\nu). \tag{4.4.19}$$

The equation of transfer of radiation in quantum-mechanics examination

Taking into account the previously examined elementary optical processes of interaction of radiation with the molecule (atom) we can write the variation of the intensity of radiation dI, propagating in the arbitrary direction dl per unit length. In order to obtain the magnitude of radiation – absorbed, forced and spontaneous – it is necessary to multiply the number of transitions of different type (equations 4.4.6–4.4.8) by the photon energy $h\nu$. It must be taken into account that the density of radiation, propagating in the direction dl in the unit solid angle, is equal to $\rho(\nu) = I_\nu/c$:

$$\frac{dI_\nu}{dl} = \frac{1}{4\pi} N_2 A_{21} f(\nu) h\nu_{21} + N_2 B_{21} \frac{I_\nu}{c} f(\nu) h\nu_{21} - N_1 B_{21} \frac{I_\nu}{c} f(\nu) h\nu_{21}. \tag{4.4.20}$$

The multiplier $1/4\pi$ in the first term of the right-hand part of equation (4.4.20) takes into account the fact that the intensity of radiation is determined as the energy of radiation propagating in the unit solid angle. In fact, the relationship (4.4.20) is the same equation of transfer of radiation but in the quantum-mechanics form. The relationship (4.4.20) may be determined in the form suitable for comparison with the transfer equation derived previously, i.e. relationship (3.4.3):

$$\frac{dI_\nu}{dl} = \frac{h\nu_{21}}{c}[N_1 B_{12} - N_2 B_{21}] I_\nu f(\nu) + \frac{h\nu_{21}}{4\pi} N_2 A_{21} f(\nu). \tag{4.4.21}$$

Quantum-mechanics equations for absorption and radiation coefficients

Comparison of the two types of equations of transfer of radiation – relationships (3.4.3) and (4.4.21) – makes it possible to write important equations for the coefficient of absorption and radiation from the viewpoint of quantum mechanics:

$$k(\nu) = \frac{h\nu_{21}}{c} N_1 B_{12} \left[1 - \frac{g_1}{g_2} \frac{N_2}{N_1} \right] f(\nu), \qquad (4.4.22)$$

$$\varepsilon(\nu) = \frac{h\nu_{21}}{4\pi} N_2 A_{21} f(\nu). \qquad (4.4.23)$$

When writing equation (4.4.22) we have used the relationship between the Einstein coefficients (4.4.15). (Here and later it is assumed, to simplify considerations, that the examined lines are narrow, i.e. it may be assumed that ν_{ij} = constant). In addition to this, we can also write the equation for the function of the source (3.4.13),

$$\tilde{B}(\nu) = \frac{\varepsilon(\nu)}{k(\nu)} = \frac{2h\nu_{21}}{c^2} \left[\frac{g_2}{g_1} \frac{N_1}{N_2} - 1 \right]^{-1}. \qquad (4.4.24)$$

If it is assumed that the medium is in the condition of thermodynamic equilibrium, and using the Boltzmann law, we obtain the following equation for the absorption coefficient:

$$k(\nu) = \frac{h\nu_{21}}{c} N_1 B_{21} \left[1 - \exp\left(-\frac{E_{21}}{k_B T} \right) \right] f(\nu). \qquad (4.4.25)$$

Intensity of lines

Knowing the expression for the coefficient of molecular absorption, one can obtain the expression for the intensity of the line corresponding to the transition 2 → 1:

$$S_{12} = \frac{h\nu_{21}}{c} N_1 B_{12} \left[1 - \exp\left(-\frac{h\nu_{21}}{k_B T} \right) \right]. \qquad (4.4.26)$$

This equation can be written in a different form, using the relationships between the Einstein coefficients:

$$S_{12} = \frac{c^2}{8\pi \nu_{21}^2} \frac{g_1}{g_2} N_1 A_{21} \left[1 - \exp\left(-\frac{h\nu_{21}}{k_B T} \right) \right]. \qquad (4.4.27)$$

In a general case, a molecule (or atom) has many states of internal energy (many levels), and the transitions between the states lead to the formation of a set of lines of absorption or emission (Fig. 4.7b). Taking this into account, the intensity of the absorption line in transition from the i-th to j-th state may be presented in the following form:

$$S_{ij} = \frac{h\nu_{ij}}{c} N_i B_{ij} \left[1 - \frac{g_i}{g_j} \frac{N_j}{N_i} \right]. \qquad (4.4.28)$$

In accordance with the Boltzmann law

$$\frac{g_i N_j}{g_j N_i} = \exp\left[-\frac{(E_j - E_i)}{k_B T} \right] = \exp\left[-\frac{h\nu_{ji}}{k_B T} \right]. \qquad (4.4.29)$$

Consequently, the intensity of the spectral line per one absorbing and emitting molecule taking into account the relationships between the Einstein coefficients may be presented in the following form:

$$S_{ij} = \frac{h\nu_{ji}}{c} \frac{N_i}{N} B_{ij} \left[1 - \exp\left(-\frac{h\nu_{ji}}{k_B T} \right) \right], \qquad (4.4.30)$$

where $N = \sum_i N_i$ is the total number of the absorbing and emitting molecules.

Using the Boltzmann law, we can write:

$$\frac{N_i}{N} = \frac{g_i \exp\left(-\frac{E_i}{k_B T} \right)}{\sum g_i \exp\left(-\frac{E_i}{k_B T} \right)} = \frac{g_i \exp\left(-\frac{E_i}{k_B T} \right)}{Q(T)}, \qquad (4.4.31)$$

where

$$Q(T) = \sum g_k \exp\left(-\frac{E_i}{k_B T} \right) \qquad (4.4.32)$$

is the statistical sum (the sum in respect of states). It should be mentioned that the energy E_i is counted from the ground state of the molecule.

Finally, we can write the following equation for the intensity of the line:

$$S_{ij} = \frac{c^2}{8\pi\nu_{ij}^2} \frac{g_j \exp\left(-\frac{E_i}{k_B T}\right)}{Q(T)} A_{ji} \left[1 - \exp\left(-\frac{h\nu_{ji}}{k_B T}\right)\right]. \qquad (4.4.33)$$

The exponential member in the square brackets in equation (4.4.33) is determined by taking into account the mechanism of induced radiation and, as shown by numerical estimates, its value for the atmospheric conditions is often very small. For example, for the vibrational band of CO_2 at 15 μm (ν = 667 cm^{-1}) this exponent is equal to 0.0082 at a temperature of T = 200 K and 0.041 at T = 300 K. Therefore, in many calculations for the conditions on the Earth this term is ignored. However, it is obvious that the simplification is not always justified. For example, when examining the molecular absorption in the atmosphere of Venus where the temperature reaches 600–700 K, the contribution of this exponential term may be very large.

Distribution of the molecules over the excited states of internal energy

The equations for the coefficient of absorption and intensity of the line contain the populations of different states of the molecules (atoms). These populations in thermodynamic equilibrium are determined by the Boltzmann law. We examine the population of different excited states in greater detail on the example of vibrational states of different molecules. Table 4.1 gives the values of the ratios of the populations of the first vibrational states to the total concentration for a number of diatomic molecules at a temperature of T = 300 K.

The data in Table 4.1 show clearly that at temperatures existing in the Earth's atmosphere it is sometimes possible to ignore the vibrational excitation of the molecules, if the required calculation accuracy is not very high. For example, for the CO molecule, the relative population of the first vibrational state N_1/N is equal to $3.4 \cdot 10^{-5}$ at T = 300 K. If the mixing ratio of CO is equal to 1 million^{-1} (ppm), then the concentration of the CO molecules in the first excited vibrational state in the normal conditions is equal to approximately $9.15 \cdot 10^9$ cm^{-3}. However, for many gases and appropriate conditions (relatively high temperatures of the medium),

Table 4.1. Equilibrium populations* of vibrational levels of different molecules at 300 K [44]

Molecule	N_1/N	N_2/N	N_3/N	N_4/N	N_5/N
CH	2.00×10^{-6}	7.45×10^{-12}	5.13×10^{-17}	6.55×10^{-22}	1.55×10^{-26}
Cl_2	6.42×10^{-2}	4.61×10^{-3}	3.43×10^{-4}	2.66×10^{-5}	2.14×10^{-6}
CN	5.52×10^{-5}	3.45×10^{-9}	2.45×10^{-13}	1.97×10^{-17}	1.80×10^{-21}
CO	3.40×10^{-5}	1.32×10^{-9}	5.79×10^{-14}	2.90×10^{-18}	1.65×10^{-22}
H_2	2.05×10^{-9}	1.25×10^{-17}	2.28×10^{-25}	1.24×10^{-32}	0
HCl	9.63×10^{-7}	1.53×10^{-12}	3.99×10^{-18}	1.72×10^{-23}	1.22×10^{-28}
N_2	1.38×10^{-5}	2.20×10^{-10}	4.01×10^{-15}	8.41×10^{-20}	2.03×10^{-24}
O_2	5.69×10^{-4}	3.64×10^{-7}	2.61×10^{-10}	2.10×10^{-13}	1.9×10^{-16}
OH	3.61×10^{-8}	2.89×10^{-15}	5.12×10^{-22}	2.01×10^{-28}	1.74×10^{-34}
NO	1.23×10^{-4}	1.72×10^{-8}	2.76×10^{-12}	5.05×10^{-16}	1.06×10^{-19}

the vibrational excitation of the atmospheric molecules should be taken into account. The absorption bands (and appropriate absorption lines), formed at transitions between the excited vibrational levels (for example, vibrational states 1 and 2) are referred to as 'hot' absorption bands, evidently because of the fact that they are manifested in the absorption spectra of gases at high temperatures. For example, the radiation spectrum of the Sun contains the 'hot' bands of CO because at a temperature of T = 5000 K the relative populations of the first and second vibrational levels are equal to 0.29 and 0.18, respectively.

When examining the distribution of molecules in respect of the rotational energies, it must be taken into account that these energies are considerably lower than the vibrational energies. This distribution is described by the same Boltzmann equation which, in this case, can be presented in the following form:

$$N_{vj} = N_v g_j \exp\left[-\frac{(E_j - E_0)}{k_B T}\right], \qquad (4.4.34)$$

where N_v is the population of the v-th vibrational level; g_j is the statistical weight of the j-th rotational level (degeneration); E_0 and E_j are the energies of the ground and excited vibrational states of the molecules.

*Evidently, in some cases the population of the levels is so small that for all practical purposes they can be regarded equal to 0. This holds for $N_m/N < 10^{-19}$.

Figure 4.8 shows the curve of the relative equilibrium population of the rotational levels of the ground state of the molecule of CO at different temperatures as a function of the rotational quantum number *j*. The figure shows that the CO molecules are distributed 'relatively uniformly' in the excited rotational states (in comparison with the distribution in respect of the vibrational states) and there is the rotational number *j* and the appropriate level of the energy for which the maximum population is found. The position of this maximum depends on the temperature of the medium and the rotational constant of the molecule B.

Statistical sums in respect of the energy states of molecule energy

In order to calculate the intensity of the spectral lines for different temperatures, it is necessary to know the statistical sums $Q(T)$. For this purpose it is necessary to know all permitted levels of the internal energy of the molecule and the degree of degeneration of the energy states (statistical weights). These energies are calculated by solving the Schrödinger equation. In this case it is necessary to specify the definited model of the molecule and the form (type) of the dependence of the potential energy of the molecule on the parameters of the molecule. For example, in the approximation of a harmonic oscillator, the solution of the Schrödinger equation gives the following simple equation [20, 24, 91] for the vibrational energies of the diatomic molecule (with rotation not taken into account):

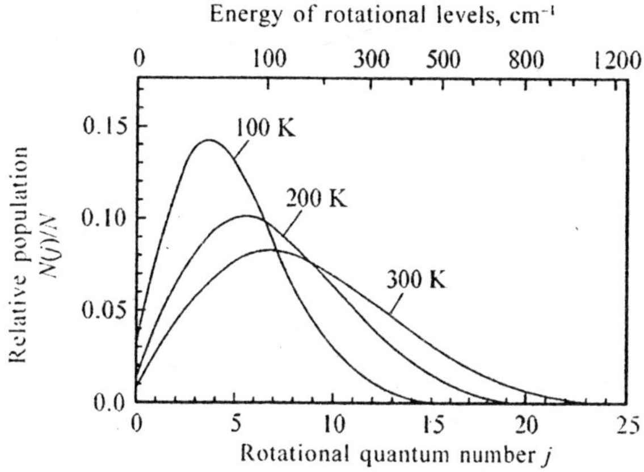

Fig. 4.8. Dependence of the relative equilibrium population of the rotational levels of the ground state of the CO molecule at T = 100, 200 and 300 K on the rotational quantum number j).

$$E(v) = h\nu_m\left(v + \frac{1}{2}\right);\ v = 0,1,2,... \tag{4.4.35}$$

where ν_m is the eigenfrequency of the vibration of the molecule; v is the vibrational quantum number. For many-atomic molecules, the relationship (4.4.35) is replaced by the appropriate the sum in respect of all eigenfrequencies of vibrations of the molecule.

After simple transformations, the following equation is derived for the vibrational statistical sum of the diatomic molecule:

$$Q_v(T) = \sum_{v=0}^{\infty} \exp\left(-\frac{h\nu_m(v+1/2)}{k_B T}\right) = \left[1 - \exp\left(-\frac{h\nu_m}{k_B T}\right)\right]^{-1}, \tag{4.4.36}$$

which takes into account that the statistical weights of such a molecule are equal to unity.

For not too high temperatures (the atmosphere of planets), the vibrational statistical sum is close to unity, because almost all molecules are in the ground state (Table 4.1). Therefore, in the majority of atmospheric applications, the vibrational statistical sum is assumed to be equal to unity. The error of this approximation is the largest for the molecules with relatively low values of natural vibrations (vibrational modes), such as CO_2, O_3, N_2O, NO_2, SO_2, COF_2 and, to the largest degree, for HNO_3. For the CO_2 molecule, this approximation should result in the error in the values of $Q(T)$ of up to 19% in the examined temperature range, and in the case of HNO_3 the error may reach 26%.

The rotating molecule is approximately regarded as a rotating solid (rigid gyroscope). For this model, all the molecules may be related to four different groups depending on the ratio between the three main moments of the inertia in relation to the three main axes of inertia of the molecules:

- linear molecules (for example CO_2, CO, N_2O, O_2, N_2);
- symmetric gyroscope (NH_3, C_2H_6, C_3H_4),
- spherical gyroscope (CH_4)
- asymmetric gyroscope (H_2O, O_3, HDO, SO_2).

For example, for a linear molecule, quantum mechanics gives the following expression for the rotational energy of the molecule:

$$E_j = \frac{h^2}{8\pi^2 J} j(j+1) = hcB_0 j(j+1), \tag{4.4.37}$$

where J is the moment of inertia of the molecule in relation to the axis of rotation, normal to the axis of the molecule, and passing through its centre of gravity; $B = h/(8\pi^2 cJ)$ is the rotational constant of the molecule; j is the rotational quantum number. The rules of selection allow the rotational transitions at $j = 0, \pm 1$. This shows that the frequencies of sucessive rotational transitions differ by $2B$. Consequently, the purely rotational spectrum of the linear molecule in the approximation of the rigid gyroscope consists of equally spaced lines (on the frequency scale):

$$\Delta E_j = h\nu_j = 2Bhc(j + 1). \qquad (4.4.38)$$

At relatively low temperatures found in the atmosphere of the planets, the rotational statistical sum can be replaced by the integral and for the linear molecules it is equal to:

$$Q_{rot}(T) = \frac{k_B T}{hcB}. \qquad (4.4.39)$$

For other molecules, having three degrees of freedom for rotational motion (and three rotational constants A, B, and C), the identical approximation of the rotational sum is given by the expression:

$$Q_{rot}(T) = \frac{k_B T^{3/2}}{hcB}. \qquad (4.4.40)$$

The examined examples relate to idealised, simplified models of the molecules. In fact, it is necessary to examine the effects of deviation from the harmonic form of the oscillations, interaction of different motions (vibrational and rotational), non-rigidity of the molecules, etc. These problems are examined in detail in lectures of molecular spectroscopy.

4.5. The shape of spectral absorption lines

The specific form of the function f_{ji} $(\nu - \nu_0)$, describing the distribution of the absorption coefficient in respect of frequency, is determined by the processes leading to the formation of the shape of the lines. These processes are often referred to as the processes of broadening of spectral lines. In the group of different broadening processes, attention will be given to the processes that are most important for the atmosphere of the planets.

Natural broadening of the spectral lines

This type of broadening may be explained on the basis of the

Heisenberg uncertainty principle:

$$\Delta E \cdot \Delta t > \hbar, \tag{4.5.1}$$

where ΔE and Δt are uncertainties of the energy and the lifetime of the excited state; $\hbar = h/2\pi$. Using (4.5.1) for molecular absorption, we set ΔE as the uncertainty in the difference of the energies of the upper E_2 and lower E_1 states of the molecules, with the transition between the states determining the examined line; t is the lifetime of the excited state of the molecule. The uncertainty of the energy shows that the frequencies of radiation (or absorption) are also not determined accurately (they may differ in the appropriate ranges). Consequently, using the relationship for the radiation quantum (4.2.1), it is easy to obtain a frequency range in which radiation or absorption of energy in the spectral line takes place:

$$\Delta \nu = \frac{\Delta E}{h}. \tag{4.5.2}$$

Namely, the value $\Delta \nu$ in (4.5.2) also characterises the natural broadening of the spectral lines. In comparison with other mechanisms of broadening of lines in the atmosphere of the planets, this type of broadening does not play any significant role and can be usually ignored but it is important because it determines the lower limit of 'narrowing' of the spectral lines in the atmosphere of the planets.

The form of the absorption line for natural broadening may be determined by solving the Schrödinger equation (quantum-mechanics approach), and also on the basis of classic examination. In the latter case, it is taken into account that the curvilinear accelerated motion of the electrons in the orbits is the reason for radiation. The emitting molecule continuously loses internal energy. The losses of energy make it necessary to introduce, into the equation of motion of the bound electron, a member describing the effect of radiation deceleration. This results in changes in respect of time (decrease) of the amplitude of the vector of electrical strength which is denoted here as A (in order to avoid confusion with energy E). Consequently, the following equation may be written:

$$A(t) = A_0 \exp(-t/\tau). \tag{4.5.3}$$

Writing the equation of the electromagnetic wave in the complex form (3.5.4) and using the moment of the start of radiation as the start of counting the time, we obtain:

$$A(t) = \begin{cases} 0, & \text{if } t < 0, \\ A_0 \exp(-t/\tau)\exp(i2\pi\nu_0 t), & \text{if } t \geq 0, \end{cases} \tag{4.5.4}$$

where ν_0 is the frequency of emission; τ is the mean lifetime of the excited state.

If the specific quantity depends on time then, as is well-known, its frequency spectrum is determined by the Fourier transformation:

$$A(\nu) = \frac{1}{\sqrt{2\pi}} \int_{-\infty}^{\infty} A(t)\exp(-2\pi i\nu t)dt =$$

$$= \frac{A_0}{\sqrt{2\pi}} \int_{0}^{\infty} \exp[-(1/\tau) - 2\pi i(\nu - \nu_0)t]dt =$$

$$= \frac{A_0}{\sqrt{2\pi}} \frac{1}{1/\tau - 2\pi i(\nu - \nu_0)}. \tag{4.5.5}$$

The energy of the electromagnetic wave at frequency ν is, according to (3.5.6),

$$E(\nu) = A(\nu)A^*(\nu) = \frac{A_0^2}{2\pi} \frac{1}{(1/\tau)^2 + 4\pi^2(\nu - \nu_0)^2}. \tag{4.5.6}$$

The value of $E(\nu)$ is maximum at $\nu = \nu_0$, i.e. at the central frequency of the radiation line. Equation (4.5.6) is the spectrum of radiation of the molecule due to the loss of energy in emission and describes the dependence of the radiation energy on frequency, i.e. the shape of the radiation line. As indicated by equation (4.4.19), the shapes of the radiation and absorption lines are identical. Consequently, the equation (4.5.6) also corresponds to the shape of the absorption line determined by the mechanism of natural broadening.

Using the normalisation relationship (4.4.4) for equation (4.5.6) one obtains

$$\int_{-\infty}^{\infty} \frac{A_0^2}{2\pi} \frac{1}{(1/\tau)^2 + 4\pi^2 x^2} dx = \frac{A_0^2}{2\pi} \frac{\tau}{2\pi} \int_{-\infty}^{\infty} \frac{dy}{1 + y^2} =$$

$$= \frac{A_0^2 \tau}{4\pi^2} \operatorname{arctg} y \Big|_{-\infty}^{\infty} = \frac{A_0^2 \tau}{4\pi}, \tag{4.5.7}$$

with the notations $x = \nu - \nu_0$, $y = 2\pi x\tau$. Separating (normalising) of the line shape (4.5.6) by the calculated integral gives

$$f(\nu - \nu_0) = \frac{2}{\tau} \frac{1}{(1/\tau)^2 + 4\pi^2(\nu - \nu_0)} = \frac{1}{\pi} \frac{1/2\pi\tau}{(1/2\pi\tau)^2 + (\nu - \nu_0)^2}$$

We introduce the half-width of the shape the spectral line α_l, i.e. the value of x $(\nu - \nu_0)$ for which $f(x) = 1/2\, f(0)$. For (4.5.8) we can easily find out (4.5.8)

$$\alpha_l = \frac{1}{2\pi\tau}. \tag{4.5.9}$$

Finally, the Lorentz shape of the radiation line (or absorption) (Lorentz or dispersion line shape) has the following form:

$$f_l(\nu - \nu_0) = \frac{1}{\pi} \frac{\alpha_l}{(\nu - \nu_0)^2 + a_l^2}. \tag{4.5.10}$$

The half-width of the line α_l in (or due to) natural broadening is inversely proportional to the lifetime of the excited state (the ratio (4.5.9)). For spontaneous transitions, the lifetime τ is inversely proportional to the Einstein coefficient for spontaneous emission A_{21}:

$$\tau \sim \frac{1}{A_{21}}. \tag{4.5.11}$$

In the real case for a molecule with many levels $A_i = \sum_k A_{ik}$, where A_{ik} are the Einstein coefficients for the spontaneous transition of the molecule from level i to different levels k.

The half-widths of the spectral lines determined by the process of natural broadening are very small and, for example, for the infrared range of the spectrum have the order of 10^{-10} cm^{-1}. The considerations and equations confirmed the previously noted impossibility of existence of monochromatic electromagnetic waves.

Broadening as a result of collisions of molecules

This type of broadening plays a significant role in the formation of the shapes of the absorption lines and radiation lines in the atmospheres of the planets. The detailed theory of broadening of the lines due to collisions is relatively complicated, and work is been carried out to develop and improve this theory, and it is described in detail in monographs concerned with molecular spectroscopy. This type of broadening is examined using both classic and quantum-mechanics approximation. In this case, examination is often carried out for the shape of the radiation line. Difference theories of broadening of the lines as a result of collisions give different, often very complicated equations for the shapes of the radiation and absorption lines. The Mickelson–Lorentz theory, one of the simplest theories, gives the same dispersion equation (4.5.10) but already with other Lorentz half-width α_L. In this theory, the molecule (or atom) is simulated by a simple harmonic oscillator with the eigenfrequency ν_0. Collisions of the molecules result in instantaneous stoppag ('break') of emission. The limited period of emission of electromagnetic energy results in the final spectral width of the radiation line and the Lorentz line shape.

It is assumed that the strength of the electrical field of an excited molecule is described as follows [44]:

$$A(t)=\begin{cases} A_0\exp[i2\pi\nu_0 t], & \text{at } 0<t<T,\\ 0, & \text{at } t\le 0 \text{ or } t\ge T, \end{cases} \tag{4.5.12}$$

where T is the lifetime of the excited state determined in the examined case by the time between collisions of the molecules. Thus, the form of (4.5.12) assumes that collisions interrupt the process of emission of the excited molecules. The corresponding Fourier transformation of the time dependence of the strength of the electrical field, giving the radiation spectrum, is equal to:

$$A(\nu,T)=\frac{1}{\sqrt{2\pi}}\int_{-\infty}^{\infty}A(t)\exp(-i2\pi\nu t)dt=$$

$$=\frac{A_0}{\sqrt{2\pi}}\int_0^T\exp(-i2\pi(\nu-\nu_0)t)dt= \tag{4.5.13}$$

$$= \frac{A_0}{\sqrt{\pi}} \left[\frac{\sin(2\pi(\nu - \nu_0)T/2)}{2\pi(\nu - \nu_0)} \right] \exp[-i2\pi(\nu - \nu_0)T/2].$$

The emission lines form as a result of the emission of the ensemble of the molecules. If τ indicates the mean lifetime for the statistical ensemble of the emitting molecules, then the time dependence of the number of these molecules may be described by the following equation:

$$N(t) = N(0)\exp(-t/\tau), \tag{4.5.14}$$

here $N(t)$ is the concentration of the excited molecules at the moment of time t. In this case, the probability of any excited molecular carrying out a transition in the time period $(t, t + dt)$ is equal to:

$$P(t)dt = -\frac{dN(t)}{N(0)} = \exp(-t/\tau)\frac{dt}{\tau}. \tag{4.5.15}$$

The observed intensity of radiation combines intensities of the emission of all excited molecules with the weights equal to the probability given previously. Thus, the spectral intensity of radiation is equal to:

$$I(\nu) = \int_0^{\infty} A(\nu,t)A^*(\nu,t)P(t)dt. \tag{4.5.16}$$

Substitution of the equations (4.5.13)-and (4.5.15) into integral (4.5.16) gives:

$$I(\nu) = \frac{\alpha A}{b^2} \int_0^{\infty} \sin^2 \alpha \left(\frac{bt}{2} \right) e^{-\alpha t} dt, \tag{4.5.17}$$

with the notations $\alpha = 1/\tau$, $b = 2\pi(\nu - \nu_0)$ and the proportionality coefficient A. Integration of equation (4.5.70) finally gives:

$$I(\nu) = \frac{A}{4\pi^2(\nu - \nu_0)^2 + (1/\tau)^2}. \tag{4.5.18}$$

The coefficient of proportionality may be expressed by means of the total radiation intensity I_0, integrating equation (4.5.18) in respect of all frequencies:

$$I_0 = \int_{-\infty}^{\infty} I(\nu)d\nu = \frac{\pi A}{a}. \qquad (4.5.19)$$

Consequently, the shape of the radiation line may be described by the following equation:

$$f_L(\nu - \nu_0) = \frac{1}{\pi} \frac{\alpha_L}{(\nu - \nu_0)^2 + \alpha_L^2}, \qquad (4.5.20)$$

where $\alpha_L = 1/2\pi\tau$ is the half-width of the line determined by collisions of the molecules. It may be seen that its shape is the same as in the case of natural broadening, i.e. Lorentz (dispersion), but its half-width (Lorentz half-width) is determined by the finite lifetime of the excited states as a result of collisions of the molecules.

Lorentz line half-widths

For the Lorentz line half-width [24, 27, 44, 52, 91] we can write the following equation:

$$\alpha_L = \frac{1}{2\pi\tau} = \frac{\tilde{\nu}}{2\pi}, \qquad (4.5.21)$$

where $\tilde{\nu}$ is the frequency of collisions (the mean number of collisions in unit time). The mean time between the collisions of the molecules at the pressure and temperature at the surface of the Earth is of the order of 10^{-10} s which gives the estimate for α_L = 0.05 cm^{-1}.

From the kinetic theory of the gases, the frequency of collisions for a one-component gas is given by the expression:

$$\tilde{\nu} = 4n\rho_k^2 \left(\frac{\pi k_B T}{m} \right)^{1/2}, \qquad (4.5.22)$$

where n and m are the concentration and mass of the colliding molecules; ρ_k is the gas-kinetic radius of the molecule ρ_k^2 is proportional to the gas-kinetic cross-section of collisions of the molecules).

Consequently, the Lorentz half-width is determined by the expression

$$\alpha_L = 2n\rho_k^2 \left(\frac{k_B T}{\pi m} \right)^{1/2} \tag{4.5.23}$$

In optical transitions, the radius of interaction (or the appropriate cross-section) is not determined by the gas-kinetic characteristics but by the perturbation of the rotational states which are characterised by other inter-molecular distances. Therefore, the quantity ρ_k in equation (4.5.23) is usually replaced by ρ_0 (the optical radius of the interaction) which is considerably higher than the gas-kinetic radius of the molecule. In the case of changes, for example of the rotational energy of the molecules, which, as a rule, is usually considerably lower than the energy of the translational movement of the molecules, the 'optical collision' is more efficient than the gas-kinetic collision.

Taking into account that the concentration of the molecules $n = p/k_B T$, the Lorentz half-width should be governed by the following dependence on pressure and temperature:

$$\alpha_L = \alpha_{0L} \left(\frac{p}{p_0} \right) \left(\frac{T_0}{T} \right)^{1/2}, \tag{4.5.24}$$

where α_{0L} is the Lorentz half-width at pressure p_0 and temperature T_0. It should be mentioned that by reason of ρ_0 also depends on temperature, expression (4.5.24) in the part of the temperature dependence of the Lorentz line half-width is approximate. Therefore, in equation (4.5.24), the temperature dependence of the line half-width is approximated in the more general form using the multiplier $(T/T_0)^{m_{ij}}$, where the exponent m_{ij} depends on the type of gas, the absorption band and the specific spectral line. As shown by the experiments and calculations carried out on the basis of the currently available theories of broadening of spectral lines as a result of collisions, exponent m_{ij} changes in the range from 0.5 to 1.0. In the mean, for all gases it may be assumed that $m = 0.7$.

In the general case of the multicomponent gas mixture, the broadening of the spectral lines takes place both as a result of collisions of the absorbing (and emitting) molecules and 'foreign' molecules (for example, the molecules of nitrogen and oxygen, i.e. the main gas components of the Earth's atmosphere). In the first case, we are talking about self-broadening, and in the second case about the broadening by 'the secondary' gas or mutual broadening. It is important that due to the efficiency of collisions of different molecules may differ, and the appropriate half-width may also differ in broadening by different gases. Therefore, the half-width of the same lines, generally speaking, differs in the atmospheres of different planets.

The broadening of the lines in collisions of different molecules may be taken into account by using the analog of the relationship (4.5.23) for the multicomponent mixture of the gases [24, 91]:

$$\alpha_L = \sum_i n_i \rho_{o,i}^2 \left[\frac{k_B T}{\pi} \left(\frac{1}{m} + \frac{1}{m_i} \right) \right]^{1/2}, \tag{4.5.25}$$

when n_i and m_i are the concentration and the mass of different interacting atmospheric molecules; $\rho_{o,i}$ are the appropriate optical radii of the collisions; m is the mass of the absorbing molecule. Equation (4.5.25) shows that, in a general case, the Lorentz half-width depends not only on the total pressure but also on the partial pressure of different atmospheric gases.

These effects of self-broadening and mutual broadening may be taken into account by introducing into equation (4.5.24) the dependence of the Lorentz half-width on the partial pressure of the absorbing p_a and broadening p_b gases in a specific form:

$$\begin{aligned} \alpha_L(p_a, p_b) &= (\alpha_a p_a + \alpha_b p_b) = \\ &= \alpha_b \left(\frac{\alpha_a}{\alpha_b} p_a + p_b \right) = \alpha_b (b p_a + p_b), \end{aligned} \tag{4.5.26}$$

where α_b and α_a are the half-widths of the lines as a result of collisions of 'foreign' (mutual broadening) and absorbing (self-broadening) molecules; b is the self-broadening coefficient. Taking into account that the total pressure is $p = p_a + p_b$, we obtain:

$$\alpha_L(p, p_a) = \alpha_b(p + (b - 1)p_a). \tag{4.5.27}$$

The self-broadening factors b may reach high values. For example, for the spectral lines of water vapour, the self-broadening factor may reach the values of 5–20 or even higher. Consequently, it is important to take into account self-broadening effects for a number of atmospheric gases, for example, water vapour in the atmosphere of the Earth.

Special attention has to given to the linear dependence of the half-width of the Lorentz line shape on atmospheric pressure. Consequently, the shape $f_L(\nu - \nu_0)$ depends strongly on pressure.

Figure 4.9 shows the shapes of the Lorentz lines for three different pressures. The graph indicates that with a decrease of pressure (increase of altitude in the atmosphere) the width of the spectral lines of absorption decreases. In this case, the value of the absorption coefficient (or radiation coefficient) in the centre of the line increases, and in the wings of the line (in the sections away from the centre) it decreases. This dependence follows from the simple analysis of equations (4.5.20) and (4.5.24). In fact, in the centre of the line at $\nu = \nu_0$, $f_L(\nu - \nu_0) = 1/\pi\alpha_L$, and in the wing of the line at $(\nu - \nu_0) \gg \alpha_L$, $f_L(\nu_0) = \dfrac{\alpha_L}{\pi(\nu - \nu_0)^2}$. The dependence of the Lorentz line shape on pressure indicates that at every altitude in the atmosphere, the lines of absorption and radiation are characterised by different shapes.

Deviations from the Lorentz line shape

The Lorentz line shape describes quite sufficiently the shape of the spectral lines in the vicinity of their centres. This is the special feature of the Lorentz line shape, and also the simple form of the Lorentz equation, and these are the reasons for its excessive application in atmospheric optics. However, in the wings of the line, the shape is often in very poor agreement with the experimental data. For the absorption lines of different molecules, the Lorentz line shape can both overestimate or underestimate the absorption

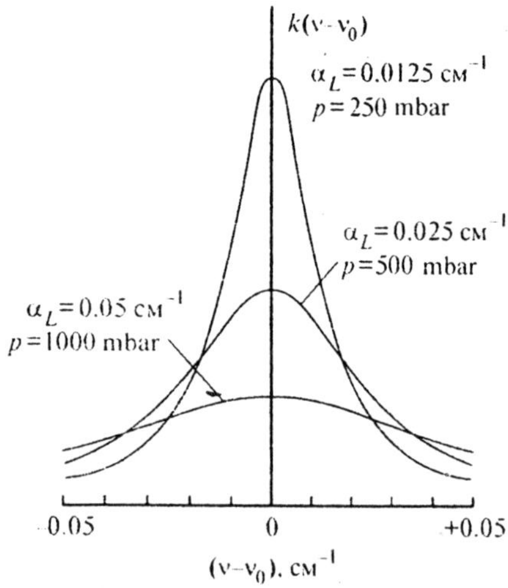

Fig. 4.9. Shapes of the Lorentz lines for three different pressures [91].

coefficient in the wings of the spectral lines. To take these deviations into account, atmospheric optics often uses correction functions $\chi(\nu)$, determined for different molecules and absorption bands, usually on the basis of the experimental data. In this case, the absorption coefficient in the spectral line is approximated by, for example, the equation

$$k(\nu) = Sf_L(\nu - \nu_0)\chi(\nu - \nu_0). \tag{4.5.28}$$

As an example, we give the correction function for absorption bands of carbon dioxide (Benedikt shape) [27, 37, 91]:

$$\chi(\nu-\nu_0)=\begin{cases}\exp[-0.46(|\nu-\nu_0|-5)], & \text{for } |\nu-\nu_0|>5\ \text{cm}^{-1},\\ 1, & \text{for } |\nu-\nu_0|\le 5\ \text{cm}^{-1}.\end{cases} \tag{4.5.29}$$

This relationship shows that the correction function decreases the absorption coefficient in the wings of the lines of carbon dioxide.

Equation (4.5.20) is valid for the infrared range of the spectrum. For the microwave region of the spectrum, where the width of the rotational lines of absorption is comparable with the values of ν_0, the theory gives the following expression for the total Lorentz line shape (Van Fleck–Weisskopf line shape) [91]:

$$f(\nu-\nu_0)=\frac{1}{\pi}\left(\frac{\nu}{\nu_0}\right)^2\times$$

$$\times\left[\frac{\alpha_L}{(\nu-\nu_0)^2+\alpha_L^2}+\frac{\alpha_L}{(\nu+\nu_0)^2+\alpha_L^2}\right]. \tag{4.5.30}$$

Thus it may easily be seen that equation (4.5.30) transforms to (4.5.20) at large values of ν and ν_0. For example, already at ν = 100 cm^{-1}, the difference between them is smaller than 1%.

Shift of the centres of spectral lines

The theory of the broadening of the spectral lines proposed by Lindholm (the theory of adiabatic collisions) gives a more general equation in comparison with the Lorentz line shape [91]:

$$f_L'(\nu-\nu_0)=\frac{1}{\pi}\frac{\alpha_L}{(\nu-\nu_0-\beta)^2+\alpha_L^2}, \tag{4.5.31}$$

where parameter β describes the shift of the centre of the line in relation to pressure. This phenomenon of the shift of the centres of the lines is detected for many gases. Like the half-width, the parameter of the shift is proportional to the frequency of collisions of the molecules and, consequently, pressure. The half-widths and the parameters of the shift of the lines for certain gases are of the same order of magnitude. For example, the half-width of the ammonia lines (self-broadening) is 0.2–0.5 cm^{-1} atm^{-1}, and the shift parameters are 0.04–0.1 cm^{-1} atm^{-1}. The effects of the shift of the spectral lines may be significant when examining the transfer of radiation in narrow spectral ranges, for example, in the absorption of laser quasi-monochromatic radiation. If this phenomenon is not taken into account this may result in large errors in both the calculations of molecular absorption and in the remote measurements, for example, characteristics of the gas composition of the atmosphere. For example, if spectral resolution of the measurements of the transmittance functions of the atmosphere better than 0.01 cm^{-1}, it is necessary to take into account, for instance, the shift of the centres of the lines for the absorption bands of HCl and HF, and also the lines of water vapour in the near-infrared region [93, 97].

Effect of line mixing

It was mentioned previously that simple summation of the absorption coefficients of the individual spectral lines is not always accurate. In the case of overlapping spectral lines there is the 'interference' of lines (line mixing). In calculations of the shapes of Lorentz line shapes, this effect is taken into account approximately using the procedure proposed by Rosencrantz [91, 103]:

$$f_R(\nu-\nu_0)=\frac{1}{\pi}\frac{\tilde{\beta}(\nu-\nu_0)+\alpha_L}{(\nu-\nu_0)^2+\alpha_L^2}, \tag{4.5.32}$$

where $\tilde{\beta}$ is the line mixing parameter which depends on temperature, the type of molecules and the examined absorption shapes. The effect of 'line mixing' of the spectral lines strongly influences the values of the absorption coefficients in the individual spectral ranges in the absorption bands of carbon dioxide, methane, ozone, etc. in the infrared range of the spectrum, in the absorption band of oxygen in the microwave range of the spectrum, when the distance between the spectral lines is small in comparison with the half-width of the lines. As an example, Fig. 4.10 compares the

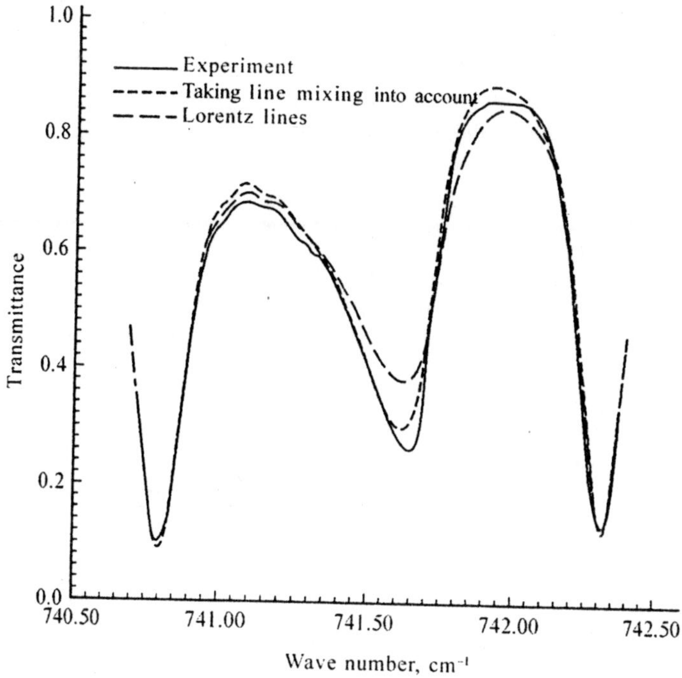

Fig. 4.10. Comparison of the experimenal transmittance functions of carbon dioxide in the spectral range 740.5–742.5 cm^{-1} with calculated values obtained with line mixing taken into account and ignored [93].

experimental functions of transmittance of carbon dioxide in the spectral range 740.5–742.5 cm^{-1} with the calculated values, obtained taking and not taking into account the line mixing effect. The graph shows clearly that if the line mixing effect is taken into account, it is possible to increase the efficiency of reproducibility of the measured results for the Q-wing where the mixing of many lines is observed (the middle part of the graph).

Doppler broadening

Even if it is assumed that there are no mechanisms of broadening such as natural broadening and broadening as a result of collisions, it is in any case likely that the absorption lines of the population of the molecules will not be infinitely narrow (monochromatic). This is associated with the motion of the molecules and the well-known Doppler effect [24, 32, 37, 91]. If the molecule has a component of velocity v in the direction of the view axis and if $v \ll c$, where c is the velocity of light, then from the viewpoint of the standing observer, the frequency of radiation (absorption) of the molecule ν_0 is replaced by the frequency:

$$\nu = \nu_0 (1 \pm v/c). \qquad (4.5.33)$$

The sign in equation (4.5.33) depends on the direction of motion of the molecule: the + corresponds to the motion in the direction to the observer, the – away from the observer.

In the case of thermodynamic equilibrium the probability of the component of velocity being in the range from v to $v + dv$, is described by the Maxwell distribution:

$$dP(v) = \left(\frac{m}{2\pi k_B T}\right)^{1/2} \exp\left(-\frac{mv^2}{2k_B T}\right) dv, \qquad (4.5.34)$$

and in this case

$$\frac{k_B}{m} = \frac{k_B N_A}{\mu} = \frac{R}{\mu}$$

Equation (4.5.34) in accordance with (4.5.33) gives the relative number of molecules emitting (absorbing) at frequency ν. Substituting the value of ν from (4.5.33) into (4.5.34), we obtain the Doppler line shape

$$f_D(\nu - \nu_0) = \frac{1}{\alpha_D \sqrt{\pi}} \exp\left[-\frac{(\nu - \nu_0)^2}{\alpha_D^2}\right], \qquad (4.5.35)$$

where

$$\alpha_D = \frac{\nu_0}{c}\sqrt{\frac{2RT}{\mu}} \qquad (4.5.36)$$

is the parameter of the Doppler width (this should not be confused with the half-width!) of the line. It should be mentioned that (4.5.35) is in fact the normal (Gauss) distribution with the dispersion α_D and, consequently, the normalisation condition (4.2.5) for this equation is fulfilled. The half-width of the Doppler line shape is:

$$\alpha'_D = \alpha_D \sqrt{\ln 2}. \qquad (4.5.37)$$

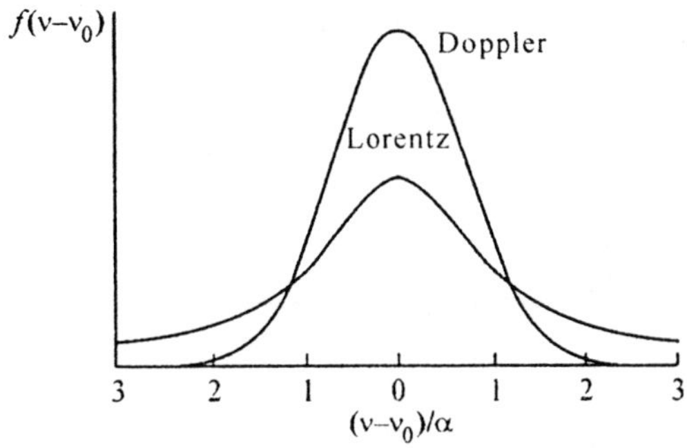

Fig. 4.11. Comparison of the Lorentz and Doppler line shapes for the same intensity and half-width of the lines [91].

Figure 4.11 compares the Lorentz and Doppler line shapes for the same intensity and half-width of the lines. If $f_D(\nu) > f_L(\nu)$ in the centre of the line, then in the wings of the line the situation is reversed: $f_D(\nu) < f_L(\nu)$. Attention should also be given to the dependence of the Doppler width on radiation frequency ν_0. Consequently, the Doppler width greatly changes in the spectrum. In transition from the visible range of the spectrum (the value of ν_0 is of the order of 20 000 cm^{-1}) to the microwave (ν_0 of the order of 10 cm^{-1}) the value of α_D changes by more than 1000 times. For example, the Doppler half-width of the line of atomic oxygen at $\lambda = 5577$ Å at $T = 300$ K is $3.3 \cdot 10^{-2}$ cm^{-1}, and the half-width of the line of the water vapour on $\lambda = 500$ μm and the same temperature is only $3.5 \cdot 10^{-5}$ cm^{-1}. Thus, the Doppler broadening is most significant for the visible and near infrared ranges of the spectrum and provide a very small contribution to the broadening of the spectral line in the microwave range. It should also be mentioned that in contrast to the Lorentz half-width, the value of α_D is independent of the pressure of air and, consequently, Doppler broadening becomes significant in the stratosphere and higher where the Doppler half-width (4.5.23) is small because of low air pressure.

Combined shape of the spectral lines (Voigt shape)

In the atmosphere of the planets, the absorption and radiation lines are determined by the two most important mechanisms of the broadening of the spectral lines – broadening as a result of collisions of the molecules and broadening as a result of the Doppler effects. These mechanisms act simultaneously and in certain conditions they must also be taken into account jointly. To consider this circumstance it is necessary to write the Lorentz line shape (4.5.31) for every molecule having the velocity v in the Maxwell distribution (4.5.34) and, consequently, for every frequency ν in the Doppler line shape (4.5.35). Here ν has already the role of ν_0 and it is necessary to find the total shape of the absorption line. Consequently, we obtain the combined (mixed) shape of the absorption (or radiation) line, i.e. the Voigt line shape [20, 32, 91]:

$$f_V(\nu-\nu_0)=\int_{-\infty}^{\infty} f_D(\nu'-\nu_0) f_L(\nu-\nu')d\nu'. \quad (4.5.38)$$

After simple transformations, the Voigt line shape maybe presented in the following form:

$$f_V(\nu-\nu_0)=\frac{1}{\pi\alpha_D}K(x,y), \quad (4.5.39)$$

where

$$K(x,y)=\frac{y}{\pi}\int_{-\infty}^{\infty}\frac{1}{y^2+(x-t)^2}\exp(-t^2)dt, \quad (4.5.40).$$

$$t=\frac{(\nu'-\nu_0)}{\alpha_D},\ x=\frac{(\nu-\nu_0)}{\alpha_D},\ y=\frac{\alpha_L}{\alpha_D}.$$

The integral $K(x, y)$ of the Voigt line shape is not expressed by means of elementary functions. However, computer calculations of the integral are not associated with any major difficulties at the present time because the effective calculation algorithms $K(x,y)$ have been developed, and different approximate equations of the Voigt line shape have also been proposed.

Deviations from the Voigt line shape – Dicke effect

When deriving the previously described Voigt line shape, we assume the independence of the two mechanisms of broadening of the spectral lines – as a result of collisions and the Doppler effect, i.e. it has been assumed that there are no changes in the velocity of the molecules during their collisions. Taking into account changes in the velocity of the molecules in collisions results in a narrowing of the width of the shape of the lines and in the increase of the absorption in the centre of the line in a specific pressure range. In certain conditions, the width of the lines may be smaller than the Doppler width. The effect of collisional narrowing of the Doppler shape of the spectral lines was predicted for the first time by Dicke and was subsequently discovered in experiments for the spectral lines of a number of atmospheric gases (for example, for hydrogen and water vapours).

The Dicke effect may be taken into account by using the Rautian or Galatri line shapes. For example, the Galatri line shape [93, 97] may be described by the following equation:

$$f(x,y,z)=\frac{1}{\alpha_D\pi}\int_0^\infty \cos(xt)\times$$

$$\times\exp\left\{-yt-\frac{1}{2z^2}(zt-1+\exp(-zt))\right\}dt, \qquad (4.5.41)$$

where the new parameter of the line shape $z = b/\alpha_D$, $b = \beta_0 p$ is the parameter of narrowing, and β_0 is a constant which differs for different gases.

The Dicke effect becomes important when examining absorption in narrow spectral ranges, for example, when examining the absorption of quasi-monochromatic radiation of a carbon dioxide laser with a wavelength of 10.6 μm over a path of ~10 km at the zenith angle of 85°, and ignoring the Dicke effect results in a 20% error in the calculation of the weakening of laser radiation.

In addition to the above mechanisms, various other phenomena may take part in the formation of the shape of the spectral lines, for example, the Stark and Zeeman effects (the effect of electrical and magnetic fields on the levels of molecular energy). These effects are quite distinctive in the upper layers of the atmosphere of the planets where the spectral lines are relatively narrow. In particular, this is manifested in the case of the lines of molecular oxygen in the microwave range of the spectrum.

'Forbidden' and induced absorption of molecules

The previously examined molecular absorption is caused by the so-called permitted transitions [20, 82, 91]. They are found for molecules with a constant electrical dipole moment, and also for molecules whose oscillations result in the formation of a dipole moment. The diatomic molecules, consisting of identical atoms, for example O_2, N_2, HO_2, etc. do not have a constant electrical dipole moment and, according to classic considerations, they should not interact with the radiation field. In fact, the interaction of molecules with the radiation field is also possible if the molecules contain a constant magnetic dipole moment or quadrupole moments of different nature (electrical or magnetic). However, the intensity of interaction of radiation with the molecules with a constant magnetic dipole moment or electric quadrupole moments, is considerably lower (by a factor of 10^5 and 10^8, respectively) in comparison with the case with the electrical dipole moment. The transitions, determined by these types of interaction, are referred to as forbidden transitions. This transitions correspond to the actual lines and absorption bands of different molecules and gases. Although in most cases the intensity of these lines is very low, they are manifested in the absorption spectra of the atmosphere in the case of a high content of these molecules in the atmospheres of the planets (as in the case of the forbidden lines of O_2 and N_2 in the atmosphere of the Earth, forbidden lines of hydrogen in the atmospheres of giant planets). From the energy viewpoint, these lines do not play any significant role in the atmosphere of the Earth, for example, in the absorption of solar radiation or the transfer of atmospheric radiation. However, these absorption bands may play a significant role in the formation of different glows of the atmosphere and also in the problems of remote sensing of the atmosphere. For example, the absorption band of oxygen in the microwave range of the spectrum, determined by the presence of a large magnetic dipole moment in the molecule, is used for temperature sounding of the atmosphere (chapter 10).

There is another type of molecular absorption, often referred to in the literature as absorption, induced by pressure (or absorption, induced by collisions) [20, 91]. The molecules with no constant dipole moments may acquire such a moment during collisions with the surrounding molecules. In the same short periods of time when the structure of the molecule is disturbed by the collisions, it may have an electrical dipole moment and, consequently, interact with the radiation field. Another explanation of induced absorption is the formation of dimers and more complicated formations of the molecules at interaction (collisions) of the molecules. This may also be accompanied by the formation of the induced electrical dipole moment in these formations and, consequently, their interaction with the radiation field (i.e. absorption and emission of electromagnetic energy) is possible. The special feature of the examined type of absorption is that the intensities of the absorption bands induced by pressure in contrast to ordinary bands vary in proportion to the pressure in the gas medium. Therefore, the volume coefficient of absorption in these bands is proportional to the concentration of the absorbing molecules n_a and the concentration of the foreign (expanding) gas n_b.

In the case of a single-component mixture, consisting of the absorbing gas, the volume coefficient is proportional to the square of concentration of this gas. It should also be mentioned that the same dependence is characteristic of the wings of the lines in the the line shape, determined by collisions of the molecules. Therefore, the pressure-induced absorption is usually characterised by the binary absorption coefficients [20, 91]:

$$k_{ab} = \frac{k_\nu}{n_a n_b} \tag{4.5.42}$$

for the mixture of two gases, and

$$k_{aa} = \frac{k_\nu}{n_a^2} \tag{4.5.43}$$

for the one-component mixture. It should be mentioned that the dimension of the coefficient k_{ab} and k_{aa} is expressed in cm^5.

4.6. Quantitative characteristics of molecular absorption

The quantitative characteristics of molecular absorption are determined experimentally and also on the basis of special calculations. The form of presentation of the quantitative characteristics of molecular absorption may be different:

– in the form of coefficients of molecular absorption as a function of the wavelength (or frequency), temperature and pressure;

– in the form of parameters of the fine structure of the spectrum of different absorption bands and different atmospheric gases (spectroscopic parameters) – position of the individual spectral lines, the intensity of these lines, half-width, etc.

– in the form of the dependence of the transmittance function for finite spectral ranges (different widths) on the content of the absorbing gas, pressure (in some cases, the pressure of the absorbing and secondary gas), temperature, the position and width of the spectral range.

Coefficients of molecular absorption

This form of presentation of the quantitative characteristics of molecular absorption is used in most cases for the absorption bands characterised by the relatively low spectral dependence of the absorption coefficient. As mentioned previously, this type of dependence in atmospheric optics is referred to as non-selective (or continuous). In fact, there is no principal difference between the selective and non-selective types of absorption. For example, selective absorption (absorption in distinctive spectral lines) transforms to non-selective absorption with the increase of pressure of either absorbing or expanding gases. Actually, the increase of pressure increases the half-width of the Lorentz lines (4.5.24) and the individual lines overlap more and more. In this case, the spectral behaviour of the total coefficient of absorption becomes smoother and smoother and the case of the non-selective absorption is realised. For example, the rotational–vibrational bands of carbon dioxide, characterised by the distinctive selective nature in the atmosphere of the Earth and Mars, have the continuous form in the lower layers of the atmosphere of Venus at high pressures and high temperatures. The non-selective nature of the spectral behaviour of the coefficient of molecular absorption is also observed in the case of relatively low atmospheric pressures (0.01–1 atm) for the absorption bands with the very closely spaced spectral lines.

Thus, the weak spectral dependence of molecular absorption is recorded in the case of the absorption spectra determined by the processes of photoionisation and photodissociation (paragraph 4.2) with the strong overlapping of the individual spectral lines in the absorption bands (electronic–vibrational–rotational absorption bands in ultraviolet, visible and near infrared ranges of the spectrum, vibrational–rotational spectra of absorption of 'heavy' molecules, such as, for example, chlorofluorocarbonates (CFC), hydrofluorocarbons (HDFC), molecules of CF_4, N_2O_5, etc. in the infrared range of the spectrum. As the example, Fig. 4.12 shows the cross-sections of molecular absorption in ultraviolet and visible ranges of the spectrum for many atmospheric molecules [54].

Parameters of the fine structure of absorption bands

Recently, special data banks have been compiled for the parameters of the fine structure, including different characteristics of the individual spectral lines of different atmospheric gases. These data

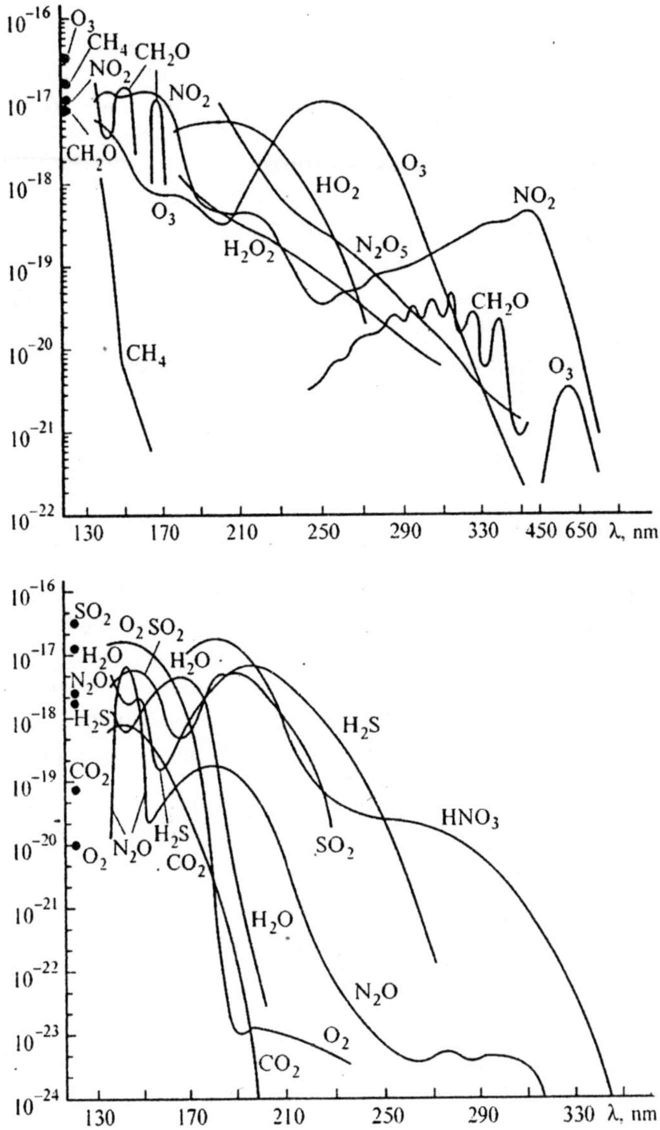

Fig. 4.12. Absorption cross-sections of molecules of gas components of the atmosphere in the ultraviolet and visible ranges of the spectrum as a function of wavelength λ [54]. The points in the left hand side of the graphs indicate the cross-sections for the wavelength of the Layman-α line (121.5 nm).

have been obtained on the basis of laboratory measurements, in most cases, with high spectral resolution, and also by means of calculations on the basis of the quantum-mechanics methods. The currently available data banks of the parameters of the fine structure include information for many atmospheric gases and their absorption bands. For the quantitative description of the molecular absorption it is necessary to use larger and larger numbers of the parameters of the fine structure. The required number of parameters for the adequate description of molecular absorption in the spectral lines depends on the requirements on the accuracy of the appropriate calculations in a specific problem of atmospheric optics. As the required accuracy increases, it is necessary to examine in greater detail different physical mechanisms, determining selective molecular absorption. In addition to this, the value of the spectral resolution (the width of the investigated spectral interval) influences to a certain degree the required level of detailed elaboration. For example, in the problems of radiation energetics (examination of wide spectral ranges) the consideration of the 'fine' effect, such as the dependence of the central frequency of the line on pressure is superfluous, but necessary, for example in the examination of the absorption of quasi-monochromatic laser radiation in the atmosphere. Finally, it should be mentioned that the realisation of new measurements and also improvement of quantum-mechanics methods of calculation result in a situation in which these data banks are constantly expanded and made more accurate.

Data banks on the parameters of the fine structure

There are several data banks on the parameters of the fine structure, for example, the data banks HITRAN, GEISA, the data banks of the Institute of Atmospheric Optics of the Siberian division of the Russian Academy of Sciences, a special databank for the spectral lines of absorption in the microwave range of the spectrum, etc.

As indicated by the above information on the relationships of molecular absorption in the individual spectral lines, the quantitative description requires definition of the entire range of the parameters. For example, the HITRAN databank of these parameters includes:

- the index of the molecule;
- the index of the isotope of the molecule;
- the central frequency (the position of the centre of the line) of the spectral line;
- the intensity of the line;

– the matrix element of the dipole moment of transition;

– the Lorentz half-width in broadening by the air (the mixture of the molecules of N_2 and O_2 in the proportion, corresponding to the atmosphere of the Earth) at a pressure of 1 atm and a temperature of 296 K;

– the Lorentz half-width in the case of self-broadening at p = 1 atm and T = 296 K;

– the energy of the lower states of the corresponding transition;

– the parameter of the temperature dependence of the half-width of the line (4.5.24);

– the parameter of the shift of the centre of the line at p = 1 atm and T = 296 K;

– the quantum numbers of the upper and lower state of the appropriate transition.

At the present time, the data banks also include the parameters of the lines for describing the line mixing phenomenon of the spectral lines. The data banks also include estimates of the errors of the tabulated parameters.

Table 4.2 shows the general information on the database HITRAN-96 (the number 96 indicates the year of issue of the database) – the examined molecules and the number of their isotopes, the number of the bands, included in the databank for the given molecule, the number of the lines in the bands and the general spectral range, including these lines.

To provide more information, Fig. 4.13 shows the spectral ranges containing the lines of absorption of different gases (index JPL for O_3, O and HO_2 indicates that the appropriate information was taken from the JPL databank (Jet Propulsion Laboratory, USA)). Attention should be given to the fact that the absorption lines of the water vapour are distributed in almost the entire examined spectral range from 0 to 10 000 cm^{-1}. It should also be mentioned that Fig. 4.13 shows the position of the CO lines in the solar radiation spectrum.

The currently available data banks of the spectroscopic parameters contain information on a vast number of the individual spectral lines. For illustration, Fig. 4.14 shows, for a relatively narrow spectral range (1200–1250 cm^{-1}), the position and intensity of the spectral lines of different atmospheric gases from the HITRAN-96 database. The figure clearly indicates the large number of the absorption lines in the examined spectral range, the 'individuality' of the spectra of different molecules, and also strong overlapping of the spectral lines of various molecules such as

Table 4.2. Gases included in the HITRAN-96 database [103]

Molecule	Number of isotopes	Number of bands	Number of lines	Spectral range, cm^{-1}
H_2O	4	137	49444	0–22657
CO_2	8	589	60802	442–9649
O_3	5	106	275133	0–4033
N_2O	5	164	26174	0–5132
CO	6	47	4477	3–6418
CH_4	3	51	48032	0–6185
O_2	3	19	6292	0–15928
NO	3	50	15331	0–3967
SO_2	2	9	38853	0–4093
NO_2	1	12	100680	0–2939
NH_3	2	40	11152	0–5295
HNO_3	1	13	165426	0–1770
OH	3	103	8676	0–9997
HF	1	6	107	41–11530
HCl	2	17	533	20–13458
HBr	2	16	576	16–9759
HI	1	9	237	12–8488
ClO	2	12	7230	0–1208
OCS	4	7	858	0–2089
H_2CO	3	10	2702	0–2999
HOCl	2	6	15565	0–3800
N_2	1	1	120	1922–2626
HCN	3	8	772	2–3422
CH_3Cl	2	8	9355	679–3173
H_2O_2	1	2	5444	0–1500
C_2H_2	2	11	1688	604–3375
C_2H_6	1	2	4749	720–3001
PH_3	1	2	2886	708–1411
COF_2	1	7	54866	725–1982
SF_6	1	1	11520	940–953
H_2S	3	15	7151	2–2892
HCOOH	1	1	3388	1060–1162
HO_2	1	4	26963	0–3676
O	1	1	2	68–159
$ClONO_2$	2	3	32199	763–798
NO^+	1	6	1206	1634–2531
HOBr	2	2	4358	0–316

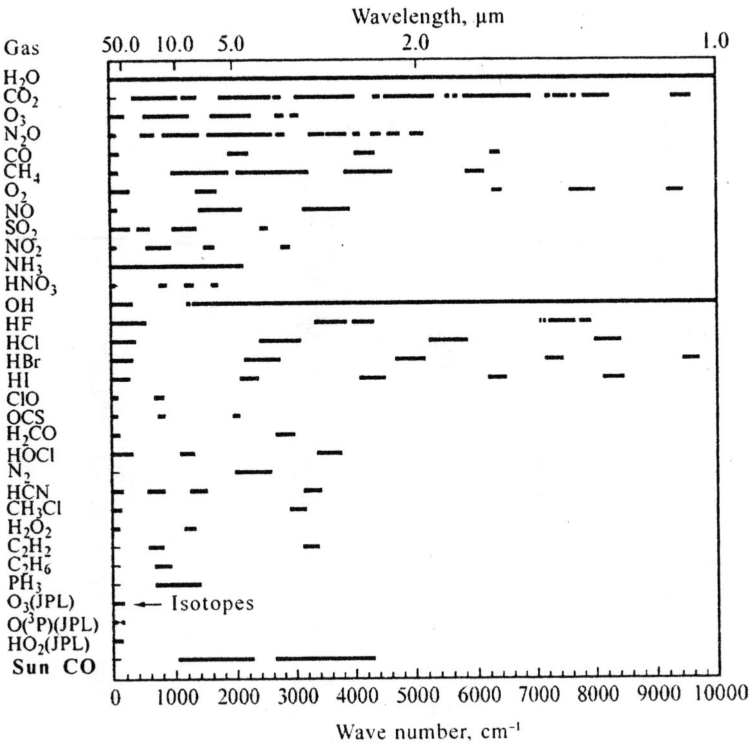

Fig. 4.13. Spectral ranges with the absorption lines of different gaes, and the position of CO lines in the solar radiation spectrum [103].

$C_2Cl_3F_3$, $C_2Cl_2F_4$ and N_2O_5.

In the data banks, for example in the HITRAN-96 database, there are (in addition to the parameters of the fine structure) also the quantitative characteristics of the continuous type of absorption in the form of the absorption coefficient for different molecules (the so-called 'spectrally unresolved' absorption bands). As an example, Fig. 4.15 shows the position of the absorption bands of this type in the spectral range 700–1800 cm^{-1}. These data, obtained in the experiments, are presented in the data banks for different temperatures, usually overlapping the range of variability of the temperatures in the atmosphere of the planets. Consequently, in the calculations of the absorption characteristics of the atmospheres of the planets it is possible to take into account the temperature dependence of the absorption coefficient. In the case of relatively low pressures the 'spectrally unresolved' absorption bands also start to show the dependence of the absorption coefficient on pressure.

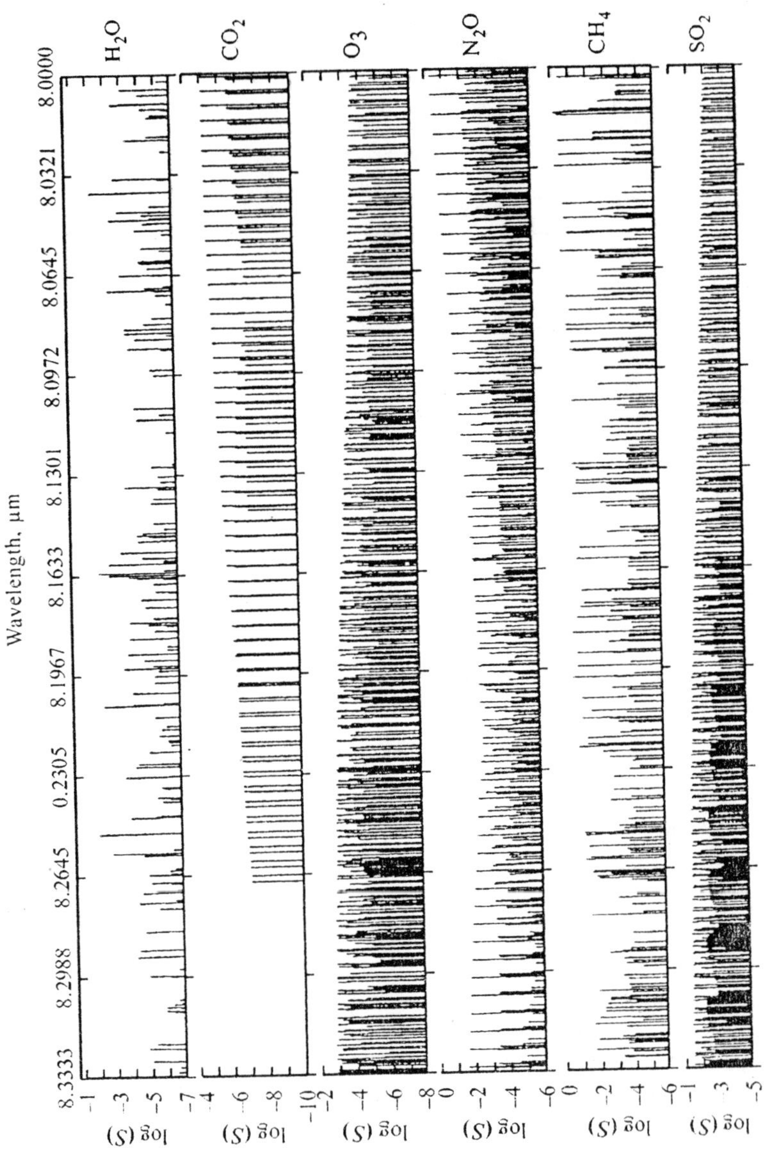
Wavelength, µm
8.0000
8.0321
8.0645
8.0972
8.1301
8.1633
8.1967
0.2305
8.2645
8.2988
8.3333
H_2O
CO_2
O_3
N_2O
CH_4
SO_2
log (S)

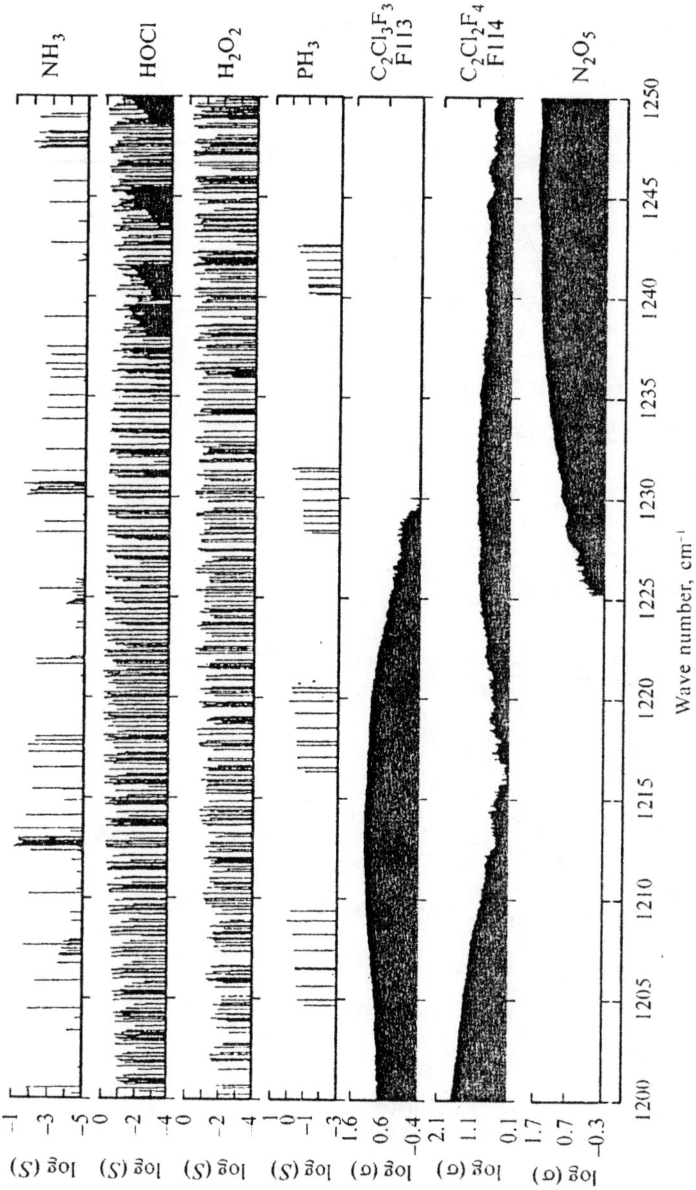

Fig. 4.14. Position and intensity of spectral lines of various atmospheric gases for the spectral range (1200–1250 cm^{-1}) as in the HITRAN-96 database [103].

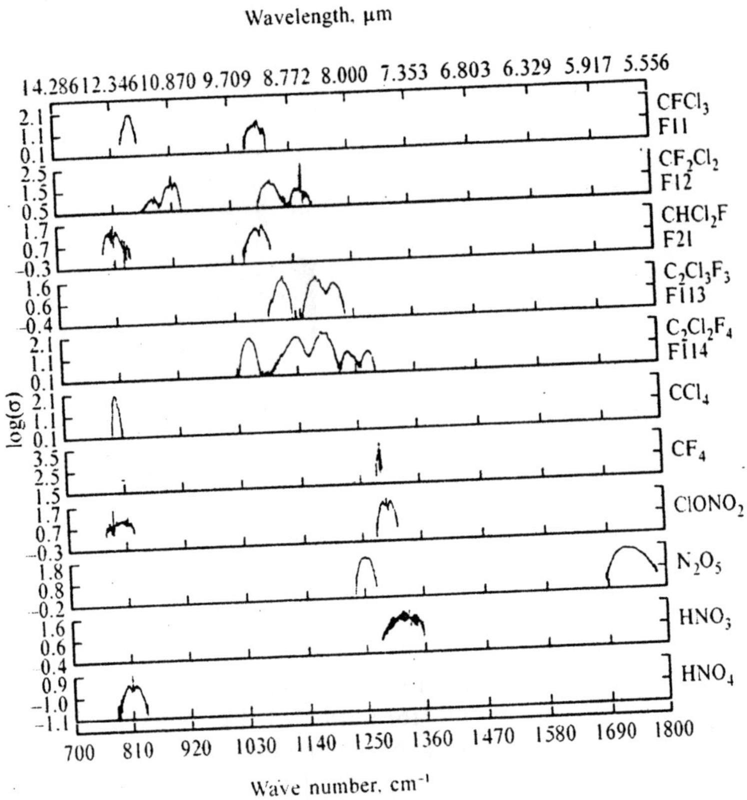

Fig. 4.15. Position of 'spectrally' unresolved' absorption bans in the range of 700–1800 cm^{-1} as in the HITRAN-96 database [103].

4.7. Molecular absorption in the Earth atmosphere

Ultraviolet and visible ranges of the spectrum

Figure 4.12 showed the absorption coefficients of a large number of different atmospheric components. The actual effect of these components on transfer of, for example, solar radiation, is determined by the content of these components in the Earth atmosphere. Table 4.3 shows the main absorption bands of different atmospheric gases (taking the content of the gases into account) in the ultraviolet and visible ranges of the spectrum.

As indicated by Table 4.3, atmospheric nitrogen and oxygen are responsible for the absorption in the range of extreme UV radiation. The molecules of oxygen and nitrogen strongly absorb

Table 4.3. Main absorption bands of the Earth atmosphere in UV and visible ranges of the spectrum

Gas	Spectral range, nm	Name of band	Absorption
N	1–100	Ionisation bands	Weak
O	1–100	"	Very strong
N_2	< 80	Ionisation continuum	Weak
	80–100	Tanaka–Worley	Very strong
	100–140	Lyman–Birge–Hopfield	Strong
O_2	< 100	Hopfield	Very strong
	100–125	–	"
	125–200	Schuman–Runge	Strong
	200–260	Hertzberg	Weak
O_3	200–300	Hartley	Strong
	300–360	Huggins	Medium
	450–700	Chappui	Weak
NO_2	400–600	–	"

solar radiation in the spectral range from 100 to 200 nm. All these components play a significant role in the absorption of solar radiation in the upper atmosphere and are responsible for its thermal regime and also for the formation of the ionosphere. The important role in the absorption of solar ultraviolet radiation is played by the ozone molecules. In particular, Hartley and Huggins bands are very important in the formation of the thermal regime of the mesosphere and stratosphere (the maximum temperature is recorded at an altitude of approximately 50 km (stratopause), determined by radiation heating), and also in protecting the surface of the Earth against the ultraviolet radiation dangerous for the biosphere. Therefore, the disruption of the ozone layer (depletion of the ozone content of the stratosphere) may have a significant effect on the climate of the Earth and the conditions of existence of the biosphere. The Shappui absorption bands O_3 and NO_2 are relatively weak in the Earth atmosphere, but the measurements taken in these bands, and also in the Hartley–Huggins bands, are used for the examination of the gas composition of the atmosphere from the satellites and from the Earth surface.

Convincing illustration of absorption in the Earth atmosphere in the ultraviolet range of the spectrum is the example shown previously in Fig. 4.2. It may be seen that in the shortwave range, solar radiation is greatly weakened already at an altitude of 60–100 km (absorption of O, N, N_2 and O_2), starting at a wavelength of 220 nm, with ozone being the main absorption agent. As indicated by Fig. 4.12, in the ultraviolet and visible ranges of the spectrum other atmospheric molecules also absorb. However, with the exception of O_2 and O_3, the absorption by other gases in the atmosphere of the Earth in the normal conditions is small (either the concentration of the corresponding gases is low or their absorption bands are located in the spectral ranges of the very strong absorption bands of O_2 and O_3 which screen them completely). Some of these bands are used for the determination of the content of the atmospheric gases by optical methods (for example, in recording solar radiation) in the ground-based and satellite experiments with special measurement geometry (for example, in twilight conditions or limb satellite experiments, see chapter 10).

The infrared range of the spectrum

The infrared range of the spectrum contains a vast number of the absorption lines of different vibrational–rotational and rotational bands of the atmospheric components. Table 4.4 gives the information on the position of the main absorption bands of different atmospheric gases in the infrared range of the spectrum and the sections of the maximum intensity of absorption – the centres of

Table 4.4. Important bands and ranges of the absorption of atmospheric gases

Gases	Band, μm	Centre of band, cm^{-1}	
CO_2	15	667	550–800
	10.6	961.0 and 1063.8	850–1100
	4.3	2349	2100–2400
H_2O	Rotational	1594.8	0–1000
	6.3		640–2800
	Continuum		400–1200
O_3	9.6	1043 and 1110	950–1200
	14.2	705	600–800
	4.8	2105	2000–2200
CH_4	7.6	1306.2	950–1650
	3.3	3018.9	2700–3300
N_2O	7.9	1285.6	1200–1350
	17.0	588.8	520–660
	4.5	2223.5	2120–2270
CO	4.7	2143.3	2000–2250
Freons	8–12		700–1300

the bands (in cases in which the maxima are easy to separate). The strongest absorption of the water vapour is recorded in the range of 6 μm and in the range of the rotational band (20–150 μm), although the rotational lines 'start' from the microwave range of the spectrum (λ = 1.35 cm) and are also situated in the 'the window of transparency' (the range of small absorption), 8–12 μm. These absorption bands of the water vapour play a significant role in the radiation heat exchange, measurements taken in these bands are used for remote (on the earth and satellites) determination of the content of H_2O vapours in the atmosphere (see chapter 10).

The molecule of CO_2 contains, in the infrared range of the spectrum, two high-intensity absorption bands, 4.3 and 15 μm. The long-wave absorption band plays a significant role in the radiation regime of the stratosphere and in the higher layers of the atmosphere. The measurements of outgoing radiation in both bands are used for the determination of the profile of the temperature of the atmosphere (for thermal sounding of the atmosphere) from space (see chapter 10).

The absorption bands of ozone are in the vicinity of 4.8, 9.6 and 15 μm. The most important absorption band is at 9.6 μm. This band plays a significant role in the radiation heat exchange in the stratosphere and mesosphere and is used for the remote determination of the vertical profile and the total content of O_3.

The important absorption bands of CH_4 and N_2O are situated in the vicinity of 7.6, 3.3 and 7.8 μm. Although the contribution of these bands to radiation heat exchange is not large, the measurements of radiation in these bands may be used for the determination of the total content of these gases. A relatively intensive band CO is found in the vicinity of 4.7 μm and has already been used for satellite mapping of the total content of carbon oxide in the global scale. The absorption bands of freons, important greenhouse and ozone-depleting gases, are situated in the transparency window of 8–12 μm.

In addition to the absorption bands of different gases in the infrared range of the spectrum, it is necessary to indicate the spectral ranges in which the intensity of atmospheric absorption is relatively small. These are determined as 'transparency windows' of the atmosphere. In addition to the already mentioned very important transparency window of 8–12 μm, there are transparency windows in the vicinity of 4.0 μm and in the near infrared range of the spectrum (the entire visible range of the spectrum is often regarded as the transparency window). The measurements of outgoing radiation in these transparency windows are used for the remote sounding of the surface of the Earth and clouds. The transparency window of 8–12 μm plays a significant role in the radiation heat exchange. In this spectral range there is the maximum of thermal radiation at the temperatures of the atmosphere existing on the Earth. In particular, the transparency window is directly responsible for the cooling of the surface of the Earth and the atmosphere as a result of radiation into space. The presence of the absorption bands of CH_4, N_2O and freons in the transparency window of 8–12 μm and the increase of the content of these gases in the atmosphere of the Earth as a result of antropogenous factors results in the additional 'greenhouse' effect and the change of the climate of the Earth: the formation of additional atmospheric absorption reduces the intensity of radiant cooling of the atmosphere and the surface.

It is also important to mention the absorption bands of O_2, H_2O and CO_2 in the near infrared range of the spectrum. These bands are relatively weak but they absorb the solar radiation arriving on the surface of the earth and, consequently, also play a significant role in the radiation energetics of the atmosphere and the surface. This is illustrated in Fig. 1.5, 4.1 and 4.3.

Microwave range of the spectrum

Although radiation in this range of the spectrum does not play any significant role in the processes of radiation heat exchange because of the small fraction of thermal radiation at a wavelength of $\lambda > 100$ μm, it is important for remote (satellite) sounding of the atmosphere and the surface of the Earth. The main absorbing agents in this region are O_2 and H_2O (Fig. 4.16). Remote measurements of the temperature profile are taken using the band of O_2 at $\lambda = 0.5$ cm (60 GHz) and the separate absorption band $\lambda = 0.25$ cm (120 GHz). The absorption bands of H_2O, situated in the vicinity of $\lambda = 1.35$ cm (22.2 GHz) and $\lambda = 0.16$ cm (187.5 GHz) are suitable for remote measurements of the profile of water vapours.

In the microwave region of the spectrum there is also a large number of rotational lines of many atmospheric gases [109], for example, ozone, ClO, NO, NO_2, HNO_3, etc. Although their intensity in the absorption spectrum of the Earth atmosphere is relatively low, at a specific geometry of measurements and in the presence of high-sensitivity devices they are used for the remote measurements of the characteristics of the gas composition of the atmosphere.

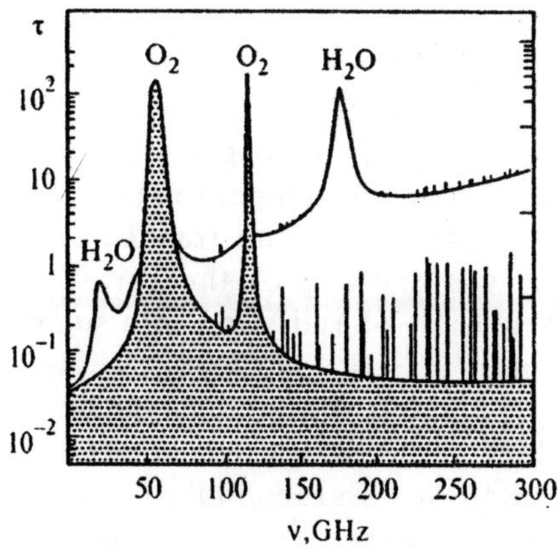

Fig. 4.16. Optical thickness of the atmosphere τ as a function of frequency ν in the microwave range. Cross-hatched area – absorption by H_2O not taken into account, upper curve – absorption by H_2O taken into account [119]

CHAPTER 5

LIGHT SCATTERING IN THE ATMOSPHERE

5.1. Molecular scattering

Modern optics uses two physical models, explaining the nature of scattering of light by gases: the scattering of light directly on molecules (atoms) of the gases (Rayleigh–Tindall theory), and the scattering of light on thermal fluctuations of the density of the gas, resulting in identical fluctuations of its refractive index (Einstein–Smolukovskii theory). The fluctuations theory of scattering is more general, and the case of molecular theory is regarded as a partial one. However, in atmospheric optics, by tradition ascending from Rayleigh, it is recommended to use the molecular theory of scattering [20, 24, 26, 32, 33, 37] and, for this reason, we shall also use his theory.

Derivation of equations for molecular scattering

We examine the interaction of an electromagnetic wave with air molecules. It is assumed that the electromagnetic wave is incident on a separate molecule. Since the dimensions of the molecule are considerably smaller than the wavelength, all points of the molecule will be found in the electromagnetic field of the same strength since the spatial variations of the strength at distances equal to the size of the molecule are ignored. Thus, the external field, acting on the molecule, may be regarded as homogeneous. Under the effect of the electrical field of the incident wave, the charges of the particles, forming the molecule (the phenomenon of polarisability of matter) are separated, and the molecule acquires its own electrical field. This field is approximated as the field of the electric dipole. The oscillations of the external field (with time) lead to identical oscillations of the dipole, i.e. its movement with acceleration and, consequently, the dipole itself becomes the secondary centre of generation of the electromagnetic wave. It is a secondary wave and radiation is scattered.

It is also assumed that an external field with the strength E_0 is incident on the molecule and induces the dipole moment of the molecule P. At the start, it is assumed that the incident wave is characterised by linear polarisation (chapter 3). Consequently, vectors E_0 and P are always found in the same plane. The directions of oscillations of the vectors are parallel. We use the formula, available in electrodynamics, for the field of a vibrating dipole E_1 in the long-range zone (i.e. at $r \gg \lambda$):

$$E_1(\theta) = \frac{1}{2c^2 r}\frac{\partial^2 P}{\partial t^2}\sin\theta, \tag{5.1.1}$$

where θ is the angle between the axis of the dipole and the direction of scattered radiation; r is the distance from the dipole to the observation point. We take into account the relationship of the dipole moment with the external field:

$$P = \tilde{\alpha} E_0,$$

here $\tilde{\alpha}$ is the polarisability of the medium (the gas in the present case). Consequently, remembering (3.1.1):

$$E_0(x,t) = E_{0,0}\cos(2\pi\nu t - \frac{2\pi}{\lambda}x + \delta),$$

$$E_1 = -\tilde{\alpha}\frac{1}{c^2 r}(2\pi\nu)^2 E_0 \sin\theta. \tag{5.1.2}$$

In final analysis, we are interested in the intensity of scattered radiation. In chapter 3 it has been confirmed that the intensity of radiation in vacuum is independent of distance. Consequently, the intensity of scattered radiation is independent of distance both directly or indirectly (through constants which depend on r). Therefore, we are completely justified to ignore, already in the initial stage, the dependence of the investigated quantities on r without waiting until r 'cancels', and write (5.1.2) in the form:

$$E_1 = -\tilde{\alpha}\frac{1}{c^2}(2\pi\nu)^2 E_0 \sin\theta. \tag{5.1.3}$$

It is now assumed that the incident wave is characterised by elliptical polarisation in a general case. Consequently, as shown in chapter 3, the vector of the electrical field of the wave maybe decomposed into two mutually normal components. We select (Fig. 5.1) the component $E_{0,\parallel}$ situated in the plane formed by the direction of the incident wave E_0 and the scattered wave $E_1(\gamma)$. This plane is referred to as the scattering plane. Component $E_{0,\perp}$ is normal to the scattering plane. However, in this case for $E_{0\perp}$, the angle between the component and the scattering direction is always $\pi/2$ and $\sin\theta = 1$. It is also taken into account that the scattering angle γ is the angle between the directions of the incident and scattered radiation, i.e. (Fig. 5.1) the angle $\gamma = \pi/2-\theta$, and consequently, $\sin\theta = \cos\gamma$. Taking this into account, from (5.1.3) we obtain:

$$E_{1,\parallel} = -\tilde{\alpha}\frac{1}{c^2}(2\pi\nu)^2 E_{0,\parallel}\cos\gamma,$$

$$E_{1,\parallel} = -\tilde{\alpha}\frac{1}{c^2}(2\pi\nu)^2 E_{0,\perp}. \tag{5.1.4}$$

The resultant equations (5.1.4) are equations (3.6.3) for the relationship between the electrical vectors of the incident and scattered waves. Consequently, the coefficients of this relationship are:

$$S_1 = -\tilde{\alpha}\frac{1}{c^2}(2\pi\nu)^2\cos\gamma,\quad S_4 = -\tilde{\alpha}\frac{1}{c^2}(2\pi\nu)^2. \tag{5.1.5}$$

The matrix and phase function of molecular scattering

According to (3.6.5), we obtain the elements of the matrix of molecular scattering

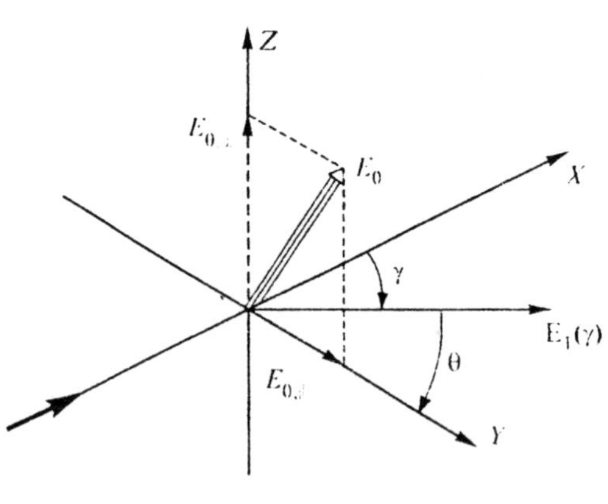

Fig. 5.1. Scattering of an electromagnetic wave on an air molecule.

$$D_{11}(\gamma)=\frac{1}{2}(S_1S_1^*+S_4S_4^*)=A(1+\cos^2\gamma),$$

$$D_{12}(\gamma)=\frac{1}{2}(S_1S_1^*+S_4S_4^*)=A(\cos^2\gamma-1)=-A\sin^2\gamma),$$

$$D_{33}(\gamma)=\frac{1}{2}(S_1^*S_4+S_1S_4^*)=2A\cos\gamma, \tag{5.1.6}$$

$$D_{34}(\gamma)=\frac{i}{2}(S_1^*S_4-S_1S_4^*)=0,$$

where (taking into account that the wavelength of light in vacuum is $\lambda = c/\nu$)

$$A=\frac{\tilde{\alpha}^2 8\pi^4\nu^4}{c^4}=\frac{\tilde{\alpha}^2 8\pi^4}{\lambda^4}. \tag{5.1.7}$$

Now, we normalise the scattering matrix, and for this purpose it should be remembered that it is necessary to separate all its elements to the integral

$$\frac{1}{4\pi}\int_{4\pi}D_{11}d\Omega=\frac{1}{2}\int_0^{\pi}D_{11}\gamma\sin\gamma d\gamma,$$

which, after calculations, gives

$$\frac{1}{2}\int_0^{\pi}A(1+\cos^2\gamma)\sin\gamma d\gamma=\frac{1}{2}A\int_{-1}^{1}(1+x^2)dx=\frac{4}{3}A. \tag{5.1.8}$$

Thus, we have obtained the normalised matrix of molecular scattering

$$\frac{3}{4}\begin{pmatrix} 1+\cos^2\gamma & -\sin^2\gamma & 0 & 0 \\ -\sin^2\gamma & 1+\cos^2\gamma & 0 & 0 \\ 0 & 0 & 2\cos\gamma & 0 \\ 0 & 0 & 0 & 2\cos\gamma \end{pmatrix}, \tag{5.1.9}$$

In the first element of this matrix is the phase function of molecular scattering

$$x(\gamma) = \frac{3}{4}(1 + \cos^2\gamma). \tag{5.1.10}$$

According to (5.1.10), molecular scattering is not isotropic. It is greater in the forward and backward directions, and smaller in the lateral direction. The phase function (5.1.0) is the Rayleigh scattering phase function (Fig. 5.2), and molecular scattering is often referred to as Rayleigh scattering.

The cross-section and volume coefficient of molecular scattering

We return to the value of the normalisation potential (5.1.8). In chapter 3 (paragraph 3.6), it was explained that the scattering matrix of a particle (a molecule in the present case) is represented in the form of the product of the scattering section and the normalised scattering matrix, i.e. the relationship (3.6.7). Consequently, the normalisation coefficient (5.1.8) is nothing else but

$$\frac{4}{3}A = \frac{1}{4\pi}C_s,$$

Here C_s is the cross-section of molecular scattering and

$$C_s = 4\pi\frac{4}{3}A = \frac{128\pi^5\tilde{\alpha}^2}{3\lambda^4}. \tag{5.1.11}$$

The expression for the polarisability of a homogeneous dielectric in a homogeneous field, known from electrostatics, is the following:

$$\tilde{\alpha} = \frac{\varepsilon - 1}{4\pi N} = \frac{n^2 - 1}{4\pi N}, \tag{5.1.12}$$

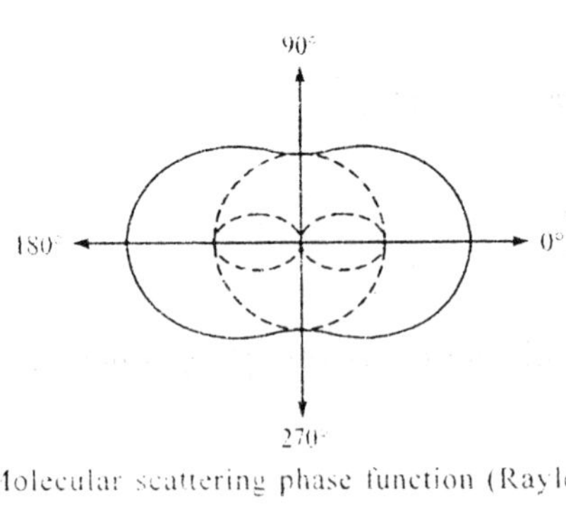

Fig. 5.2. Molecular scattering phase function (Rayleigh phase function).

where $\varepsilon = n^2$ is dielectric permittivity, associated with the refractive index of the gas n; N is the number of gas molecules in the unit volume, i.e. the countable concentration of the gas. Finally, we obtain the following equation for the molecular scattering cross-section

$$C_s = \frac{8\pi^3(n^2-1)^2}{3N\lambda^4}. \tag{5.1.13}$$

Assuming that all molecules interact independently with the radiation, we obtain the following equation for the volume coefficient of molecular scattering $\sigma = NC_s$

$$\sigma = \frac{8\pi^3(n^2-1)^2}{3N\lambda^4}. \tag{5.1.14}$$

The relationships (5.1.13) and (5.1.14) are slightly confusing because the concentration of the particles N is in the denominator of the equations, i.e. it would appear that as the concentration N decreases, the intensity of molecular scattering should increase which, evidently, contradicts the physics of the process. However, it should be mentioned that the polarisability $\tilde{\alpha}$ in the initial equation (5.1.11) is the characteristic of the molecule of matter, i.e. it is independent of concentration. However, the dependence on N is reflected in the refractive index of the gas because, according to (5.1.12), (n^2-1) is proportional to N. Therefore, in complete correspondence with (5.1.11) and the physical meaning, the section of molecular scattering (5.1.13) is independent of the concentration of the particles and the volume coefficient of molecular scattering (5.1.14) is directly proportional to concentration.

Rayleigh law

According to (5.1.14), the volume coefficient of molecular scattering is inversely proportional to the fourth power of the wavelength of light. This claim is referred to as the Rayleigh law. According to this law, blue and grey light rays (approximately 0.45 μm) are scattered in the air to a considerably greater extent (by almost a factor of 4) in comparison with orange and red rays (0.65 μm). This also explains the blue colour of the cloudless sky determined by the scattered solar radiation with the blue and grey rays being dominant.

The extinction of the intensity of light as a result of molecular scattering is also stronger for the blue and grey rays in comparison with the orange and red ones, as directly indicated by the Bouguer law (3.4.5). This also explains the red colour of the sunset and also the red colour of the setting Sun and the Moon: in the case of high zenith angle θ the power of the exponent increases and the extinction in the atmosphere becomes considerable and, consequently, in the spectrum of direct and scattered radiation, mainly red and orange rays are retained..

Volume coefficients and optical thicknesses of Rayleigh scattering

The quantitative characterisation of Rayleigh scattering and extinction may be carried out using Table 5.1 which gives the volume coefficients of molecular scattering σ at pressure p = 1 atm and T = 15 °C and the appropriate optical thicknesses (along the vertical) $\tau(0, \infty)$ of the entire atmosphere of the earth.

Table 5.1 shows clearly the strong spectral dependence of the coefficient of Rayleigh scattering and optical thickness of the entire of atmosphere. If in the case of the wavelength of 0.30 μm the optical thickness of the atmosphere is higher than 1, in the near infrared range it does not exceed fractions of unity. This indicates the important role played by Rayleigh scattering in the extinction of, for example, solar radiation in the ultraviolet range of the spectrum and its low value in the infrared and even more so microwave spectral ranges. Usually, the molecular scattering in these longwave spectral ranges is ignored when solving different

Table 5.1. Molecular scattering coefficients and optical thickness of the atmosphere along the vertical [24]

Wavelength, μm	σ, km⁻¹	τ(0,∞)	Wavelength, μm	σ, km⁻¹	τ(0,∞)
0.30	1.446×10^{-1}	1.2237	0.65	5.893×10^{-3}	0.0499
0.32	1.098×10^{-1}	0.9290	0.70	4.364×10^{-3}	0.0369
0.34	8.494×10^{-2}	0.7188	0.80	2.545×10^{-3}	0.0215
0.36	6.680×10^{-2}	0.5653	0.90	1.583×10^{-3}	0.0134
0.38	5.327×10^{-2}	0.4508	1.06	8.458×10^{-4}	0.0072
0.40	4.303×10^{-2}	0.3641	1.26	4.076×10^{-4}	0.0034
0.45	2.644×10^{-2}	0.2238	1.67	1.327×10^{-4}	0.0011
0.50	1.726×10^{-2}	0.1452	2.17	4.586×10^{-5}	0.0004
0.55	1.162×10^{-2}	0.0984	3.50	6.830×10^{-6}	0.0001
0.60	8.157×10^{-3}	0.0690	4.00	4.002×10^{-6}	0.0000

atmospheric optics problems. The strong effect of molecular scattering is found in the Venus atmosphere where the optical thickness of molecular scattering may reach hundreds of units.

Light polarisation in molecular scattering

Having the scattering matrix (5.1.9), it is quite easy to calculate the polarisation of light during Rayleigh scattering. We convert the right upper square of the 2 × 2 matrix to the diagonal form using the equations (3.6.12)–(3.6.13):

$$\begin{pmatrix} \frac{1}{2}(1+\cos^2\gamma-\sin^2\gamma) & 0 \\ 0 & \frac{1}{2}(1+\cos^2\gamma-\sin^2\gamma) \end{pmatrix} = \\ = \begin{pmatrix} \cos^2\gamma & 0 \\ 0 & 1 \end{pmatrix}. \quad (5.1.15)$$

As explained in section 3 .6, the elements of the matrix (5.1.15) have the meaning of the coefficients of transformation of the components of the intensity of scattering radiation in the scattering plane $I_{0,\parallel}$ and in the perpendicular direction $I_{0,\perp}$. In many cases, the coefficients of the matrix (5.1.15) are referred to as the phase function of light scattering with linear polarisation in the scattering plane ($x(\gamma) = \cos^2\gamma$) and the linear polarisation in the plane normal to the scattering plane ($x(\gamma) = 1$). Attention should be given to the fact that in the latter case isotropic scattering takes place. These scattering phase functions are indicated by the dashed line in Fig. 5.2.

Let us assume that the incident light with intensity I_0 is not polarised. Consequently, according to the transformation (3.6.13), both mutually perpendicular components for the light are equal to $I_{0,\parallel} = I_{0,\perp} = 1/2\ I_0$. We should remember the definition of the degree of linear polarisation (3.5.21) as the ratio of the difference of the maximum and minimum intensities to their sum. After scattering the maximum intensity will be $I_{\perp} = 1\cdot\frac{1}{2}I_0$, and the minimum intensity $I_{\parallel} = 1\cdot\frac{1}{2}I_0\cos^2\gamma$ (because in all cases $\cos^2\gamma \leq 1$). Consequently, the degree of linear polarisation of scattered light (3.5.21) is:

$$P_l = \frac{1-\cos^2\gamma}{1+\cos^2\gamma} = \frac{\sin^2\gamma}{1+\cos^2\gamma}. \tag{5.1.16}$$

Thus, the degree of linear polarisation in Rayleigh scattering is equal to zero at $\gamma = 0$ and $\gamma = 180°$ and equals 100% at $\gamma = 90°$, i.e. in the directions normal to incident light, the scattered light is completely linearly polarised.

Polarisation of scattered radiation of the cloudless sky

As discussed in detail in chapter 8, the scattering in the atmosphere of the Earth and planets is multiple and, consequently, equation (5.1.16) is valid only for the first scattering. Also, in addition to molecular scattering, there is also aerosol scattering. Therefore, the pattern of polarisation of the scattered light of the cloudless sky, obtained in measurements, differs from the ideal schema described previously. Firstly, the polarisation is actually maximum at points under the angle of 90° in relation to the Sun but the actual values of the degree of linear polarisation at these points are approximately 60–70%. Secondly, the zero value of polarisation is not obtained at the points 0 and 180° but at a 'distance' of approximately 15–20° from them. The upper point of zero polarisation (20° above the Sun) is referred to as the Babinet point, the lower point (20° below the Sun) as the Brewster point. At a low Sun, there is the third point of zero polarisation, i.e. the Arago point, which is 20° above the antisolar point (scattering angle 160°).

Clarification of molecular scattering theory

In derivation of the equations of molecular scattering, the molecules were regarded as ideal spheres. However, because of the anisotropy in the structure of the molecules, the exact theory provides a correction for the resultant relationships in the form of a multiplier which depends on the depolarisation factor δ [20]. It appears that in molecular scattering, the degree of linear polarisation (5.1.16) for the angle of 90° is theoretically not equal to unity, and is lower by the value of δ in comparison with unity. The value of δ depends on the type of gas and for air $\delta = 0.035$. The value of δ is used to express the correction multiplier in the equations of molecular scattering:

for the volume coefficient:

$$\sigma = \frac{8\pi^3(n^2-1)^2}{3N\lambda^4}\frac{6+3\delta}{6-7\delta}, \qquad (5.1.17)$$

and for the phase function

$$x(\gamma) = \frac{3}{4+2\delta}(1+\delta+(1-\delta)\cos^2\gamma).$$

It should be mentioned that for air $\frac{6+3\delta}{6-7\delta} = 1.06$ and, consequently, equation (5.1.17) is 6% more accurate than equation (5.1.14).

Molecular scattering in the presence of absorption

Previously, we examined molecular scattering in the pure form. However, it may be accompanied by radiation absorption. We examine again a molecule of air and an electromagnetic wave with the strength E_0 incident on on the molecule. It is assumed that the absorption of the molecule is determined by the transitions between the energy levels of only one electron. Remaining in the framework of classic electrodynamics, we examine this electron as an attenuating harmonic oscillator oscillating under the effect of the external field E_0 [62]. According to the second Newton law, the equation of motion of the electron may be presented in the form:

$$m\frac{d^2x}{dt^2} = -kx - q\frac{dx}{dt} + eE_0, \qquad (5.1.18)$$

where m is the mass, e is the charge of the electron, $-kx$ is the quasi-elastic return force, trying to restore the electron to the equilibrium position; $q\frac{dx}{dt}$ is the force, identical to the friction force and introduced to take into account light absorption. Dividing by m, equation (5.1.18) is reduced to the form:

$$\frac{d^2x}{dt^2} + \gamma\frac{dx}{dt} + \omega_0^2 x = \frac{e}{m}E_0, \qquad (5.1.19)$$

where $\gamma = q/m$ is the extinction coefficient; $\omega_0^2 = k/m$ is the natural cyclic frequency of oscillations of the electron $\omega_0 = \pi 2\nu_0$.

It is now assumed that the field of incident radiation is again represented by a linearly polarised wave, and the spatial variation of the amplitude of the wave on the scales of the molecule is ignored (as previously). The molecule is again regarded as a vibrating dipole carrying out forced oscillations with the frequency of the external field of $\omega_0 = \pi 2\nu$, consequently, using the complex form of writing (3.5.4) we obtain

$$x = x_0 \exp(i2\pi\nu t). \tag{5.1.20}$$

substitution of (5.1.20) into (5.1.19) gives

$$-4\pi^2\nu^2 x + 2\pi i\nu x + 4\pi^2\nu_0^2\chi = \frac{e}{m}E_0,$$

and consequently

$$x = \frac{e/m}{4\pi^2(\nu_0^2 - \nu^2) + 2\pi i\nu\gamma}E_0.$$

According to the definition of the dipole, moment $P = ex$, on the other hand, $P = \tilde{\alpha}E_0$, and consequently we obtain for the polarisability of a single molecule:

$$\tilde{\alpha} = \frac{e^2/m}{4\pi^2(\nu_0^2 - \nu^2) + 2\pi i\nu\gamma}. \tag{5.1.21}$$

Further, we may use, as previously, the resultant value of $\tilde{\alpha}$ in all considerations and, consequently, the normalised matrix and the scattering phase function show no changes, and the equation for the scattering cross-section (5.1.11) changes to the form:

$$C_s = \frac{128\pi^5\tilde{\alpha}\tilde{\alpha}^*}{3\lambda^4}. \tag{5.1.22}$$

It should be mentioned that the dependence of the cross-section on the wavelength (5.1.22) is now complicated because $\tilde{\alpha}$ depends on $\nu = c/\lambda$.

Taking into account the relationship of the polarisability and the refractive index (5.1.12), and using the complex refraction index (CRI), introduced in chapter 3, $n-i\kappa$ may be written in the following form:

$$(n-i\kappa)^2 = 1 + 4\pi N\tilde{\alpha} = 1 + \frac{2Ne^2}{m}\frac{1}{2\pi(\nu_0^2 - \nu^2) + i\nu\gamma}. \tag{5.1.23}$$

Separating from (5.1.23) the real n and imaginary κ parts of the CRI, we may easily obtain for them explicit expressions but they are very cumbersome. We shall use a simpler procedure: it is taken into account that CRI for these gases is very close to unity and, consequently, it is possible to write:

$$(n - i\kappa)^2 - 1 = (n - i\kappa - 1)(n - i\kappa + 1) \approx 2(n - i\kappa - 1).$$

Consequently, separating the real and imaginary parts of the fraction of CRI (5.1.23) we obtain

$$n - i\kappa = 1 + \frac{Ne^2}{m} \frac{2\pi(\nu_0^2 - \nu^2) - i\nu\gamma}{4\pi^2(\nu_0^2 - \nu^2)^2 + \nu^2\gamma^2}. \tag{5.1.24}$$

The absorption in the model examined here is significant only in the vicinity of resonant frequency ν_0. Consequently, we set $\nu \approx \nu_0$. Therefore, $\nu_0^2 - \nu^2 = (\nu_0 - \nu)(\nu_0 + \nu) \sim 2\nu_0(\nu_0 - \nu)$. Consequently, we obtain the following equations for the real and imaginary parts of the CRI:

$$n = 1 + \frac{4\pi Ne^2}{m\nu_0} \frac{\nu_0 - \nu}{16\pi^2(\nu_0 - \nu)^2 + \gamma^2},$$

$$\kappa = \frac{Ne^2}{m\nu_0} \frac{\gamma}{16\pi^2(\nu_0 - \nu)^2 + \gamma^2}. \tag{5.1.25}$$

Attention should be given to the equation (5.1.25) for the imaginary part of the CRI κ. Its dependence on frequency is identical to the Lorentz shape of the spectral line. The volume coefficient of absorption (denoted by β_ν in this case) in the spectral line with the Lorentz shape is written in the form:

$$\beta_\nu = k_\nu N = NSf_L(\nu - \nu_0) = NS \frac{1}{\pi} \frac{\alpha_L}{(\nu - \nu_0)^2 + \alpha_L^2}.$$

On the other hand, in chapter 3 we determined the relationship between the volume coefficient of absorption and the imaginary part of the CRI (3.5.12):

$$\kappa = \beta_\nu \frac{c}{4\pi\nu} = \frac{NS_c}{4\pi^2\nu} \frac{\alpha_L}{(\nu - \nu_0)^2 + \alpha_L^2}. \tag{5.1.26}$$

Comparing (5.1.26) with (5.1.25) and taking into account $\nu \approx \nu_0$, we formally obtain:

$$\gamma = 4\pi\alpha_L, \quad \frac{e^2}{m} = \frac{cS}{\pi}. \tag{5.1.27}$$

Substituting (5.1.27) into (5.1.25), we obtain the expression for the CRI of the gas through the spectroscopic parameters:

$$n = 1 + N\frac{Sc}{4\pi^2\nu_0}\frac{\nu_0 - \nu}{(\nu_0 - \nu)^2 + \alpha_0^2},$$

$$\kappa = N\frac{Sc}{4\pi^2\nu_0}\frac{\alpha_L}{(\nu - \nu_2)^2 + \alpha_L^2}. \tag{5.1.28}$$

We have derived relationships (5.1.28) within the framework of classical mechanics. However, in the condition of fulfilling the previously mentioned approximations and the equivalence of the shapes of the lines of absorption and radiation, discussed in chapter 4, quantum mechanics leads to the same expressions, only with the addition of summation in respect of all lines within the limits of some range close to ν_0. Thus, the knowledge of the spectroscopic parameters of the gases makes it possible to calculate the volume coefficient of absorption and the volume coefficient of scattering.*

Away from the absorption line ν_0, the equations (5.1.28) are no longer applicable. However, in this case, we have the condition opposite to the condition used previously, $|\nu_0^2 - \nu^2| = |(\nu_0 - \nu)(\nu_0 + \nu)| \gg \nu$, and in equation (5.1.23) we can ignore the term $i\nu\gamma$ in the denominator and write immediately already for the real refractive index n:

$$n^2 = 1 + N\frac{e^2}{\pi m}\frac{1}{\nu_0^2 - \nu^2}$$

or, taking into account $n + 1 \approx 2$,

$$n = 1 + N\frac{e^2}{2\pi m}\frac{1}{\nu_0^2 - \nu^2}, \tag{5.1.29}$$

where it is again necessary to carry out summation in respect of all the absorption lines.

*The relationships indicate that the spectral dependences of the real and imaginary parts of the CRI are interconnected. In optics (by other way and with deriving some other final formulae – Kramer–Kronig relationships) it is proved that this claim holds for any substance [6].

The refractive index of air

The calculations of the CRI using equation (5.1.28) are relatively complicated and time-consuming. However, we take into account the fact that the molecular scattering, as reported previously, is significant only in the ultraviolet and visible ranges of the spectrum. Therefore, in the practical calculations we do not take into account the effect on the refractive index of air of molecular absorption in the infrared region ignoring consequently the error resulting from this procedure. In the ultraviolet and visible ranges we take into account absorption by air in the long-range ultraviolet range (Fig. 4.14) and different modifications of the equation (5.1.29). In particular, the approximate empirical relationship [20] is sufficiently popular

$$(n_0 - 1) = 10^{-6}\left(64.328 + \frac{29498.1}{146 - \lambda^{-2}} + \frac{255.4}{41 - \lambda^{-2}}\right), \qquad (5.1.30)$$

where λ is the wavelength in μm; n_0 is the refractive index at a pressure of p_0 = 1000 mbar, a temperature of T_0 = 15 °C and zero humidity.

Finally, taking into account (5.1.28) and (5.1.29), it is possible to use different relationships for the dependence of the refractive index of air on the concentration of molecules N. The simplest formula is [24]

$$n - 1 = (n_0 - 1)\frac{\rho}{\rho_0}, \qquad (5.1.31)$$

where ρ is the density of air; ρ_0 is the density of dry air at p_0 and T_0 (ρ_0 = 1.20903 · 10^{-3} g·cm^{-3}).

5.2. Scattering and absorption on aerosol particles

Aerosols optics

To determine the characteristics of the interaction of aerosol particles with radiation, they are mathematically modelled by bodies of a specific geometrical form and, consequently, it is possible to solve for these bodies the problem of diffraction of electromagnetic waves on them. Thus, the optics of aerosols in the theoretical plan is closely linked with classic electrodynamics.

The main difficulty in the theoretical analysis of scattering on the aerosol particles is that, in a general case, the dimensions of the particles are no longer small in comparison with the wavelength of incident radiation (chapter 2 – the characteristic dimensions of the aerosols). Therefore, it is not possible, as in molecular scattering, to ignore the variations of the vector of the electrical strength of the incident wave on the particle surface. It is therefore necessary to calculate the characteristics of a heterogeneous electromagnetic field inside the particle which, taking into account the boundary conditions on the surface of the particle, is associated with the field of scattered radiation in which we are interested [6]. For an exact solution of this problem, it is necessary to solve the Maxwell equations which, even in the simplest cases, results in time-consuming operations. After writing Maxwell equations and the boundary conditions, the solution of the equations is transformed into a purely mathematical problem. The solution itself, as will be shown, results in a very complicated dependence of the characteristics of scattering on the initial parameters and it is very difficult to understand the 'physical meaning' of the results.

However, in aerosol optics, we can use approximations which make it often possible to obtain simple solutions of the problem of scattering [5]. For example, the Rayleigh–Hans–Jeans approximation is based on the assumption that the field inside the particle is homogeneous and is formed by dipoles with the same orientation; consequently, the external field can be determined by the superposition of the fields of all dipoles. This approximation is fulfilled quite satisfactorily for the particles with the dimensions considerably smaller than the wavelength. In the approximation of the 'soft' particles according to van de Hulst, it is assumed that the internal field of the particle coincides with the external field of the incident wave. This is fulfilled for the particles with the refractive index close to unity, in particular for the water particles. Examination using the approximations makes it possible to carry out physical analysis of the scattering processes.

Mie theory

The simplest case for which the general solution of the diffraction problem has been obtained is light scattering by a homogeneous sphere. The solution is referred to as the Mie theory (according to German scientist Gustav Mie, who proposed this solution in 1908). We shall present the Mie equations in the form of final results, omitting the explanation because it is time consuming.

Briefly, the method of the deriving the Mie equations may be described as follows [6]. We write the Maxwell equations for the incident wave, the scattered wave, and the wave passed through inside the particle, and the boundary conditions for them. Subsequently, using the well-known method of theoretical electrodynamics – the introduction of the scalar and vector potentials – the system of equations is transformed from the vector to scalar form. Because of the spherical symmetry of the particle, the solution is sought in the form of expansion into a series in respect of spherical functions, and the incident wave and boundary conditions are transformed to the same type. Consequently, the variables in the equations are separated, the equations are reduced to the cases with known solutions and for the coefficients of the series we obtain systems of easy solvable linear algebraic equations. The results are expressed through the Bessell functions with the half-integer index and Legendre polynomials. All the mathematical operations in the derivation of the Mie equations are quite simple but are accompanied by cumbersome transformations. This derivation is described in greater detail in [6].

The absorption and scattering of light by a homogeneous spherical particle is characterised by three dimensionless parameters: the ratio $x = 2\pi r/\lambda$, where r is the radius of the particle, and m is the complex refractive index of the material of the particle* (the CRI is the pair of the numbers and, consequently, there are three parameters).

The equations for the characteristics of interaction are constructed on the basis of the coefficients of complex series a_n and b_n:

$$a_n = \frac{m\psi_n(mx)\psi'_n(x) - \psi_n(x)\psi'_n(mx)}{m\psi_n(mx)\xi'_n(x) - \xi_n(x)\psi'_n(mx)},$$

$$b_n = \frac{\psi_n(mx)\psi'_n(x) - m\psi_n(x)\psi'_n(mx)}{\psi_n(mx)\xi'_n(x) - m\xi_n(x)\psi'_n(mx)}, \tag{5.2.1}$$

here $\psi_n(z)$ and $\xi_n(z)$ are the Riccati–Bessell functions in the general case from the complex argument; $\psi'_n(z)$ and $\xi'_n(z)$ are the

*More accurately, if the wavelength of light is the wavelength in vacuum, then m is the ratio of the CRI of the particle to the real part of the refractive index of the medium. For particles in the atmosphere, the real part of the refractive index of the medium is always assumed to be equal to unity but, for example, for particles in water (hydrosols) one should use the ratio.

derivatives of these functions. The recurrent equations for calculating the Riccati–Bessell functions will be presented below.

In aerosol optics, in addition to the extinction, scattering and absorption cross-sections, it is also necessary to introduce the factors of extinction, scattering and absorption Q_e, Q_s, Q_a which are determined as the ratios of the cross-sections to the area of projection of the particle, perpendicular to the incident wave. For the sphere, this area is πr^2 and, consequently $Q_e = C_e/\pi r^2$, $Q_s = Q_s/\pi r^2$, $Q_a = C_a/\pi r^2$. The factors are dimensional quantities and, consequently, make it possible to compare the relative characteristics of the interaction of particles of different dimensions (this possibility is the reason for introducting the factors). The Mie theory for the scattering $Q_s = Q_s/\pi r^2$ and extinction $Q_e = C_e/\pi r^2$ factors gives

$$Q_s = \frac{2}{x^2}\sum_{n=1}^{\infty}(2n+1)(|a_n|^2 + |b_n|^2),$$

$$Q_e = \frac{2}{x^2}\sum_{n=1}^{\infty}(2n+1)\,\mathrm{Re}(a_n + b_n). \tag{5.2.2}$$

The absorption factor is $Q_e - Q_s$.

The matrix of scattering on a homogeneous spherical particle has the form (3.6.4). Its elements depend only on the scattering angle γ. The matrix is calculated using equation (3.6.3) for the complex coefficients S_1 and S_4 for which, as in molecular scattering, the vector $E_{\parallel}$ is found in the scattering plane and the dependence of these coefficients on γ is associated with the 'angular' functions $\pi_n(\gamma)$ and $\pi_n(\gamma)$:

$$S_1(\gamma) = \sum_{n=1}^{\infty}\frac{2n+1}{n(n+1)}(a_n\tau_n(\gamma) + b_n\pi_n(\gamma)),$$

$$S_4(\gamma) = \sum_{n=1}^{\infty}\frac{2n+1}{n(n+1)}(a_n\pi_n(\gamma) + b_n\tau_n(\gamma)). \tag{5.2.3}$$

Since we have decided to use only the normalised scattering matrices, the constant multipliers in equation (5.2.3) are ignored. It is now only necessary to introduce the recurrent equations for the calculation of the Riccati–Bessell functions and angular functions $\pi_n(\gamma)$ and $\tau_n(\gamma)$:

$$\psi_{n+1}(z)=\frac{2n+1}{z}\psi_n(z)-\psi_{n-1}(z),\ \ \psi_{-1}=\cos z,\ \ \psi_0=\sin z;$$

$$\xi_{n-1}(z)=\frac{2n+1}{z}\chi_n(z)-\chi_{n-1}(z),\ \ \chi_{-1}=-\sin z,\ \ \chi_0=\cos z;$$

$$\xi_n(z)=\psi_n(z)+i\chi_n(z); \tag{5.2.4}$$

$$\pi_n(\gamma)=\frac{(2n-1)\cos\gamma}{n-1}\pi_{n-1}(\gamma)-\frac{n}{n-1}\pi_{n-2}(\gamma),\ \ \pi_0(\gamma=0),\ \ \pi_1(\gamma)=1;$$

$$\tau_n(\gamma)=n\cos\gamma\pi_n(\gamma)-(n+1)\pi_{n-1}(\gamma).$$

As already mentioned, the Mie equations (5.2.1)–(5.2.4) were obtained as a purely mathematical solution of the diffraction problem on a homogeneous sphere and do not make it possible to examine the physics of the process. Below, we present several results of calculations obtained using these equations but for the moment we continue theoretical examination of aerosol scattering and examine the limiting cases which would enable us to determine some important physical relationships.

Small particles

Let us assume that the dimensions of the particles which, as previously, are regarded as homogeneous spheres, are considerably smaller than the radiation wavelength. In this case, we can, as in the case of molecular scattering, ignore the inhomogeneity of the external field incident on the particle. In addition to this, it is assumed that the matter of the particle is a dielectric material, i.e. the conductivity of matter is either non-existent or insignificantly small. Because of the phenomenon of polarisability of the dielectric (the matter of the particle), the charges induced by the external field start to appear on the surface of the particle. Because of the spherical symmetry of the particle, the homogeneity of the external field and the absence of conductivity, the positive and negative charges gather in different hemispheres of the particle and are distributed strictly symmetrically in relation to each other. The distribution of the charges means that we have an emitting dipole [6]. Further, it is necessary to repeat all the considerations, used in the derivation of the equations of molecular scattering and, consequently, all the relationships (5.1.1)–(5.1.11) are valid for the small aerosol particles.

To determine the cross-section of scattering by a small particle, we use the expression available in electrostatics for polarisability of a homogeneous sphere in a homogeneous field:

$$\tilde{\alpha} = r^3 \frac{\varepsilon - 1}{\varepsilon + 2} = r^3 \frac{n^2 - 1}{n^2 + 2}, \qquad (5.2.5)$$

where r is the radius of the sphere; n is the refractive index of the material of the sphere. Substituting (5.2.5) into (5.1.11) we obtain

$$C_s = \frac{128\pi^5}{3} \frac{r^6}{\lambda^4} \left(\frac{n^2 - 1}{n^2 + 2} \right)^2 . \qquad (5.2.6)$$

The above equation gives the same inverse proportionality of the scattering cross-section of the fourth degree of the length of the wave as molecular scattering. This means that the Rayleigh law is valid for the small aerosol particles. In addition to this, previously we have mentioned that the phase function and the scattering matrix of the small particles and, consequently, their polarisation characteristics, are also identical to the molecular characteristics. Consequently, we are talking about Rayleigh aerosol particles (for which the above approximation is fulfilled), the Rayleigh phase function of scattering (the phase function $x(\gamma) = \frac{3}{4}(1 + \cos^2 \gamma)$, Fig. 5.2), the region of Rayleigh scattering (the range of wavelengths and size of the particles in which the Rayleigh approximation is fulfilled). Since the boundaries of approximation of the Rayleigh scattering depend on the wavelength of light, in the long-range infrared and microwave ranges Rayleigh scattering also takes place in the case of verylarge particles of clouds and precipitation.

It should be mentioned that the coincidence between the relationships of scattering for the molecules and small particles is not accidental. Its physical basis is the approximation of the homogeneity of the field, incident on the particle. This approximation is referred to in the literature as dipole approximation, or electrostatic approximation, Rayleigh approximation, or Rayleigh-Hans–Jeans approximation. Since the fluctuations of the density of air in the normal conditions are also small in comparison with the wavelength of light, this approximation is also fulfilled for them. This is the physical reason for the coincidence of the formulae of molecular and fluctuation theories of scattering by gases.

Small particles as a limiting case of Mie theory

The previously made conclusions on the scattering by small particles should evidently be confirmed by a general theory, Mie theory. It is assumed that $|mx| \ll 1$, i.e. $\left|\frac{2\pi mr}{\lambda}\right| \ll 1$. Since the value of the CRI for the aerosol substances does not greatly differ from unity, the examined condition transfers to $r \ll \lambda$, i.e. we examine the particles small in comparison with the wavelength of light. However, this coincides with the previously examined Rayleigh approximation.

Substituting, into equation (5.2.4) the expansion of the sinuses and cosines into the series in the vicinity of zero

$$\sin z = z - \frac{z^3}{6} + \frac{z^5}{120} - O(z^7),$$

$$\cos z = 1 - \frac{z^2}{2} + \frac{z^4}{24} - O(z^6),$$

we obtain, for the first two Riccati–Bessell functions, the following asymptotics:

$$\psi_1(z) = \frac{z^2}{3} - \frac{z^4}{30} + O(z^6), \quad \psi_2(z) = \frac{z^3}{15} + O(z^5);$$

$$\chi_1(z) = \frac{1}{z} + \frac{z}{2} - \frac{z^3}{8} + O(z^5), \tag{5.2.7}$$

$$\chi_2(z) = \frac{3}{z^2} + \frac{1}{2} + \frac{z^2}{8} + O(z^4).$$

Subsequently, substituting (5.2.7) into (5.2.1) we obtain for α_1:

$$a_1 = \frac{\frac{2}{9}x^3\left(m^2-1\right) + \frac{2}{90}x^5\left(1-m^4\right) + O\left(x^7\right)}{-\frac{i}{3}\left(m^2+2\right) + \frac{i}{30}x^2\left(m^2-1\right)\left(m^2+10\right) + \frac{2}{9}x^3\left(m^2-1\right) + O\left(x^4\right)}$$

Multiplying the numerator and denominator by the value, complex conjugate to the numerator, and neglecting the small values in the denominator that are independent of x, we obtain

$$a_1 = \frac{2i}{3}x^3\frac{m^2-1}{m^2+2}+$$

$$+\frac{2i}{5}x^5\frac{(m^2-2)(m^2-1)}{(m^2+2)^2}+\frac{4}{9}x^6\left(\frac{m^2-1}{m^2+2}\right)^2+O(x^7). \qquad (5.2.8)$$

After carrying out identical procedures for the coefficients b_1, a_2 and b_2 and ignoring the terms of the order x^7, we finally obtain the asymptotics:

$$b_1 = \frac{i}{45}x^5(m^2-1)+O(x^7),$$

$$a_2 = \frac{i}{15}x^5\frac{m^2-1}{2m^2+3}+O(x^7),\ b_2 = O(x^7). \qquad (5.2.9)$$

For the scattering characteristics we should calculate the squares of the moduli of the coefficients. In this case, taking into account $x \to 0$, we can ignore the terms with a higher order of smallness in comparison with x^3 in a_1. Substituting a_1 into (5.2.2) we obtain

$$Q_s = \frac{2}{x^2}3|a_1|^2 = \frac{8}{3}x^4\left|\frac{m^2-1}{m^2+2}\right|^2.$$

Taking into account that $x = \frac{2\pi r}{\lambda}$ and $C_s = \pi r^2 Q_s$, we obtain the following equation for the scattering cross-section:

$$C_s = \frac{128}{3}\pi^5\frac{r^6}{\lambda^4}\left|\frac{m^2-1}{m^2+2}\right|^2. \qquad (5.2.10)$$

Equation (5.2.10) fully coincides with (5.2.6) if the refractive index m is regarded as purely real (this was also assumed when deriving (5.2.6)). Thus, the Rayleigh approximation is justified as a limiting case of Mie's theory. On the other hand, from Mie's theory we have obtained the general equation (5.2.10), which holds true for particles with any CRI, including conductors.

The agreement between the scattering matrix and Rayleigh approximation will now be verified. For complex vectors from

(5.2.3), taking into account $\pi_1(\gamma) = 1$ and $\tau_1(\gamma) = \cos\gamma$, we have $S_1 = \frac{3}{2}a_1\cos\gamma$, $S_4 = \frac{3}{2}a_1$. In subsequent stages, no calculations are required because the angular dependences of S_1 and S_4 are completely identical (5.1.5), and in normalisation of the matrix we obtain (5.1.9). Thus, both the special features of scattering and polarisation of small spherical particles (with any CRI) are equivalent to the case of molecular scattering.

Although Rayleigh approximation has a clear physical meaning, in the framework of the general theory, purely mathematically, we can obtain the results which no longer follow from simple considerations. We examine the extinction factor of small particles Q_e. For the factor, we substitute asymptotic relationships (5.2.8), (5.2.9) into (5.2.2), but now it is necessary to ensure that the final equation does not convert to zero for real m, and the single member with the non-zero real part in (5.2.8), (5.2.9) – the member with x^6 – to a_1. Consequently, for Q_e it is necessary to take into account all the terms to the sixth degree x exclusively:

$$Q_e = \frac{2}{x}(3\,\mathrm{Re}(a_1) + 3\,\mathrm{Re}(b_1) + 5\,\mathrm{Re}(b_2)) =$$

$$= -\mathrm{Im}\left(4x\frac{m^2-1}{m^2+2} + \frac{12}{5}x^3\frac{(m^2-2)(m^2-1)}{(m^2+2)^2} + \frac{2}{15}x^3(m^2-1) + \right.$$

$$\left. + \frac{2}{3}x^3\frac{m^2-1}{2m^2+3}\right) + \frac{8}{3}x^4\,\mathrm{Re}\left(\left(\frac{m^2-1}{m^2+2}\right)^2\right) = -\mathrm{Im}\left(4x\frac{m^2-1}{m^2+2}\times\right.$$

$$\left.\times\left(1 + x^2\frac{18(m^2-2)(2m^2+3)(m^2+2)^2(2m^2+3)+5(m^2+2)^2}{30(m^2+2)(2m^2+3)}\right)\right) +$$

$$+\frac{8}{3}x^4\,\mathrm{Re}\left(\left(\frac{m^2-1}{m^2+2}\right)^2\right).$$

Transforming the denominator at the term with x^2

$$2(9(m^2-2)(2m^2+3) + (m^2+2)^2(m^2+4)) =$$

$$= 2(9(m^2-2)(2m^2+3) + (m^2+2)^2(2m^2+3) -$$

$$-(m^2 + 2)^2 (m^2 - 1))=$$

$$=2((2m^2 + 3)(m^4 + 13m^2 - 14)-(m^2 + 2)^2(m^2 - 1))=$$

$$=2((2m^2 + 3)(m^2 - 1)(m^2 + 14) - (m^2 + 2)^2(m^2-1)),$$

we finally obtain

$$Q_e = -4x\,\mathrm{Im}\left(\frac{m^2-1}{m^2+2}\left(1+\frac{x^2}{15}\left(\frac{m^2-1}{m^2+2}\right)\times\right.\right.$$

$$\left.\left.\times\frac{m^4+27m^2+38}{2m^2+3}\right)\right)+\frac{8}{3}x^4\,\mathrm{Re}\left(\left(\frac{m^2-1}{m^2+2}\right)^2\right). \qquad (5.2.11)$$

For the real part complex refractive index the first member in (5.2.11) is evidently zero and, consequently, in a general case it is the absorption factor Q_a. The sign – in front of the imaginary part in (5.2.11) cannot cause misunderstanding. Because in optics (chapter 3) we use the complex refractive index in the form $m = n - i\kappa$, it may easily be shown that Q_a is positive. The second term in equation (5.2.11) is not interpreted so unambiguously. Like the entire extinction factor, for the real CRI this term coincides with the scattering factor (5.2.10). In a general case of the absorbing particles, there is no longer any agreement between the second term in (5.2.11) and the scattering factor (5.2.10). Consequently, the second term describes the contribution to extinction of both scattering and absorption. It should be mentioned that for the absorbing particles the absorption factor Q_a is inversely proportional to the first degree of the wavelength, whereas the scattering factors Q_s is proportional (as previously) to the fourth degree. Therefore, with increasing wavelength the absorption properties of the aerosol particles become more and more significant in comparison with the scattering properties.

Large particles

We now examine an opposite case – the case of particles considerably larger than the wavelength of light, i.e. the particles for which $x = \frac{2\pi r}{\lambda} \gg 1$. Evidently, for these particles geometrical optics should hold in the limit. Within the framework of the geometrical optics it may be assumed that the particle scatters and absorb the entire amount of light incident on the surface. Consequently, according to the definition of the extinction cross-section, it is equal to πr^2. Correspondingly, for the extinction factor we should have $Q_e = 1$. However, in fact this result is not accurate. Actually, in the framework of the geometrical optics we cannot take into account the additional extinction caused by diffraction (scattering) at the edges of the particle of beams passing away from the particle. This diffraction takes place because of the wave nature of light.

Consequently, the real extinction cross-section should be equal to the sum of the geometrical and some additional scattering. In order to obtain the quantitative value of this additional diffraction section, it is necessary to use the optical theory proposed by Babinet as resulting, for example, from the Fresnel–Kirchhoff diffraction equation [7, 62]. According to the Babinet theorem, the scattering cross-section as a result of diffraction on large particles is equal to its geometrical cross-section. Adding up this cross-section with the geometrical cross-section, we find that for any particle with the dimensions considerably larger than the wavelength, the extinction cross-section is equal to the double area of projection of the particle on the plane normal to the light rays (more accurately, asymptotically tending to the double area when the size of the area tends to infinity). In particular, for spherical particles we obtain $S_e = 2\pi r^2$ ($Q_e = 2$). Since diffraction is associated exclusively with light scattering, the following simple consequence follows from the above considerations: the absorption cross-section of the large particle cannot exceed its scattering cross-section or, in other words, the amount of radiation energy absorbed by the large particle is not larger than the amount of energy scattered by the particle.

Extinction paradox

Thus, a large particle (of any shape) receives from the light beam the amount of energy which is double the amount incident on its surface. This surprising fact, contradicting the geometrical optics and the common sense based on it, is referred to as the 'extinction paradox'. The presence of this paradox stresses that, generally speaking, geometrical optics cannot be applied to the processes of scattering of radiation and any results obtained in geometrical optics require verification within the framework of wave optics. To reconcile the extinction paradox and the common sense, we use a simple explanation proposed by Van de Hulst. The paradox is associated with the following: when examining rigorously the diffraction phenomenon, we assume that any rays, including those scattered at extremely small angles, change direction and, consequently, they are removed from the beam of transmitted light. Consequently, the diffraction pattern can be examined and the extinction cross-section be measured at a very large (in the limit– at infinite) distance from the particle, where we can take into account the contribution of these small angles. Therefore, observing the shadow from a stone placed on the window of a room we cannot separate the radiation passed by the stone at the radiation scattered hrough very small angles and in accordance with the geometrical optics and rational sense we obtain for it the extinction factor $Q_e = 1$. However, if the stone is a meteorite located hundreds of millions kilometres from the Earth, then the light beams, scattered through very small angles, already pass away from our device and we record the extinction factor close to $Q_e = 2$. Returning to the Mie equations (5.2.2) it may be asserted that the following relationship should be fulfilled

$$\lim_{x \to \infty} Q_e(x, m) = 2. \qquad (5.2.12)$$

The above relationship is confirmed in the numerical calculations and used for testing the algorithm and computer programs used in computations by Mie theory.

Anomalous diffraction

In the framework of geometrical optics other, relatively simple approximate equations were obtained for the optical characteristics of aerosol particles. In scattering theory, developed by Van de Hulst, it is assumed that

$$m - 1 \ll 1 \text{ and } x = 2\pi r/\lambda \gg 1. \qquad (5.2.13)$$

Van de Hulst referred to this approximation as anomalous diffraction. Strictly speaking, the first condition indicates that $m \to 1$, but it has been found that the resultant equations can also be used for the case with $m \to 2$. The second condition means that the radius of the particles may be several times greater than the wavelength. Since the radius of the particles of fog and clouds is greater than several microns, for the visible range of the spectrum particles of this type may be used as an example of the application of the approximation of anomalous diffraction in atmospheric optics. For non-absorbing particles, we examine the scattering factor Q_s.

for the absorbing particles Q_e. The main parameter in the theory of anomalous diffraction is the parameter:

$$\rho = 2x(m-1) = \frac{4\pi r}{\lambda}(m-1), \qquad (5.2.14)$$

which combines three main scattering parameters.

Taking into account these assumptions, Van de Hulst obtained the equation for the scattering factor from the Huygens–Fresnel principle:

$$Q_s = 2 - \frac{4}{\rho}\sin\rho + \frac{4}{\rho^2}(1-\cos\rho). \qquad (5.2.15)$$

As already mentioned, in practice this equation describes the main properties of the scattering factor (and extinction factor for non-absorbing particles) not only at $m \approx 1$ but also for the case $m \approx 2$.

The additional parameter b is introduced for the absorbing particles. The parameter is determined by the relationship

$$\operatorname{tg} b = \frac{\kappa}{(m-1)}, \qquad (5.2.16)$$

where κ is the imaginary part of the refraction index.

The extinction factor is defined by the following equation:

$$Q_e = 2 - 4\exp(-\rho tgB)\frac{\cos b}{\rho}\sin(\rho - b) -$$

$$-4\exp(-\rho tgb)\left(\frac{\cos b}{\rho}\right)^2 \cos(\rho - 2b) + 4\left(\frac{\cos b}{\rho}\right)^2 \cos 2b. \qquad (5.2.17)$$

This expression is reduced to equation (5.2.15) at $b = 0$ which corresponds to $\kappa = 0$, i.e., to the non-absorbing particles. Later, empirical correction multipliers were proposed for equation (5.2.17) which increase the accuracy of calculating the optical characteristics of aerosol particles in the approximation of anomalous diffraction.

Complex refractive indexes

When calculating optical characteristics of aerosol particles it is necessary to have the information on the complex refractive indexes of the material of the particles. Taking into account the differences in the physical–chemical properties of atmospheric aerosol and its temporal and spatial variability, the determination of these data is a relatively cumbersome and time-consuming task. Measurements have been taken of the refractive indexes of many substances, entering into the composition of the atmospheric aerosols. In particular, a large amount of data on the spectral behaviour of the real and imaginary parts of refractive indexes of different substances have been published in monographs, handbooks and databases. As an example, Table 5.2 characterises the information of the refractive indexes of different substances present in the HITRAN-96 database (chapter 4).

As indicated by Table 5.2 and a large number of other data, the refractive indexes are available for wide ranges of the spectrum so that it possible to calculate different optical characteristics of aerosol particles in the appropriate regions. For example, the refractive index of ice is given in the range from 0.04 to $8 \cdot 10^6$ μm, i.e. from 40 nm to 8 m. In a general case, the refractive indexes are functions of temperature and, consequently, the HITRAN-96 database gives the refractive indexes of several substances for a number of temperatures.

As an example, Fig. 5.3 gives the spectral behaviour of the real and imaginary parts of the refractive index of water and ice – the aerosol substances prevailing in the Earth atmosphere. The graph demonstrates the spectral variability of n and κ, which is especially strong for the imaginary part of the refractive index.

The comparison of the imaginary part of the refractive index in a wide range of the spectrum from 0.1 to 100 μm for different atmospheric aerosols is shown in Fig. 5.4. It should be noted that the imaginary part determines the absorbing properties of atmospheric aerosol. Figure 5.4 shows that the imaginary parts of the refractive index for different substances differ by orders of magnitude – from the values close to unity to 10^{-6} or less. In addition to this, for the individual substances, the distinguishing feature is the strong spectral dependence $\kappa(\lambda)$. The lower part of the graph shows the spectral regions where the imaginary part of the refractive index for water, quartz, ammonium sulphate and sodium chloride is lower than 10^{-6}. For example, for water this range includes part of the ultraviolet range, the visible range and part of the near infrared range. On the other hand, water is characterised by high values of κ in the infrared range of the spectrum.

Table 5.2. Data on the refractive index of various substances in the HITRAN-96 database [109]

Substance	Comment
Water (liquid)	Spectral range 0.65–1000 μm
Water (ice)	0.04–8·10^6 μm
H_2SO_4 /H_2O solution	Room temperature, 0.35–25 μm, 25, 38, 50, 75, 84.5, 95.6% solutions of H_2SO_4 by weight; 6.4–13 μm, 90% H_2SO_4 by weight
HNO_3/H_2O solution	Room temperature, 2–32 μm, 3, 12, 22, 40, 70% HNO_3 solution by weight
NAT ($HNO_3 \cdot 3H_2O$)	1.4–20 μm, trihydrate of nitric acid at 181 and 196 K
NAD ($HNO_3 \cdot 2H_2O$)	1.4–20 μm, dihydrate of nitric acid at 841 K
NAM ($HNO_3 \cdot H_2O$)	1.4–20 μm, monohydrate of nitric acid at 179 K
Water (ice)	Ice sheet at 163 K
NAT	Solid film of solution of amorphous NAT at 153 K
NAD	Solid film of solution of amorphous NAD at 153 K
NAM	Solid film of solution of amorphous NAM at 153 K
NaCl	Room temperature, 0.2–30000 μm
Sea salt	Room temperature, 0.2–30000 μm
Ammonium sulphate	Room temperature, 0.2–40 μm
Carbon material	Room temperature, 0.2–40 μm
Volcanic dust	Room temperature, 0.2–40 μm
Meteor dust	Room temperature, 0.2–40 μm
Quartz	Room temperature, 0.2–300 μm
Iron oxide	Room temperature, 0.2–300 μm
Sand	Room temperature, 0.2–300 μm

Examples of calculations using Mie theory

When discussing the calculations using the Mie equations, it should be mentioned that the success of Mie theory and, generally, of the optics of aerosols started with the appearance of computers because the Mie series slowly converge and, in a general case, it is necessary to take into account in tens, hundred and in some cases thousands of terms. This was not possible in the era of manual calculations. In developing the calculation algorithms for computers, it was necessary to transform equations (5.2.1) to a slightly different form and develop algorithms stable in relation to the accumulation of machine errors. We shall not discuss this problem

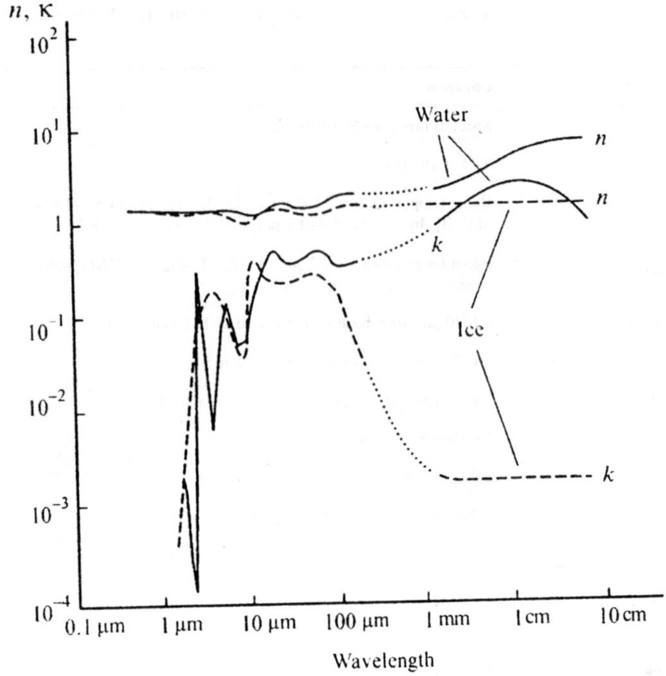

Fig. 5.3. Spectral dependence of the real and imaginary parts of the refractive index of water and ice [114].

and we only mention that at present there are algorithms and programs which operate in a stable manner in the entire range of variations of x and m that are possible in the problems of atmospheric optics [4, 6, 10].

We analyze the behaviour of the characteristics of extinction, absorption and scattering of aerosol particles on the example of the results of calculations carried out using Mie theory. Figure 5.5 shows the dependence of the extinction factor on parameter x for three substances: water ($m=1.33-i\cdot 1\cdot 10^{-9}$), quartz ($m = 1.535-i\cdot 5\cdot 10^{-4}$) and soot ($m = 1.82-i\cdot 0.74$) under the condition of constant CRI. The last condition physically corresponds to the variation of only the radius of the particle at a fixed wavelength. The dependences $Q_e(x)$ for water and quartz have the nature of damping oscillations converging to the asymptotic value 2. The periodicity of the dependence follows from the definition of the Riccati–Bessell functions (5.2.4) through sinuses and cosines of the parameters mx and x, the distinctive period of oscillations does not form, and this is explained by the superposition in summation of the

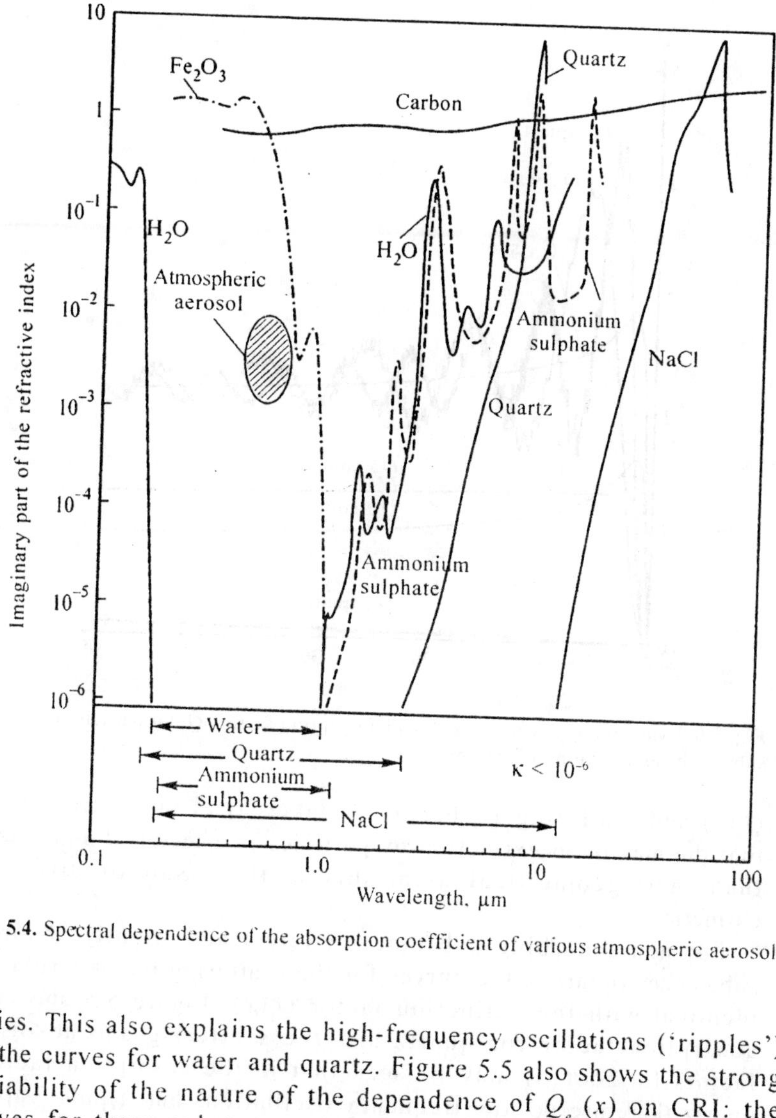

Fig. 5.4. Spectral dependence of the absorption coefficient of various atmospheric aerosols [6].

series. This also explains the high-frequency oscillations ('ripples') on the curves for water and quartz. Figure 5.5 also shows the strong variability of the nature of the dependence of $Q_e(x)$ on CRI: the curves for three substances greatly differ. In particular for soot, because of strong absorption, the vibrational nature of the dependence is generally not evident, and the $Q_e(x)$ dependence is smooth since the large imaginary part of the CRI in the calculations of the sinuses and cosines of the complex argument in (5.2.4) results in an exponential multiplier 'extinguishing' vibration terms. Consequently, it may be concluded that as the imaginary part of the CRI increases, the oscillations of the extinction factors of

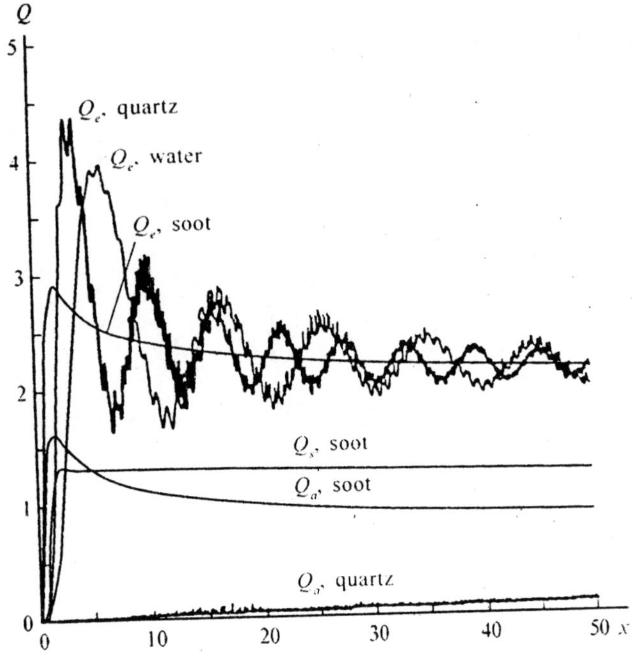

Fig. 5.5. Extinction Q_e, scattering Q_s and absorption factors Q_a for particles of different substances as functions of parameter x.

$Q_e(x)$ and scattering Q_s become weaker. It is interesting to note that the maximum extinction by particles is a factor of 3–4.5 larger than their geometrical area; this is the result of diffraction extinction.

For non-absorbing substances (water) and slightly absorbing substances (quartz), the curves for the scattering factors are almost identical with the extinction factor $Q_e(x)$. Figure 5.5 shows the absorption factor for quartz and the scattering and absorption factors for soot. It may be seen that in the absorption factor of quartz there are no low-frequency oscillations but 'ripples' appear. In soot, the scattering and absorption factors have the same smooth form as in the case of $Q_e(x)$.

The scattering phase functions in relation to x for water, soot and quartz are presented in Fig. 5.6a. For small particles, as shown previously, the phase function is close to the Rayleigh phase function. With increase of the particle size the phase function is extended in the forward direction. This is caused by the fact that scattered emission is the diffracted emission. The diffraction theory

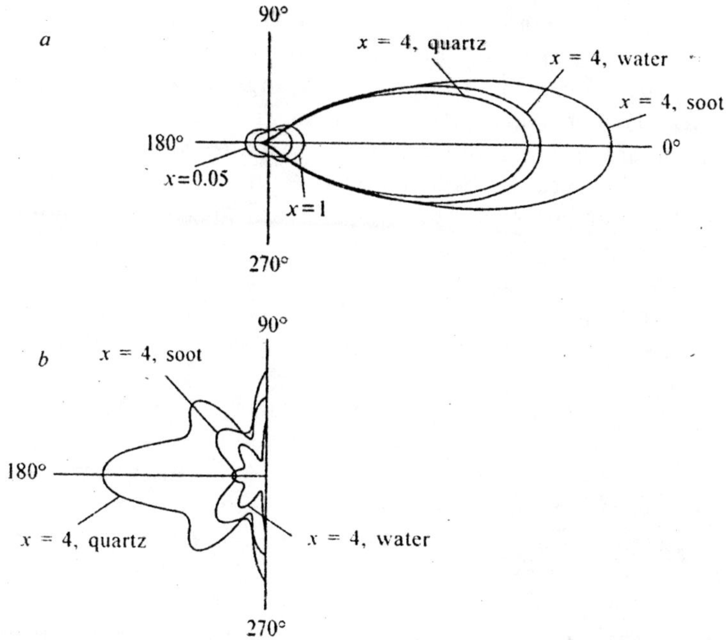

Fig. 5.6. Scattering phase functions of various substances as a function of parameter x. a) general view, b) range of back scattering.

shows that as the size of the obstacle (particle size) increases the 'concentration' of diffracted radiation increases in the smaller solid angle in the vicinity of the direction of propagation of initial radiation. This also follows from equation (5.2.4): the cosine of the scattering angle in the region of zero is maximum, consequently, the function τ_n is maximum, and in the summation of the Mie series these maxima are added together. Therefore, the phase functions of the aerosol particles are elongated and the extent of elongation increases with increase of the particle size. As these phase functions are obtained in the calculations using the Mie theory, they are referred to (in contrast to Rayleigh phase functions) as Mie phase functions. As the opposite of Rayleigh scattering we use the term Mie scattering (i.e. the scattering which can be described using calculations based on general theory).

For relatively small particles, the form of the phase function is almost independent of the CRI. For example, in Fig. 5.6a, the phase functions at $x = 1$ almost completely coincide. With increase of x secondary maxima start to appear in the scattering phase function: the maximum of backscattering at the angle of 180 °C and lateral maxima. The magnitude and form of these maxima already depend strongly on the type of substance. This is indicated by Fig. 5.6a and also 5.6b where the range of the angles 90–180 °C is shown separately. For quartz, the maximum of backscattering and lateral maxima are very distinctive, in the case of water and soot they are considerably weaker. But in this case, the phase function of water has the strongest minimum. The lateral maxima for water are of special interest because they are 'responsible' for various atmospheric phenomena, such as rainbow, glory, etc. Later, we shall return to them. However, it should be mentioned that, as indicated by Fig. 5.6b, the rainbow phenomena are not the exclusive privilege of water droplets, in principle, they accompany scattering on particles of any substance, simply this has not been detected in nature because of the comparatively low concentration of large non-water particles. For the scattering matrix (for the elements different from the phase function), in particular for the characteristics of polarisation of aerosol scattering, the dependences on x and CRI are complicated and varied.

5.3. Aerosol scattering and absorption in the atmosphere

Optical characteristics of ensembles of aerosol particles

In the previous section, we examined in considerable detail the interaction of radiation with single aerosol particles. However, this does not exhaust the problem of describing aerosol scattering in the atmosphere, at least owing to the fact that the real atmospheric aerosols content of different substances and have different dimensions (chapter 2). We start with the last circumstance. As already mentioned in chapter 2, the characteristic of the difference in the dimensions, dispersion of the aerosols, is the size distribution function of the particles $n(r)$, where n is the concentration of the particles, r is the radius of the particles. According to the definition of the distribution function, the number of particles dN with the dimensions from r to $r + dr$ in the unit volume is $dN = n(r)dr$. As shown in chapter 3, the volume coefficient of aerosol extinction α_a for the particles in the given radius range, according to (3.3.20) is:

$$\alpha_a(r) = dNC_e(r) = n(r)C_e(r)dr.$$

To obtain the total volume coefficient of all aerosol particles we should, in accordance with (3.3.21), add up all $\alpha_a(r)$ and consequently we obtain:

$$\alpha_a = \int_0^\infty n(r) C_e(r) dr. \tag{5.3.1}$$

As previously, the aerosol particles will be simulated by homogeneous spheres. Consequently, equation (5.3.1) can be written in a more convenient form. It is taken into account that $C_e(r) = \pi r^2 Q_e(r, m)$, and the distribution function may be written in the form $n(r) = Nf(r)$, where N is the total concentration of the aerosol particles (of all sizes); $f(r)$ is respectively the normalised function of distribution (chapter 2). As a result, (5.3.1) is presented in the following form:

$$\alpha_a = N \int_0^\infty \pi r^2 Q_e(r,m) f(r) dr. \tag{5.3.2}$$

Equation (3.3.22) shows that the relationships, completely identical with (5.3.2), are also obtained for the volume coefficients of aerosol scattering and absorption:

$$\sigma_a = N \int_0^\infty \pi r^2 Q_s(r,m) f(r) dr,$$
$$k_a = N \int_0^\infty \pi r^2 Q_a(r,m) f(r) dr. \tag{5.3.3}$$

As explained in chapter 3, for the phase function of aerosol scattering it is necessary to add up the products of the phase functions by the volume scattering coefficients. In the range from r to $r+dr$ it is $n(r)\,\sigma(r)\,x(r, \gamma)\,dr = Nf(r)\,\sigma(r)\,x(r, \gamma)dr$. Now, for the phase function we use 'the rules of composition' (3.4.34) and consequently

$$x_a(\gamma) = \frac{\int_0^\infty \pi r^2 Q_s(r,m) x(r,m,\gamma) f(r) dr}{\int_0^\infty \pi r^2 Q_s(r,m) f(r) dr}, \tag{5.3.4}$$

where $x_a(\gamma)$ is the total aerosol phase function of scattering; $x(r,m,\gamma)$ is the phase function of the particle radius r.

The aerosol particles in the atmosphere taking into account their dispersion are referred to as the ensemble of the particles. The resultant equations (5.3.1)–(5.3.4) make it possible to solve completely the problem of definition of the optical characteristics of the ensemble of the spherical homogeneous particles: it is necessary to know the concentration of the particles, the function of the size distribution of the particles and the complex refractive index (CRI) of the material of the particles (and also to be able to take into account the characteristics of interaction of radiation with a single particle, for example, according to Mie's theory). If the atmosphere contains aerosol particles of several types (with different $f(r)$ and CRI) it is always possible to calculate the optical characteristics separately for every type and, subsequently, use the rules of composition (3.3.21), (3.3.22), (3.4.34).

Angström formula. Aerosol optical effects

The effects of aerosol particles on the transfer of radiation is more significant in the visible range of the spectrum where there are almost no absorption bands of the atmospheric gases and the controlling mechanism of transformation of solar radiation in the atmosphere is molecular scattering and aerosol extinction. In the visible range, the CRI of the main aerosol substances shows only small changes over the spectrum and they can be regarded as constants. We approximate the function of size distribution of the aerosol particles by the Junge distribution (2.4.7), i.e. it is assumed that $f(r) = cr^{-b}$, where b is the distribution parameter, D is a constant coefficient. Consequently, equation (5.3.2) has the following form:

$$\alpha_a = \pi ND \int_0^\infty r^{2-b} Q_e(r)\,dr.$$

According to the Mie theory, Q_e depends on $\frac{2\pi r}{\lambda}$, where λ is the wavelength of light. We introduce into the integral the substitution $x = \frac{2\pi r}{\lambda}$ and, consequently $r = \frac{\lambda x}{2\pi}$, $dr = \frac{\lambda}{2\pi}dx$, and

$$\alpha_a = \frac{\pi ND}{2\pi^{2-b}} \int_0^\infty \lambda^{2-b} x^{2-b} Q_e \frac{\lambda}{2\pi} dx,$$

and therefore

$$\alpha_a = A\lambda^{-\beta}, \tag{5.3.5}$$

where A is a constant which is independent of λ, and the exponent is linked with the Junge distribution parameter by $\beta = b-3$ (in many cases, the Junge distribution is written in the form $f(r) = Dr^{-1-b'}$ and, consequently, $\beta = b'-2$).

Equation (5.3.5) is the Angström formula [46]. According to this equation, aerosol extinction is inversely proportional to some degree of the wavelength of light. By this, it is similar to molecular scattering for which $\beta = 4$. As already explained in the previous section, the same parameter of proportionality is typical for the extinction cross-section and, according to (5.3.1), for the volume coefficient of extinction of small (Rayleigh) particles. Since the case of Rayleigh particles is a limiting case, for the atmospheric aerosol $\beta < 4$. The relationship of β with Junge's distribution shows that as the parameter b decreases, i.e., as the number of large particles increases, the exponent β decreases. Thus, the presence of large particles makes the spectral dependence of aerosol extinction less marked, up to its absence ($\beta = 0$ at $b = 3$).

In the visible range of the spectrum, aerosol absorption is considerably less intensive than aerosol scattering (the imaginary parts of the CRI of many aerosol substances are close to zero). Therefore, all the considerations regarding the dependence of β on the size of the particles also relate to aerosol scattering. Thus, in contrast to molecular scattering, aerosol scattering decreases with increasing wavelength at a lower rate, and in the case of large particles it does not change at all, and the light of different wavelengths is scattered by the aerosol to the approximately same degree. This explains the neutral colour of the aerosol formations: clouds, mist, dust accumulations, smoke are white, grey, black, whereas the blue colour of the sky is associated, as explained in paragraph 5.1, with $\beta = 0$ for molecular scattering.

If β in (5.3 .5) is sufficiently high, i.e., the fine aerosol particles are dominant, the blue rays are attenuated to a greater extent than the red ones. Therefore, when examining a glowing object, in particular, the Sun and the Moon, the objects will appear as redden. Similar effects have already been discussed for molecular scattering in section 5.1, but in the case of the aerosols the difference is that the extinction is directly proportional to their concentration and, therefore, if the concentration of fine particles in the atmosphere is high, the reddening effects are detected not only at sunrise an sunset but also at a relatively high position of the Sun and the Moon above the horizon.

When deriving Angström equation (5.3.5), two assumptions were made: the CRI is independent of wavelength and the distribution function corresponds to the Junge equation. These assumptions impose very strict restrictions on the practical application of (5.3.5): it is not valid in the infrared range, where CRI greatly changes with a wavelength; for the distributions differing from the Junge distribution; finally, in the presence of aerosols of different nature with different CRI in the atmosphere. Below, when examining the results of computer calculations of the optical characteristics of the aerosol ensembles we shall slightly improve the accuracy of our knowledge regarding the spectral dependence of the volume coefficient of aerosol extinction. Here, it should be mentioned that, regardless of all the assumptions, the Angström formula is used quite often in the processing of the results of field measurements. Actually, measuring the spectral dependence of the volume coefficient of aerosol extinction in the visible range of the spectrum (at least at two wavelengths) and approximating it by (5.3.5), it is easy to determine parameter b of the Junge distribution, i.e. in fact, obtain information on the size distribution function of the aerosol particles.

Calculations of the optical characteristics of atmospheric aerosols. Special and anomalous aerosol optical effects

The special features of the optical characteristics of polydisperse aerosols are determined by the smoothing of the oscillations of the factors of extinction, scattering and scattering phase functions (mentioned in the previous chapter) as a result of integration in respect of the ensemble of aerosol particles. Therefore, the dependence of the optical aerosol characteristics on the wavelength of light is relatively smooth which is also indirectly confirmed by the analytically derived Angström formula (5.3.5). The intensity of smoothing increases with increase of the spread of the particle dimensions, i.e. the dispersion of the distribution function of the particles. In cases in which the particles of a specific size are dominant, smoothing is weak, and the optical characteristics of the aerosol ensembles start to show special features, typical of the individual particles. This results in 'unique' optical effects: rainbow, halo, etc. They will be discussed in detail in section 5.6. Here, it should only be mentioned that the water droplets show small dispersion during rain and, consequently, they clearly show a rainbow, sometimes even a double rainbow. Small dispersion is also typical of the icy particles of cirrus and cirrus–cirrostratus clouds, and, consequently, a halo forms in these clouds. Thus, examination of the unique optical phenomena in the clouds is a confirmation of the low dispersion of the particles forming the clouds and, consequently, even purely visual examination may be used for making conclusions regarding the dimensions of cloud particles.

The Angström equation (5.3.5) was derived using the Junge distribution. However, the real samples of aerosol particles are characterised by a specific maximum of the size and this is followed by a decrease both in the range of large and small particles. A suitable example of such a more adequate distribution is the log-normal distribution (2.4.6). It is no longer possible to obtain the dependence of aerosol extinction on the wavelength in the explicit analytical form for distributions of this type, but the results of numerical calculations make it possible to introduce certain additions to the analysis carried out on the basis of the Angström equation.

The characteristic feature of the spectral dependence of aerosol extinction is the presence of a maximum determined by the parameters of the distribution function. This is shown in Fig. 5.7 which gives three spectral curves of the volume coefficient of aerosol extinction for dust ($m = 1.5 - i \cdot 0.008$). The log-normal distribution (2.4.6) with $\sigma = 0.7$ and different values of r_0: 0.1, 0.25 and 0.5 μm, was used. To facilitate analysis, the relative curves are given (normalised to the value at 0.5 μm). The presence of the maximum of the curve with $r_0 = 0.25$ is clearly visible. The identical maxima are also recorded for other curves, but they are outside the graph: at $r_0 = 0.1$ in the ultraviolet region, at $r_0 = 0.5$ in the infrared range. Thus, if the large particles prevail over fine particles, aerosol extinction may increase from short waves to long ones. These phenomena, sometimes detected in measurements in the atmosphere, are referred to as the 'the anomalous spectral dependence of aerosol extinction'.

If the concentration of aerosol is sufficiently high and the aerosol extinction is more intensive than molecular scattering, the anomalous spectral dependence may result in interesting optical effects. For example, on photographs of the landscape, obtained during landing of the American Viking space station on Mars, the Martian sky was yellow instead of blue or black [47]. This may easily be explained if it is taken into account that because of the very thin atmosphere (Table 1.3) the intensity molecular scattering on Mars is low but the sandy surface of Martian deserts and periodic sand storms result in a high concentration of aerosol particles. According to

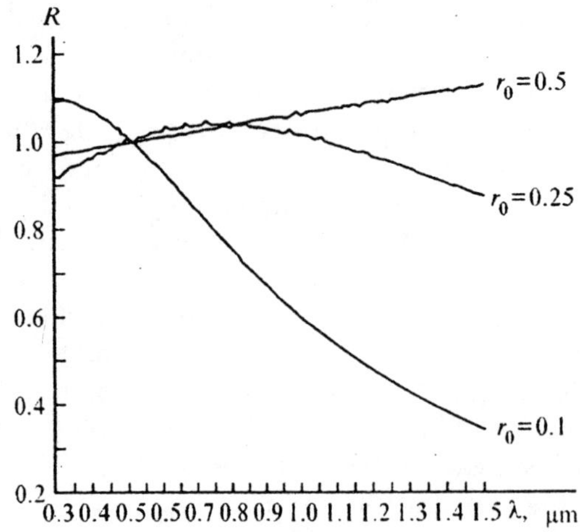

Fig. 5.7. Dependence of the spectral course of the relative volume coefficient of aerosol extinction R = ()/ (0.5 μm) on the parameter r_0 (μm) as a function of thelog-normal distribution of aerosol particles.

measurements taken in deserts on the Earth, in the sand aerosols the particles with the sizes causing the anomalous spectral course of aerosol scattering are dominant and result in the maximum scattering of yellow and red rays. On the Earth, these effects are suppressed by strong molecular scattering but in the deserts, the colour of the sky and the Sun is not so saturated with colours. In the case of anomalous aerosol extinction, if the concentration of the large particles is relatively high, the red rays can be attenuated to a greater extent than the blue ones and this results in the extremely rare phenomenon of 'blueing' of glowing objects [6].

Previously we discussed mainly aerosol scattering. This is not accidental. Although the aerosol absorbs radiation, this process is not important: in the visible range of the spectrum, the aerosol absorption is low, and in the infrared region it is usually ignored in comparison with gas absorption. However, it should be mentioned that all this is valid only for the 'standard' conditions on the Earth. Powerful emissions of volcanoes, the arrival of large meteorites, large forest fires, wars, etc. may result in the ejection into the atmosphere of a very large quantity of the aerosols absorbing radiation (mainly, soot and ash) resulting in extensive changes of the optical properties of the Earth's atmosphere.

Other examples of the polydisperse optical characteristics for the case of cumulus clouds are presented in Figs. 5.8 and 5.9. Figure

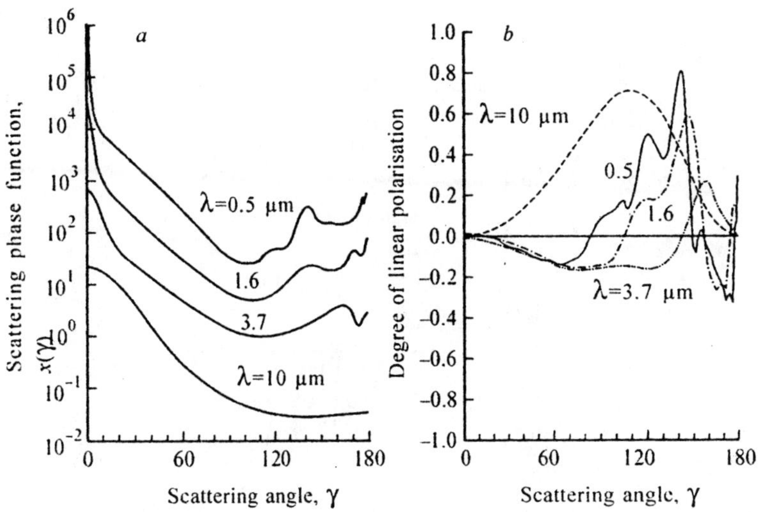

Fig. 5.8. Scattering phase functions $x(\gamma)$ and degree of linear polarisation (b) for the case of cumulus clouds.

5.8a and 5.8b show the scattering phase functions x (γ) and the degree of linear polarisation of this type of cloud for different wavelengths in the visible and infrared ranges of the spectrum. In the visible range of the spectrum at λ = 0.5 μm there is extensive scattering in the forward direction (high values of the parameter $x = 2\pi r/\lambda$). In the infrared range (λ = 10 μm), the relative contribution to the 'forward' scattering is considerably smaller. As indicated by Fig. 5.8b, the degree of linear polarisation in scattering in cumulus clouds may reach 60–80% at specific scattering angles.

Figure 5.9 shows, for the same cumulus clouds, the spectral dependence of the normalised extinction coefficient σ_e (0.5), the albedo of single scattering $\tilde{\omega}_0 = \sigma_a / \alpha_a$, and the asymmetry parameter g (this parameter will be discussed in greater detail in the following section). It should be mentioned that in the mean, the extinction coefficient of the cumulus clouds increases in transition from the visible to the infrared range of the spectrum (approximately up to λ = 6 μm) and subsequently decreases. The spectral behaviour of the albedo is determined mainly by the spectral course of the imaginary part of the refractive index of water. In particular, the spectral course of $\tilde{\omega}_0$ contains three minima at the wavelength of approximately 3, 6 and 10 μm. The asymmetry parameter g is approximately constant (0.82–0.86) for wavelengths up to ~10 μm (with the exception of ~3–6 μm, where the diffraction

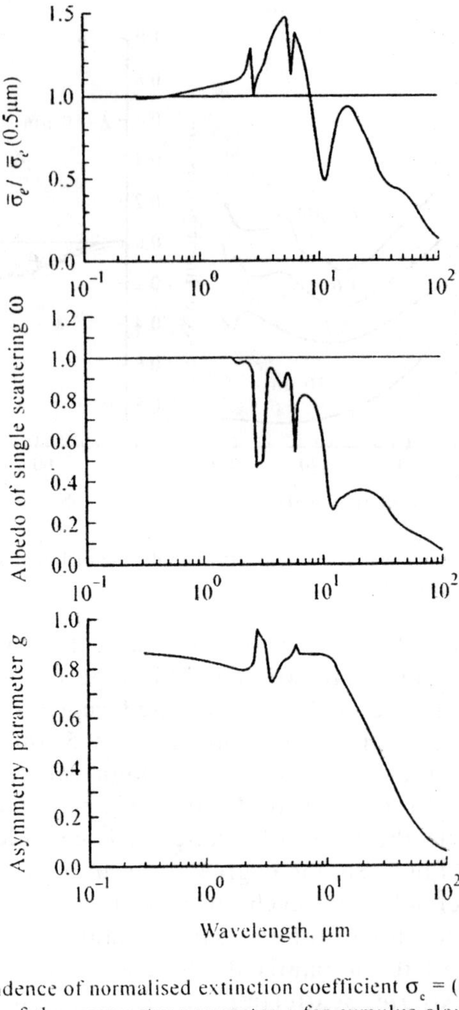

Fig. 5.9. Spectral dependence of normalised extinction coefficient σ_e = (0.5), albedo of single scattering ω and of the asymmetry parameter g for cumulus clouds [103].

effect, leading to strong forward scattering, is dominant). With a further increase of the wavelength a decrease of the scattering parameter x results in a large decrease of the asymmetry parameter.

Approximation of scattering phase functions

The scattering phase functions as a function of the scattering angle are often approximated using polynomial expansions, in particular in the solution of problems of transfer of solar radiation in the atmosphere of planets (chapter 8). Usually, the orthogonal basis in the scattering theory is represented by the Legendre polynomials:

$$x(\cos\gamma) = \sum_{l=0}^{N} x_l P_l(\cos\lambda), \qquad (5.3.6)$$

where P_l (cos γ) are the Legendre polynomials of the first order, and x_l is the expansion coefficient. The Legendre polynomial is selected as the orthogonal basis because of the fact that, in particular, the zero Legendre polynomial is equal to unity, and the first one to cos γ. Thus, the zero and second terms of the expansion make it possible to take into account the contribution of molecular (Rayleigh) scattering to the scattering phase function.

Table 5.3 gives examples of the simplest expansions of the scattering phase function and the first four of their expansion coefficients. One of the expansion coefficients for two analytical functions is presented in the form of a special coefficient, i.e. the asymmetry parameter g = 1/3 x_1. It should be mentioned that g = 1 corresponds to the case of complete forward scattering (scattering angle 0°), g = −1 to the case of complete scattering backwards (scattering angle 180°), and at g = 0 there is isotropic or symmetric scattering (for example, Rayleigh scattering).

It should be mentioned that as the particle size increases and the height of the peak of the phase function in forward scattering becomes larger, the number of the terms of expansion in equation (5.3.6) which must be considered to obtain the acceptable accuracy of approximation increases.

Table 5.3. Examples of approximation of the scattering phase functions and their expansion coefficients (values $\delta^+ = 1$ at $\gamma = 0$ deg and zero at other angles, $\delta^- = 1$ at $\gamma = 180$ deg and zero at other angles)

Type of scattering	Formula for phase function	Expansion coefficients			
		x_0	x_1	x_2	g
Isotropic	1	1	0	0	0
Rayleigh	$\frac{3}{4}\left(1+\cos^2\gamma\right)$	1	0	1/2	0
Henyey–Greenstein	$\frac{1-g^2}{\left(1+g^2-2g\cos\gamma\right)^{\frac{3}{2}}}$	1	$3g$	$5g^2$	g
Forward and back scattering	$(1+g)\delta^+ + (1-g)\delta^-$	1	$3g$	$5g^2$	g

Aerosol optical models

The volume coefficient of aerosol scattering and absorption and the phase function (matrix) of aerosol scattering, according to the transfer equation (3.4.35), are used as parameters when calculating the radiation field in the atmosphere. It is necessary to determine the numerical values of these components. Using (5.3.2)–(5.3.4), the problem is solved by calculating the optical characteristics of the aerosols in respect of the given complex refractive index, the size distribution function of the particles and the particle concentration, and they are in turn determined from experimental measurements. The set of these characteristics and of the optical characteristics, calculated from them, is referred to as the aerosol optical model [26].

Problem of nonsphericity in aerosol optics

Completing discussion of aerosol scattering, we return to the starting point. In the previous section, it was shown that to solve the problem of diffraction of light on aerosol particles, they are simulated by solids of a specific geometrical form. However, in all cases, we discussed only homogeneous spherical particles. As already mentioned in chapter 2, the form of the particles in the atmosphere of the planets may greatly differ. The form close to spherical is shown only by liquid aerosols. Even this form, for example, in the case of large liquid cloud particles, can be simulated more efficiently by ellipsoids of revolution (spheroids). The dust and soot particles and icy crystals of the clouds, have the form which greatly differs from spherical.

The description of the scattering of radiation on non-spherical particles leads, in a general case, to very complicated equations of electrodynamics, and consequently, solutions of these equations have been obtained only for several models. For example, there are the equations similar to the Mie equations, for calculating the optical characteristics of two-layer spheres ('an envelope in a shell') [6]. They are used quite widely for describing the optical properties of flooded aerosols in the atmosphere [11, 26] (in condensation of liquid water on aerosol particles at a high relative humidity of air). There are solutions for the ellipsoids of revolution (spheroids) [76, 87, 91] which can be used for simulating large liquid particles in clouds and precipitation. Figure 5.10 shows examples of calculations of the factors of the efficiency of scattering for prolate and flattened spheroids. The graph shows the form of these particles, characterised by two parameters, a and b, and their orientation in relation to the direction of propagation of radiation. In this case, the normalisation of the efficiency factor of scattering is carried out in relation to the areas b^2 and a^2. The curves match different ratios a/b. The optical characteristics of the spheroids are similar to the appropriate characteristics for the spheres. In the limit of small spheroids, we obtain the Rayleigh type of scattering. For large spheroids, the interference and resonant special features in scattering are typical. The spheroids are sometimes used in modelling of the optical properties of icy particles ('needles' and 'discs').

It should be mentioned that in passing to the non-spherical particles, the difficulties of describing the ensembles of the particles become more severe because, in this case, it is necessary to integrate not only in respect of the radius of the spheres but also a number of the geometrical parameters of the form, taking into account the orientation of the particles in relation to the direction of propagation of radiation. In this case, the scattering phase function of the ensemble of these particles already depends not only on the angle but also the scattering azimuth. This greatly complicates the already complicated equation of radiation transfer when taking into account the multiple scattering for these particles

Icy particles of cirrus clouds are often simulated by hexagonal particles. In this case, the dimensions of the hexagonal crystals are usually of the order of several hundreds of micrometers. In this case, since the dimensions of the particles are considerably greater than the wavelength of radiation, we can use the approximation of geometrical optics. In this method we examine the propagation of the individual light rays taking into account the processes of reflection and refraction of light on the crystal faces. The rays, leaving the crystals in different directions, are characterised by different amplitudes and phases. In the case of multiple reflections and refraction, it is necessary to use the Snell's law. The energy of the reflected and refracted fields is determined by the Fresnel equation (for more details, see chapter 6).

The scattering on non-spherical particles is characterised by its specific angular and polarisation special features. The pattern of polarisation for the non-spherical particles greatly differs from the polarisation recorded for the spheres. Large differences in the polarisation characteristics of scattered radiation for different cloud particles are the physical basis of remote methods of examining the form of cloud particles.

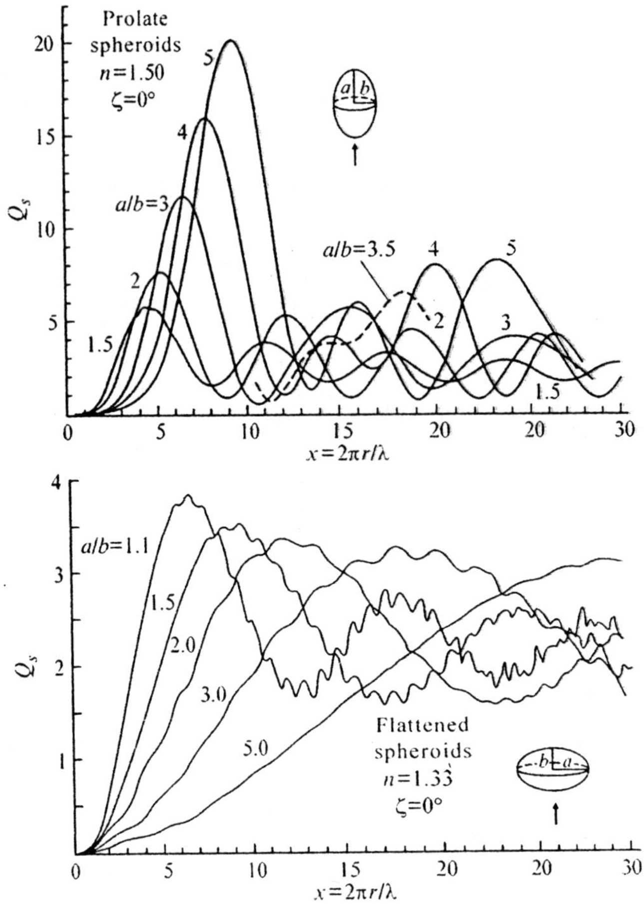

Fig. 5.10. Examples of calculations of the efficiency factor of scattering for prolate and flattened spheroids [51].

Approximations in the theory of scattering in the atmosphere

The scattering in the atmosphere depends on many parameters:

- the number of particles;
- the size and shape of the particles (distribution functions);
- the refractive index of the particles (its real and imaginary parts);
- the wavelength of radiation.

Since the scattering on the spherical particles depends on the parameter $x = 2\pi r/\lambda$, the behaviour of particles in different regions of the spectrum differs. This fact is clearly indicated in Fig. 5.11 which shows the spectral regions of the applicability of the approximations of geometrical optics, Mie scattering, Rayleigh scattering and the negligible small scattering for different particles. The vertical axis gives the dimensions of the particles, from rain droplets to the air molecules, the horizontal axis gives the wavelength. Evidently, depending on the wavelength, the particles of atmospheric mist may also be Mie particles and Rayleigh particles. In the microwave range, these particles practically do not scatter.

Variability of aerosol extinction

Because of the extremely large variations of the characteristics of the aerosol particles, i.e. their concentration, shape, the function the size distribution of the particles, etc, the optical characteristics of the aerosols may change greatly.

Figure 5.12 shows the spectral dependence of the extinction coefficients for different types of atmospheric aerosol. It may be seen that the maximum extinction (~10^{-2} m^{-1}) is characteristic of the cumulus clouds. It is important to note the relatively weak spectral dependence of the coefficient of extinction for this type of clouds in the examined spectral range, 0.1–10 μm.

The coefficient of extinction of high layer clouds is an order of magnitude smaller and greatly varies in the spectral range 7–12 μm.

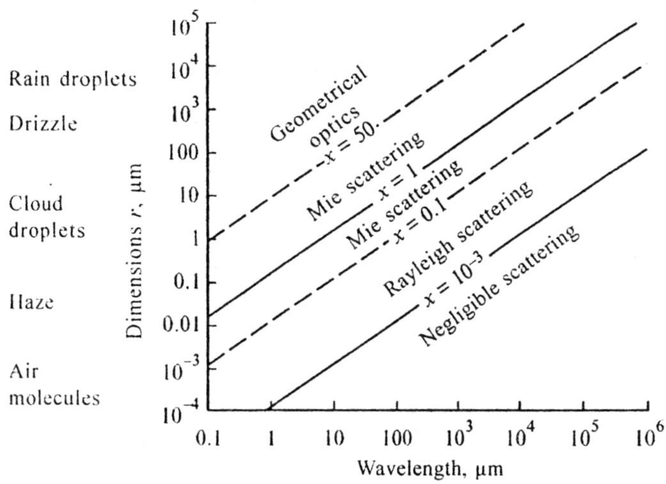

Fig. 5.11. Spectral ranges of applicability of various approximations for different particles [101].

The coefficients of extinction of haze of different types change in a wide range, from ~10^{-4} to $5 \cdot 10^{-7}$ m^{-1}. For comparison, Fig. 5.12 also shows the curve of the spectral dependence of the coefficient of Rayleigh molecular extinction. It may be seen that at $\lambda > 3$–5 μm, Rayleigh extinction in the atmosphere of the Earth may be ignored.

5.4. Scattering of radiation with redistribution in respect of frequency

The previously examined mechanisms of scattering of radiation in the atmosphere of the planets (molecular and aerosol) are characterised by the fact that scattered radiation has the same wavelength (frequency) as incident radiation. This type of scattering is referred to as the elastic or coherent scattering. However, there are also other types of scattering known in optics, at which the

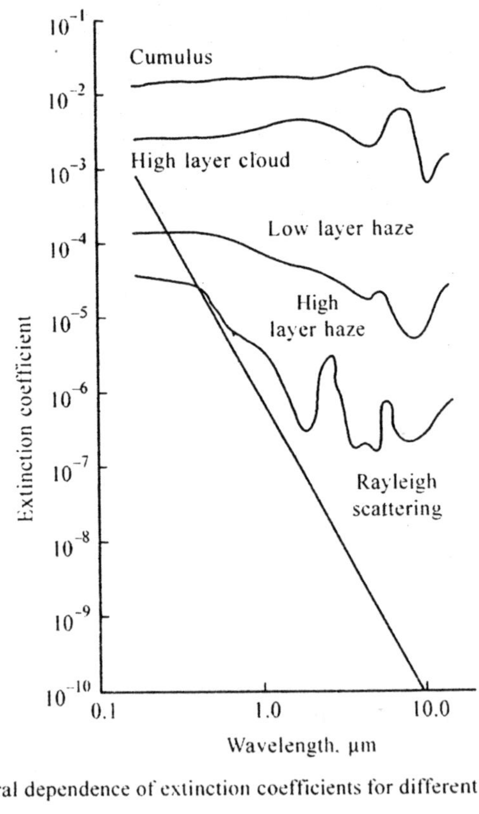

Fig. 5.12. Spectral dependence of extinction coefficients for different types of atmospheric aerosol [44].

frequency of radiation after scattering changes. Such scattering in the theory of radiation transfer is referred to as the scattering of radiation with redistribution in respect of frequency. In a number of textbooks and monographs, this type of scattering was previously often referred to as inelastic scattering. The terminology 'elastic–inelastic' scattering is associated with the interpretation of scattering as collisions of photons with molecules, examined in chapter 4. This term cannot be regarded as suitable because, for example, elastic collisions of the particles are not accompanied by the transfer of the energy of translational movement to the internal degrees of freedom. At the same time, in aerosol scattering in cases in which the imaginary part of the refractive index is not equal to zero, part of radiation energy is absorbed, although scattering takes place at the same frequency and may be referred to as elastic. This can also be said in a general case about molecular scattering.

It is more accurate to refer to scattering with the variation of the frequency of scattered radiation as non-coherent scattering, where the disruption of coherence is represented by the change of the frequency of radiation. If in previous examination of molecular and aerosol scattering it was sufficient to use classic electrodynamics, then the non-coherent types of scattering require the application of quantum mechanics representations.

Raman scattering

The most widely known type of non-coherent scattering is Raman scattering. It is assumed that a molecule has a system of internal energy levels, shown in Fig. 5.13, and consisting of two electronic levels split into vibrational sublevels. In particular, Fig. 5.13a shows schematically the process of Rayleigh scattering, when the frequency of incident radiation differs from the frequencies corresponding to the transition of the molecule from the lower to upper states. In this case, as mentioned previously, the frequency in Rayleigh scattering remains the same, i.e. $\nu = \nu_0$. This is illustrated in the central part of Fig. 5.13a in the graph – Spectral manifestation.

In addition to the process of Rayleigh scattering, there may be another type of scattering, shown in Fig. 5.13b. It appears that if a molecule is in the radiation field with frequency ν_0, which does not coincide with the frequencies corresponding to the transitions of the molecule from the lower to upper state of the molecule, the so-called virtual transition may take place, i.e. the transition to the virtual state of the internal energy of the molecule. It does not coincide with the allowed quantum states, shown in Fig. 5.13. (The virtual level of the molecule in Fig. 5.13b is indicated by the broken line). Regardless of the fact that the frequency of incident radiation does not correspond to the distance between the allowed quantum states of the molecule, there is a slight interaction between the incident radiation and the investigated molecule. In this case, part of the radiation energy may be transferred to the internal energy of the molecule, and the molecule, 'leaving' the virtual level, may emit at a frequency of $\nu_1 < \nu_0$. Under certain conditions, the molecule may pass to the ground electronic state, but to the level, corresponding to the radiation frequency of $\nu_2 > \nu_0$. Naturally, part of the internal energy of the molecule is transformed to radiation energy.

The described process of interaction of radiation with the molecule is referred to as the ordinary combination (Raman) scattering and leads to the formation in the spectrum of scattered radiation, in addition to the line with frequency ν_0, of two additional lines, the so-called Stokes and anti-Stokes components. These components are situated on different 'sides' of the ground line of scattering with frequency ν_0. It is important that the position of these two additional lines in the scattered radiation in relation to the frequency of incident radiation with frequency ν_0 is fully determined by the system of the levels of the internal energy of the molecule. In the case examined in Fig. 5.13b, this is determined by the difference of the energy of the vibrational half-levels of the ground electronic state of the molecule. At the same time, the Raman spectrum of scattered radiation is individual for every specific molecule. In other words, the 'shifts' of the Stokes and anti-Stokes components in relation to the frequency of incident radiation are individual for every molecule.

The efficiency of the process of interaction of radiation with the molecule, leading to the process of ordinary Raman scattering and shown schematically in Fig. 5.13b, is small. In other words, the probability of transition of the molecule to the virtual state is low. Consequently, the cross-section of ordinary Raman scattering is smaller than, for example, the cross-section of molecular (Rayleigh) scattering by three orders of magnitude. The intensity of Raman scattering is proportional to the number of the molecules in the initial state, and the transitions from this state to the virtual state generate the given scattering line. This is the physical basis of the

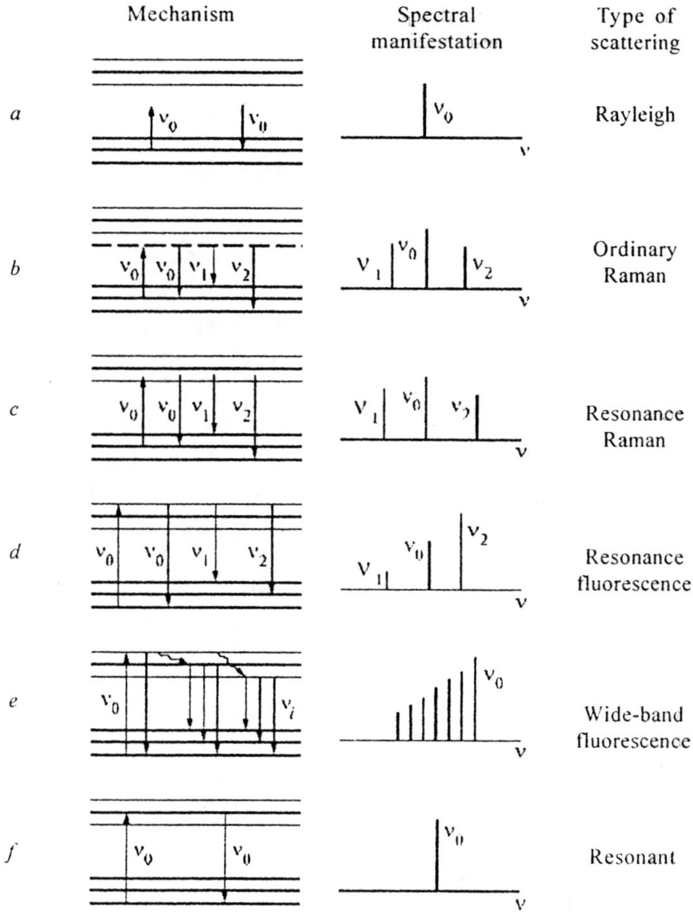

Fig. 5.13. Types of scattering. a) Rayleigh, b) ordinary Raman, c) resonant Raman; d) resonance fluorescence, e) wide-band fluorescence; f) resonance scattering [38].

determination of the concentration in the atmosphere of different molecules in the spectrum of conventional Raman scattering.

When the frequency of incident radiation is very close to the frequencies corresponding to the transitions of the molecules from the ground to excited states, the intensity of interaction of radiation with the molecule greatly increases. In other words, the situation corresponds to cases in which the virtual level of the molecules is close to the allowed quantum states of the molecules. This type of scattering is referred to as resonance combinational scattering (Fig.

5.13c). This is accompanied by the increase of the scattering cross-section by 3–6 orders of magnitude in comparison with the cross-section of ordinary (extra resonance) Raman scattering.

Fluorescence

Another important process of interaction of radiation with the molecule is shown in Fig. 5.13d. The frequency of incident radiation ν_0 corresponds to the difference of the energies of the ground and excited states. This is accompanied by the absorption of incident radiation by the molecule, and the molecule changes to the excited state for a specific period of time. If this period is not characterised by emission-free transitions as a result of collisions of the molecules ('quenching' of excitation), the molecule may pass to the ground electronic state, re-emitting the energy. Here, because of the presence of vibrational sub-levels of the ground state, these transitions may provide radiation at different frequencies ν_i. These frequencies are evidently determined by the system of these sub-levels. Consequently, the spectrum of scattered radiation (this spectrum may also be referred to as the spectrum of emission of the molecule) contains many lines. The described processes are referred to as fluorescence (in some cases, resonance fluorescence).

In many cases, another process of interaction of radiation with a molecule is mentioned – wide-band fluorescence (Fig. 5.13e). The process differs from simple fluorescence by the fact that during the time in which the molecule is in the excited state the collisions of the molecules result in emission-free transitions on different vibrational sub-levels of the excited electronic state of the molecules. After these emission-free processes and the 'distribution' of the molecule over the additional vibrational sub-levels of the excited electronic state, the molecule may change to the ground electronic state (again on different sub-levels), and the scattering (emission) spectrum contains an even larger number of the scattering lines in comparison, for example, with the case of resonance fluorescence. This set of the large number of the scattering lines occupies a wide spectral range and the spectrum itself, when recorded by equipment with low spectral resolution, has the form of a wide scattering band (radiation).

In optics, fluorescence is usually regarded as a set of two single-photon processes, i.e. as the two-stage interaction of radiation with the molecule, consisting of the absorption of an individual photon with frequency ν_0 followed by spontaneous emission of a photon with another frequency ν_i. The difference $\nu_i - \nu_0$ is determined by the position of the sub-levels of the ground and excited state of the molecule and depends on the structure of the molecule. At the same time, it is evident that the fluorescence spectra are strongly individual for every molecule and enable the type of molecule to be identified.

Examining the processes of fluorescence as two single-photon processes, it is important to pay attention to the fact that there is a specific time period between the acts of absorption and emission of the photon. Therefore, the intensity of fluorescence after irradiation is a function of time, in particular, at low pressure fluorescence decreases exponentially with time.

All types of fluorescence in the atmosphere are subjected to quenching (without forming the photon) caused by inelastic collisions with the surrounding molecules. Therefore, the intensity of fluorescence at a low pressure (in the upper layers of the atmosphere), i.e. at a small number of collisions of the molecules, is several orders of magnitude higher in comparison with that at high-pressure (in the lower layers of the atmosphere). Nevertheless, the cross-section of the 'quenched' fluorescence is usually larger than the cross-section of Rayleigh or ordinary Raman scattering. This is associated with the resonance nature of the interaction of radiation with the molecule because the frequency of incident radiation ν_0 corresponds to the difference of the energies of the internal states of the molecule. It should be mentioned that the processes of collisions of the molecules result not only in the quenching of fluorescence but also in the previously discussed phenomena of broadening of the spectral lines (Chapter 4).

As already mentioned, the fluorescence processes may be regarded as two one-photon processes. On the other hand, the scattering on an isolated (individual) atom or molecule (for example, Rayleigh or Raman scattering) is usually regarded as a two-photon process, described by the single-stage interaction. This process also leads to the disappearance of the photon with the frequency ν_0 and the appearance of a photon with the same frequency ν_0 or with different frequency ν. All the experimental data indicate that the quenching collisional processes, examined in connection with fluorescence, do not play any role for the two-photon scattering because this process is almost completely instantaneous.

Figure 5.13f shows another process, consisting of the excitation of the molecule by resonant radiation with frequency ν_0 (resonant in the sense of fulfilment of the condition $h\nu_0 = E_2 - E_1$, where E_1

and E_2 are the energies of the ground and excited states of the molecule) and the almost simultaneous emission of a photon with the frequency equal to or very close to frequency ν_0. In scientific literature, this process is also sometimes referred to as resonance fluorescence. Since we have examined fluorescence as a combination of two one-photon processes, this term is not suitable in the given case. It is more logical to referred to the process as resonance scattering.

Since the frequency of radiation of resonance scattering by molecules is identical with the frequency of aerosol scattering or close to it, as in the case of resonance fluorescence, it is necessary to solve the problem of separation of these different components of scattering. This separation is not complicated in the upper layers of the atmosphere where the contribution of aerosol scattering is very small. It should be mentioned that although according to the formal features of the Rayleigh and resonant scattering (Fig. 5.13a and 5.13f) these two types of scattering are similar (the frequency of scattering is equal to frequency of incident light), Rayleigh scattering differs from resonance scattering by the fact that the frequency of incident radiation ν_0 'does not coincide' with the differences of the internal energies of the molecule. According to our definition, these two types of scattering belong to coherent scattering*.

Comparison of different types of scattering

Although we have already discussed the intensities and duration of different processes of scattering and the times of occurrence of these processes, it is useful to examine Table 5.4 which gives, for the visible range of the spectrum, characteristic values of the differential backscattering cross-sections and the duration of the processes for different acts of scattering.

Table 5.4 shows clearly the many previously mentioned special features of different types of scattering – a wide range of the values of differential cross-sections of scattering for aerosol scattering, and also resonance Raman scattering and resonance fluorescence, almost the 'instantaneous' nature of processes such as a Rayleigh, aerosol, and ordinary Raman scattering, the relative continuance of other scattering processes.

Role of non-coherent scattering in atmospheric optics

The non-coherent types of scattering play a very small role in the

Table 5.4. Characteristics of different types of scattering [38]

Type of scattering	Differential backscattering cross-section, $cm^2 \cdot sr^{-1}$	Process duration, s
Rayleigh	10^{-27}	10^{-14}
Aerosol	10^{-27}–10^{-8}	10^{-14}
Ordinary Raman	10^{-30}–10^{-29}	10^{-14}
Resonant Raman	10^{-30}–10^{-23}	10^{-14}–10^{-8}
Resonant fluorescence	10^{-23}–10^{-16}	10^{-8}–10^{-1}
Wide-band fluorescence	10^{-16}	10^{-8}
Resonant scattering	10^{-27}–10^{-20}	10^{-14}–10^{-6}

*Generally speaking, both Rayleigh and resonance scattering are characterised by a narrow spectrum of scattering as a result of, for example, thermal motion of the molecules – one should remember the Doppler effect and the Doppler line shape (chapter 4).

radiation energetics of the lower layers of the atmosphere of the planets. This is due to the fact that a large number of the collisions of the molecules and intensive 'quenching' of the excited states result in the negligible role of different types of fluorescence. Raman scattering also does not play any significant role in the majority of cases. This is because solar radiation is characterised by a wide spectrum. Therefore, the redistribution of radiation over frequency, taking place during non-coherent scattering, occurs 'continuously' through the spectrum (i.e., for every frequency of solar radiation). The effect of this redistribution may be detected in the case of the presence in the spectrum of solar radiation or in the spectrum of absorption of the atmosphere of considerable spectral heterogeneities (minima or maxima of radiation or absorption). In fact, in the presence of, for example, Fraunhofer lines in the solar spectrum, experiments showed the presence of the so-called Ring effect*, leading to the smoothing of the spectral lines of absorption in the spectrum of the radiation, scattered by the atmosphere of the Earth. It may be assumed that the effects of non-coherent scattering may be significant when calculating the fields of scattered solar radiation in ultraviolet and visible ranges of the

*This effect should really be referred to as the Shefov–Ring effect because it was detected for the first time separately by two scientists, and Shefov's study was published earlier than that Ring).

spectrum characterised by the presence of a large number of Fraunhofer lines. The same effect may also be found in the scattering spectra of the atmosphere in the region of strong lines and absorption bands of the atmospheric gases.

In the upper atmospheres of the planets, different types of non-coherent scattering play a significant role. In particular, the processes of formation of atmospheric radiation in the case of breakdowns of the local thermodynamic equilibrium and the interaction of solar radiation with the upper layers of the atmosphere are closely linked with the previously examined processes of fluorescence (see also chapter 8).

The processes of non-coherent scattering of radiation are very important in the interpretation of the results of optical measurements of the characteristics of the natural and artificial fields of radiation. Different non-coherent processes of scattering play a special role in the laser remote sensing of the atmosphere where they can be used efficiently for obtaining different information on the parameters of the physical state of the atmosphere. For example, Raman scattering represents the basis of a number of laser methods of determination of the concentration of different atmospheric gases and temperature of the atmosphere.

5.5. Atmospheric refraction

The refraction phenomenon

The spatial inhomogeneities in the values of the refractive index of atmospheric air, caused by spatial changes of the physical parameters of the air, result in deviations in the straight propagation of light. This phenomenon is referred to as refraction, i.e. the distortion of the trajectories of the beams of light in the inhomogeneous atmosphere [18, 43, 47].

Refraction is divided into a number of types:

–astronomical refraction – the phenomenon of the variation of the visible position of extraterrestrial sources of light in relation to the true position of the sources on the skies sphere;

–the Earth (atmospheric) refraction – phenomenon associated with the variations of the visible position of the source of light (or an object) situated in the atmosphere, when examined from the Earth's surface or from another point in the atmosphere;

–cosmic refraction – the effect of variation of the position of light sources when examined from the space through the atmosphere of the Earth.

In the literature, there are also definitions of regular (normal) or random refraction. Regular refraction is caused by the smooth changes of the parameters of the atmosphere and, correspondingly, by small changes of the refractive index. Random refraction is caused by a relatively small-scale spatial variation of the parameters of the atmosphere and the refractive index. These variations are characterised by different spatial scales, from centimetres to tens of metres. They are caused by, for example, turbulence in the atmosphere. Random refraction results in the well-known phenomenon of the scintillation of .light sources, for example, the flickering of stars when viewed from the Earth's surface. Finally, one should mention the phenomenon of anomalous refraction – stable, long-term (up to several hours) deviations of the refractive index of air from its mean values. Anomalous refraction is responsible for various phenomena such as mirages, which will be discussed later.

The refraction equation

The refraction phenomenon may be explained by the refraction of light at the boundaries of layers with different optical properties. We examine the propagation of light from an extraterrestrial source (Fig. 5.14). The atmosphere is divided into several concentric layers, relatively thin, so that they can be regarded as homogeneous, with a constant refractive index. The refractive indexes, corresponding to these layers, will be denoted by n_1, n_2, n_3, etc. According to (5.1.31), the refractive index is associated with the density of air which decreases with increasing altitude and, consequently, $n_1 < n_2 < n_3 <$.... The incidence angle θ and refraction angle ψ at the boundary of two adjacent layers are linked by the Snell's law:

$$\frac{\sin\theta_1}{\sin\psi_1} = \frac{n_2}{n_1}, \frac{\sin\theta_2}{\sin\psi_2} = \frac{n_3}{n_2}, \frac{\sin\theta_3}{\sin\psi_3} = \frac{n_4}{n_3}, \ldots \qquad (5.5.1)$$

The triangle $1O2$, according to the sinus theorem, shows that

$$\frac{r_1}{\sin(\pi - \theta_2)} = \frac{r_2}{\sin\psi_1},$$

where r_1 and r_2 ar4 distances from the points 1 and 2 to point O (the centre of the Earth). Similarly, from the triangles $2O3$, $3O4$,

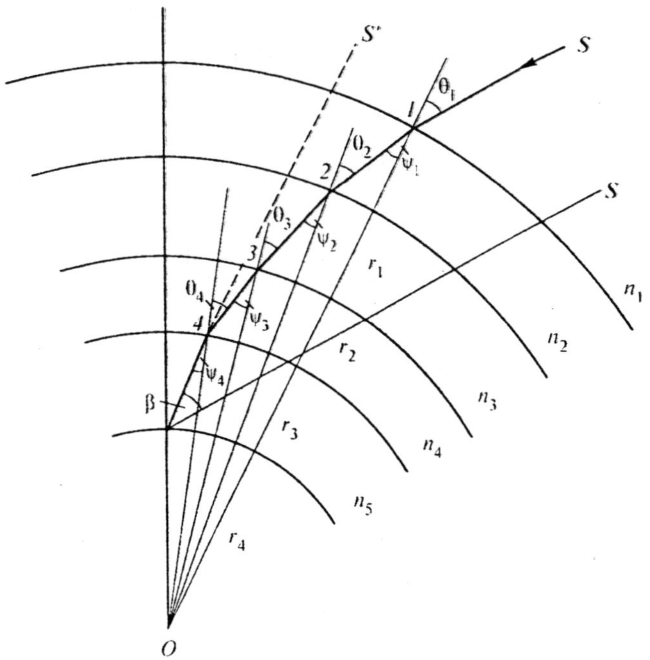

Fig.5.14. Derivation of the refraction equation in the atmosphere [43].

etc., it follows that

$$\frac{\sin\psi_1}{\sin\theta_2}=\frac{r_2}{r_1},\frac{\sin\psi_2}{\sin\theta_3}=\frac{r_3}{r_2},\frac{\sin\psi_3}{\sin\theta_4}=\frac{r_4}{r_3},\ldots \tag{5.5.2}$$

Multiplying the equalities (5.5.1) and (5.5.2) in pairs, we obtain

$$\frac{\sin\psi_1}{\sin\theta_2}=\frac{n_2r_2}{n_1r_1},\frac{\sin\psi_2}{\sin\theta_3}=\frac{n_3r_3}{n_2r_2},\frac{\sin\psi_3}{\sin\theta_4}=\frac{n_4r_4}{n_3r_3},\ldots$$

Consequently

$$n_1r_1\sin\theta_1 = n_2r_2\sin\theta_2 = n_3r_3\sin\theta_3 = \ldots$$

Thus, at any point of the trajectory of the ray, the following relationship is fulfilled:

$$n(r)r\sin\theta = \text{const}, \tag{5.5.3}$$

where r is the distance from the centre of the earth; $n(r)$ is the refractive index of air; θ is the zenith angle of the light ray. Equation (5.5.3) is also the equation of the trajectory of the light ray in the atmosphere or the refraction equation. The constant in (5.5.3) is evidently equal to $r_0 \sin \theta_0$, where r_0 is the distance from the centre of the earth to the upper boundary of the atmosphere (where $n \equiv 1$); θ_0 is the angle of incidence of the beam on the upper boundary.

Astronomical refraction

As a result of astronomical refraction, all the extraterrestrial sources of light, i.e. the Sun, the planets, stars, appear to be lifted to some angle above the horizon. An important characteristic is the angle of astronomical refraction β, i.e. the angle between the true S and visible S' directions to the light source (Fig. 5.14). Table 5.5 gives the calculated values of the angle of astronomical refraction for the visible altitudes of the source of radiation h above the horizon.

The maximum angles of astronomical refraction are obtained at the moments of rising and setting of heavenly bodies and for small negative altitude angles. For the mean atmospheric conditions, they reach the values of 35', but at low temperatures and high pressure at the surface of the Earth, the variations of the refractive index of air may become large and the refraction angles increase to 2–3°. Consequently, this phenomenon increases the duration of the day (the light time of the day). In the case of high latitudes, this increase may reach hours and days. For example, at the pole, the duration of the polar days (when the Sun does not descend below the horizon) is more than 14 days longer than the duration of the polar night.

Astronomical refraction also results in the flattening of the Sun disc (or the Moon) during its rising and setting, because the rays from the upper and lower edges of the Sun are characterised by different angles of refraction. The flattening effect reaches 7 min at the angular dimensions of the Sun of 32 min. It should be mentioned that according to (5.1.30), the refractive index of air is characterised by a spectral dependence and, therefore, the refraction angle for different ranges of the spectrum slightly differs. This is associated with a very rare and beautiful phenomenon of the green beam when, usually at sunset, the last

Table 5.5. Astronomical refraction angles β (in arc minutes) for different visible altitudes h (in degrees) of heavenly bodies above the horizon [18]

h, deg	−0.10	0.00	0.10	0.30	1.0	3.0	10.0	30.0	90.0
β, min	36.8	34.3	32.3	28.7	24.3	14.3	5.3	1.7	0

visible point of the Sun disc changes to green. The green beam can be detected only in a very quiet atmosphere above the ideally flat edge of the horizon: it is usually the sea in the conditions of complete calm.

Earth refraction, mirages

The light rays from the ground-based objects also propagate along curvilinear trajectories. The Earth refraction angle is the angle between the direction to the visible and actual position of an object. The values of this angle depend on the distance to the examined object and the temperature stratification of the layer of air above the Earth. Depending on the nature of the vertical gradient of temperature which, according to (5.1 .31), determines the gradient of the refractive index, the layer of the atmosphere above the surface of the Earth may be characterised by lifting and expansion or lowering and narrowing of the visible horizon. This effect results in the increase (during expansion) or decrease (during narrowing) of the geometrical range of visibility of the objects.

It is assumed that the atmosphere is horizontally homogeneous, i.e. pressure p and temperature T change only with the altitude in the atmosphere. To a first approximation (disregarding the humidity of the atmosphere), equation (5.1.31) may be presented in the following form:

$$n-1=(n_0-1)\frac{\rho}{\rho_0}=(n_0-1)\frac{T_0 p(z)}{T(z)p_0}. \tag{5.5.4}$$

Differentiating (5.5.4) in respect of altitude, we obtain

$$\frac{dn}{dz}=(n_0-1)\frac{dp(z)}{dz}\frac{T_0}{T(z)p_0}-(n_0-1)\frac{dT(z)}{dz}\frac{T_0 p(z)}{T^2(z)p_0}.$$

We denote $\gamma = -dT(z)/dz$ is the vertical temperature gradient. Subsequently, using the equation of hydrostatics (2.2.1) obtain

$$\frac{dp}{dz} = -\rho \frac{\mu g}{RT(z)},$$

we finally obtain

$$\frac{dn}{dz} = (n_0 - 1)\frac{\rho(z)}{\rho_0 T(z)}(\gamma - \gamma_0), \tag{5.5.5}$$

where $\gamma_0 = \mu g/R = 34.2$ K/km is a constant. We analyze equation (5.5.5) in different conditions in the layer of the atmosphere just above the surface of the Earth.

It is assumed that the temperature gradient is positive but smaller than γ_0. In these conditions $dn/dz < 0$ – the refractive index decreases with altitude, the trajectory of the light ray faces the Earth's surface by the concave side. Consequently, the horizon is lifted (in comparison with the true horizon) by a certain angle. This refraction is referred to as positive. In the mean meteorological conditions expansion of the horizon reaches 6–7%, the radius of curvature of the trajectory of propagation of radiation is approximately 6 times greater than the Earth radius. In the case of temperature increasing (not decreasing) with altitude, the density of air and its refractive index may rapidly decrease and the radius of curvature of the ray becomes equal to the Earth radius. The visible horizon appears to lift up and the Earth is 'straightened' and becomes flat. In the case of an even larger increase of temperature the curvature of the beam may become smaller than the curvature of the Earth. The visible horizon is lifted even more and to the observer it may appear that he/she is at the bottom of a huge basin. Because of the visible horizon the objects situated far away behind the 'current' horizon, appear to be lifted and become visible. For example, from the Canadian shores through the Smith Bay (approximately 70 km) one can observe sometimes the shoreline of Greenland with all details of the relief and buildings.

Similar conditions result in the formation of upper mirages when the visible image of an object, which is often magnified and distorted, is situated above the object, and the object itself is often not visible because it is far away behind the horizon. The upper mirages are usually found in the polar regions where the air near the surface is colder in comparison with the temperature of air at a low altitude in the atmosphere.

If the temperature gradient γ in (5.5.5) is equal to γ_0, then $dn/dz = 0$ and refraction evidently does not occur.

Let us finally assume that $\gamma > \gamma_0$ in particular, the temperature rapidly decreases with the altitude and in this case $dn/dz > 0$. The trajectory of the ray faces the Earth's surface by the conved side and this refraction is referred to as negative. This temperature distribution is found in most cases in the steppes and deserts during the daytime hours in the summer when the Earth's surface is overheated by the Sun. Lower mirages form in similar conditions. The formation of lower mirages is characterised by the formation of the images of 'lakes': it appears that the water surface is situated below the actual object and the latter is reflected in the water surface. Lower mirages are easily found in hot summer days above tarmac roads when it appears that the completely dry tarmac is covered with paddles over long distances in which the images of vehicles are visible.

Cosmic refraction

The development of cosmic methods of the measurement of the parameters of the atmosphere requires examination of the refraction phenomena when examining extraterrestrial sources through the atmosphere from the space (Fig. 5.15). Although the atmosphere is a three-dimensional heterogeneous medium, in examination of the many problems of refraction, it is necessary to use the model of the local spherical symmetric atmosphere, for example, in the vicinity of the perigee (the point closest to the Earth) of the trajectory of the propagation of radiation. It is taken into account that the refractive index n changes with the altitude at a considerably higher rate in comparison with the rate of change on the horizontal and, consequently, ignoring the horizontal heterogeneities, n may be regarded as a function of only the distance from the centre of the Earth r.

We find the angle of cosmic refraction β equal to the angle of deflection of the ray, passed through the atmosphere, θ_2 (point 2 in Fig. 5.15) from the initial direction θ_1 (point 1). It is determined by the equation

$$\beta = \int_1^2 \frac{dl}{R(l)}, \qquad (5.5.6)$$

where integration is carried out along the path of the ray from point 1 to point 2, and R is the radius of curvature of the ray. Equation (5.5.6) is evident if we remember the determination of the radius of curvature of the curves: R is the radius of the circle, which is

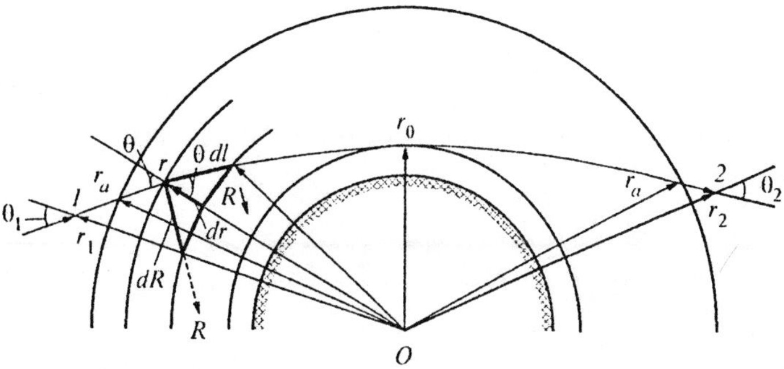

Fig.5.15. Sounding of the atmopshere in examination from space [18]. Points r_1 and r_2 correspond to the source and radiation receiver.

in contact with the curve at the given point, or in other words, the curve is regarded as a set of elementary pieces of the circles; consequently, ratio dl/R is the plane angle of rotation of the trajectory of the light rays, and to determine β it is necessary to add up the angles of turn along the entire trajectory.

The problem of determination of the radius of curvature R may be solved strictly (on the basis of the well-known Fermi principle), but the solution is very time-consuming. We obtain the value of R, using a simple mechanical analogy [62]. The light is regarded as a flux of photons – the particles with the mass m, moving with the velocity $v = c/n$ (n is the refractive index of air). The variation of the direction of motion of the photon, according to the second Newton law, is the results of the effect of some force on the photon. Formally, we can write

$$ma_R = F_R,$$

where the index R indicates the relevant components of the acceleration a and force F, acting along the radius to the instantaneous circle R. As indicated by the mechanics, in uniform rotation along the circle $a_R = mv^2/R$, and the force is the variation of potential energy U and, consequently:

$$\frac{mv^2}{R} = \frac{\partial U}{\partial R}. \qquad (5.5.7)$$

According to the law of conservation of energy $mv^2/2 + U =$ constant. Consequently, after differentiation, we obtain

$$-\frac{\partial U}{\partial R} = mv\frac{\partial v}{\partial R}. \tag{5.5.8}$$

Substituting (5.5.8) into (5.5.7) and taking into account $v = c/n$, we obtain

$$\frac{1}{R} = -\frac{1}{v}\frac{\partial v}{\partial R} = \frac{1}{n}\frac{\partial n}{\partial R} = \frac{\partial}{\partial R}\ln n(r),$$

and, consequently

$$\frac{1}{R} = \frac{d}{dr}(\ln n(r))\frac{\partial r}{\partial R}.$$

The derivative $\partial r/\partial R$ can be easily determined by examining in the differential manner the small sections of the trajectory dl. The radius of the instantaneous circle is normal to dl, and other facts are evident from Fig. 5.15: $dR = dr/\cos(\pi/2-\theta)$ and consequently $dr = dR \sin\theta$, where θ is the zenith angle of the light ray at the point r. Finally,

$$\frac{1}{R} = \frac{d}{dr}(\ln n(r))\sin\theta. \tag{5.5.9}$$

The required integral (5.5.6) is divided into two integrals, using the perigee point of the beam r_0:

$$\beta = \int_{r0}^{r1}\frac{dl}{R(l)} + \int_{r0}^{r2}\frac{dl}{r(l)}.$$

In examination from the space both points 1 and 2 are situated outside the limits of the atmosphere. Consequently, the expression for β is presented in the symmetric form:

$$\beta = \int_{r0}^{ra}\frac{dl}{R(l)},$$

where r_a is the height of the upper boundary of the atmosphere. Subsequently, taking into account that $dl = dr/\cos\theta$ and, according to the refraction equation (5.5.3)

$$\sin\theta = \frac{r_0 n(r_0)}{rn(r)}$$

(since $\sin\theta_0 = 1$), we finally obtain

$$\beta = 2\int_{r_0}^{r_a}\frac{d}{dr}(\ln n(r))\frac{\sin\theta}{\cos\theta}dr =$$

$$= 2n(r_0)r_0\int_{r_0}^{r_a}\frac{d}{dr}(\ln n(r))\frac{dr}{\sqrt{n^2(r)r^2 - n^2(r_0)r_0^2}}.$$

An important effect of cosmic refraction is the refraction elongation of the element of the ray dl. In fact, as defined previously, in the presence of refraction

$$dl = \frac{dr}{\cos\theta} = dr\frac{rn(r)}{\sqrt{n^2(r)r^2 - n^2(r_0)r_0^2}},$$

and in the absence of refraction, formally assuming $n(r)$ = const, we have

$$dl = dr\frac{r}{\sqrt{r^2 - r_0^2}}.$$

Table 5.6 shows the calculated values of the refraction elongation of the element of the length of the rayas the ratio dl in the presence and absence of refraction. The altitudes H_0 in Table 5.6 are counted from the surface of the Earth, i.e. $H_0 = r - R_0$, where R_0 is the radius of the Earth. The data, presented in Table 5.6, show that at low values of H_0 the refraction elongation may reach 5–15% which obviously must be taken into account when solving different atmospheric optics problems.

In examinations through the atmosphere of the Sun disc or the Moon, the variation of the angle of refraction β with the height of the ray r results in refraction divergence, i.e. the variation of the angle between the rays, emitted from different edges of the disc [18]. This variation may be quite significant at a sufficiently large distance of the examination point (space equipment) from the perigee of the rays, propagating through the atmosphere. In this case, the atmosphere may act as a scattering lens resulting in a visible decrease of the brightness of the Sun disc (Moon), i.e. the phenomenon of refraction extinction. Reversed situations of refraction amplification when the atmosphere acts as a collecting lens, reducing the angular dimensions of the Sun (Moon) are also possible. In particular, the phenomena are very strong in the case of low values of r_0, i.e. in examination through the lower layers of

Table 5.6. Dependence of refraction elongation on the altitude of the perigee H_0 and current altitude H [18]

H_0, km	$H - H_0$, km				
	0	5	10	15	20
0	1.159	1.108	1.079	1.060	1.048
5	1.073	1.051	1.038	1.030	1.024
10	1.035	1.025	1.019	1.015	1.012
15	1.018	1.013	1.010	1.008	1.006
20	1.009	1.007	1.005	1.004	1.003

the atmosphere. These regions may be characterised by the occurrence of different distortions of the images of the Sun and the Moon, including also their 'ruptures'.

It should be mentioned that on the basis of the measurements of the angle of refraction, it is possible to determine the vertical profile of the refractive index of air and from this value the density of air. In the radiowave range, the refractive index depends strongly on the partial pressure of water vapour so that it is possible to determine the humidity of the atmosphere from space.

In examination of the radiation passing through the atmosphere from the space, scintillation is also detected from point sources. This phenomenon has been studied in detail by scientists of the Institute of Atmospheric Physics of the Russian Academy of Sciences (Moscow) and cosmonauts under the supervision of A.S. Gurvich in a series of experiments onboard the long-life orbital stations (LOS) Salyut and Mir [21]. Examination from the space through the atmosphere (during rising or setting of a star under the horizon of a planet) differs from the ground-based sounding by the fact that, firstly, the examiner is far away from the atmosphere (hundreds and thousands of kilometres) and, secondly, the effect of denser layers of the atmosphere below the perigee of the ray is automatically excluded. The first circumstance results in considerable increase of scintillations in the propagation of radiation in the free space behind the atmosphere. Consequently, the observed fluctuations of the radiation flux from the star becomes sufficiently strong even in the presence of very small disturbances of air density. The second circumstance also supports the examination of the inhomogeneities of the density of air at large altitudes, inaccessible for sensing from the surface of the Earth because of the screening effect of the denser lower turbulised layers of the atmosphere.

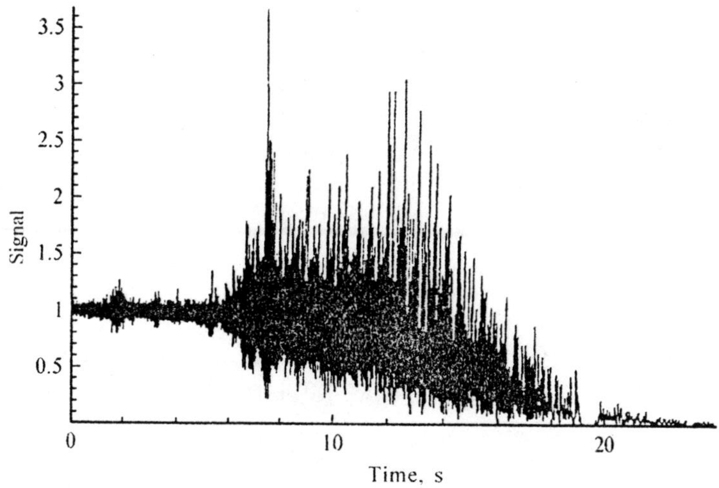

Fig. 5.16. Time dependence of the radiation flux of a star (in relative units) passing through the Earth's atmosphere [21]

As an example, Fig. 5.16 shows the results of cosmic experiments with the measurement of the radiation of a star, passing through the atmosphere of the Earth [21]. The figure shows the time variation of the radiation flux of the star (in the relative units). The increase of time corresponds to the 'immersion' of the star in lower and lower dense layers of the atmosphere. Figure 5.16 shows that on the background of the smooth decrease of the signal from the star, determined by Rayleigh and aerosol extinction in the atmosphere, there are sharp fluctuations of the radiation, reaching hundreds of percent. The phenomenon of extinction of the radiation of the star, refraction characteristics, and statistical properties of scintillation are used for examining different parameters of the atmosphere of the Earth and other planets.

The transfer equation taking refraction into account

The refractive divergence must be taken into account not only for the Sun disc (the Moon) but also for the intensity of the point sources. In fact, when deriving the transfer equation in chapter 3, we examine the energy passing through the faces of the elementary volume $dl\ dS$, but now part of the energy may additionally leave the volume (or arrive in the volume) as a result of refraction caused by the variation of the refractive index n along the path dl [47]. We obtain the transfer equation for the given case, using again mechanical analogy.

Let us assume that we have a flux of photons with the mass m and velocity $v = c/n$. Evidently, intensity I, as the measure of radiation energy (see chapter 3), is proportional to the total number of the photons, passing through the face dS, and consequently, to their total mass. However, the energy of the photon is $E = mv^2/2$ and consequently

$$E = I\frac{v^2}{2} = \frac{c^2}{2}\frac{I}{n^2}.$$

According to the definition of the volume extinction coefficient α, the extinction of the energy in the volume is

$$dE_e = \alpha E dl.$$

Remembering the definition of the volume coefficient of radiation (chapter 3) and its the relationship with the intensity $dI = \varepsilon dl$, it is concluded that the variation of the energy of the radiation of the volume is

$$dE_r = dI\frac{v^2}{2}\varepsilon dl = \frac{v^2}{2}\varepsilon dl = \frac{c^2}{2}\frac{\varepsilon}{n^2}dl.$$

Now, as in chapter 3, we can write the law of conservation of energy

$$dE = -dE_e + dE_r,$$

$$\frac{c^2}{2}d\left(\frac{I}{n^2}\right) = -\frac{c^2}{2}\alpha\frac{I}{n^2}dl + \frac{c^2}{2}\frac{\varepsilon}{n^2}dl.$$

Finally, we obtain the differential transfer equation with refraction taken into account

$$n^2\frac{d}{dl}\left(\frac{I}{n^2}\right) = -\alpha I + \varepsilon. \qquad (5.5.10)$$

It should be mentioned that (5.5.10) changes to the 'conventional' differential transfer equation, if we set n = const, i.e. (5.5.10) is the generalisation of the 'conventional' equation (3.4.3).

The influence of the refraction effects on the transfer radiation has been examined in detail by Minin [47]. He obtained a simple parameter which makes it possible to evaluate the contribution of refraction in the transfer of scattered radiation. In accordance with Minin, we introduce the length of the free path of the photon in molecular scattering $l_m = 1/\sigma_m$, where σ_m is the volume coefficient of molecular scattering. The meaning of l_m follows from the Bouguer law: it is the distance over which the intensity weakens as a result of molecular scattering by *e* times. Approximating the length of the free path of the photon in movement in the atmosphere taking into account the refraction by the arc of the circle with the radius of curvature determined by the equation (5.5.9), we obtain the following equation for the angle of deviation

$$b = \frac{l_m}{R} = \frac{1}{\sigma_m}\frac{d}{dr}(\ln n(r))\sin\theta.$$

Since we are interested in the evaluation of the maximum contribution of refraction, it is assumed that $\sin\theta$ is maximum, i.e. $\sin\theta = 1$. Finally

$$b = \frac{1}{\sigma_m}\frac{d}{dr}(\ln n(r)). \tag{5.5.11}$$

Parameter b gives the angle of turn of the beam along the free path length of the photon. If the value of b is small, the light between the acts of interaction with the atmosphere propagates almost in a straight line and the refraction effects may be ignored. Calculations carried out using equation (5.5.11) show that in the standard conditions in the atmosphere of the Earth the value of b for scattered radiation is 3–4 min of the arc and, consequently, in calculations of the field of scattered radiation refraction may always be ignored.

5.6. Optical phenomena in the atmosphere

Twilight

Twilight is the entire set of optical phenomena occurring in the atmosphere during the sunrises and sunsets. Large changes of the illumination of the surface of the Earth (approximately by a factor of a billion) take place twice every day resulting in the qualitative rearrangement of the radiation regime of the Earth's surface and of the surrounding atmosphere. The atmosphere 'softens' this transition because of scattering. The transition is not instantaneous and is spread over a more or less long period, referred to as twilight.

The explanation of the twilight phenomenon is very simple (Fig. 5.17). After sunset, the altitude h of the Sun becomes negative and the Sun rays do not illuminate the surface of the Earth (point A). However, they illuminate the atmosphere at some altitude H from the surface – point B. The radiation at point B and higher is scattered (in all directions) and part of the radiation reaches the surface. To determine the relationship of the height H with the angle h, the Sun is approximated by a point source, illuminating the Earth by a parallel flux of rays (since the distance from the Earth to the Sun is considerably greater than the radius of the Earth, the divergence of the beams may be disregarded), and refraction is ignored. Figure 5.17 shows clearly that $(R + H) = R/\cos h$, where R is the radius of the Earth, and consequently

$$H = R\left(\frac{1}{\cos h} - 1\right). \tag{5.6.1}$$

According to (5.6.1), as the Sun descends, the strength of illumination of the upper and, consequently, less dense layers of the atmosphere is higher and, therefore, the scattered radiation reaching the surface weakens. This is also associated with the smooth transition from day to night on the Earth. For analysis of the twilight the point B – zenith – was selected in order to simplify the explanation of the effect of scattered light at point A; in fact, this scattering takes place along the entire path of the rays of the Sun through the atmosphere, and the lowest layers of the atmosphere correspond to the direction to the horizon.

It should be mentioned that, as explained in the previous paragraph, refraction may bring significant corrections to equation (5.6.1). For example, in polar regions at a strong positive refraction, one can observe false sunrises when during the polar night the Sun appears for a short period of time above the horizon, although realistically, its altitude is less than zero.

The biological, geophysical, social and economic significance of twilight may be indicated by the scale of this phenomenon [61]. If we look at the globe from the space, it appears to be surrounded by a wide band of twilight half shadows, covering 20–25% of the Earth's surface, depending on the condition of the atmosphere. On one side of the surface, to 42–45% of the area of the globe, there is the day, and on the other side, 33–35% of the Earth's surface is in the night condition. The duration of twilight in the tropics, where the Sun rapidly descends to the horizon, is 10–15% of the

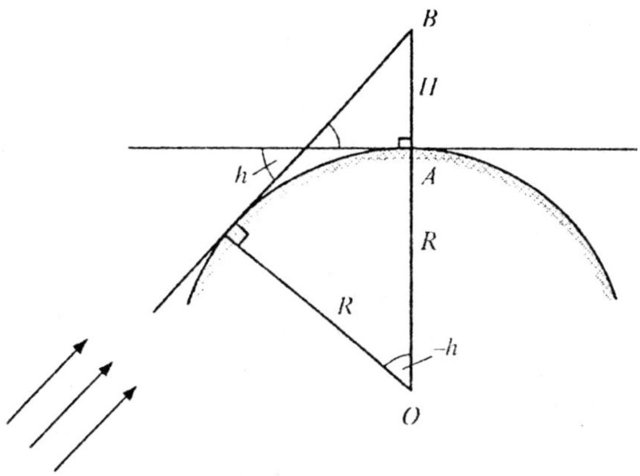

Fig. 5.17. Explanation of the twilight phenomenon.

duration of the year, whereas at high latitudes, it increases to 30–40%, and in the polar regions in the spring and autumn periods, the continuous twilight – white knights – lasts for weeks.

The main factor, determining the course of twilight is the scattering of the solar light in the Earth atmosphere and the associated attenuation of direct Sun rays. When the Sun approaches the horizon, the path of its rays through the atmosphere increases and the brightness of the rays decreases. This results in a decrease of the degree of illumination of the Earth's surface both with the direct light of the Sun and the light scattered by the atmosphere. During the day, this dependence of illumination on the altitude of the Sun is not strong. However, when the Sun descends 5–10° below the horizon, the process of decrease of the strength of illumination is rapidly accelerated. The start of accelerated decrease of illumination is regarded as the start of twilight. This is accompanied by a decrease of the relative role of the rapidly attenuated straight rays of the Sun in the illumination of the Earth surface. From the moment of descent of the Sun beyond the horizon the atmosphere is the only source of light and, as reported previously, the gradual decrease of the radiation, scattered by the atmosphere, determines the decrease of the degree of illumination of the Earth's surface. When the Sun descends by approximately 10–15° below the horizon, scattering becomes so small that the glow, generated by the upper layers of the atmosphere (chapter 8) already becomes evident, together with the radiation of the stars: illumination gradually approaches the night illumination. The transition to the night is usually completed at when the centre of the Sun descents 17–19° below the horizon but in some cases at a higher concentration of the aerosols in the stratosphere or in the presence of silvery clouds, it is extended to 22–23°. Thus, the boundaries of the twilight period are greatly smudgy and depend on the atmospheric conditions.

Taking into account the practical requirement to evaluate the conditions of visibility in different periods of the day, there has been a tendency to divide the twilight into three stages, depending on the level of illumination. The brightest part of the twilight, when the natural illumination in the open area makes it possible, for example, to read, is referred to as public twilight. This twilight corresponds to the Sun descending 6–8° below the horizon. This is followed by the sea or so-called navigation twilight, during which the darkness already covers small details, but the silhouettes of large objects, for example, the shoreline, are quite clearly visible. Their boundaries are represented by the angles of descent of the Sun of 6 and 12°. Finally, the sea twilight is replaced by astronomical twilight (when the stars become visible) continuing up to the moment when the Sun descents to 18° below the horizon , i.e. to the start of the night.

'Special' optical phenomena in the atmosphere

The 'special' optical phenomena in the atmosphere usually include a number of phenomena requiring the presence of special conditions (special state of the atmosphere). These are rainbows, halo, false suns, mirages. With the exception of the mirages, examined in the previous paragraph, these phenomena are associated with the scattering of visible light by water droplets and icy particles. They can be regarded as part of the aerosol optics [6]. For example, for a rainbow, as already mentioned in section 5.2, the approximation of the spherical particles and the Mie theory enable both the illustration of this phenomenon and determination of its quantitative characteristics.

The optical phenomena may be examined by the geometrical optics because the wavelength of the visible light is considerably smaller than the dimensions of the water droplets and icy crystals in the atmosphere. As shown in section 5.2, geometrical optics cannot be used for determining the quantitative characteristics of scattering, but in practice for the rainbows and halo the values of the intensity are usually not required and it is sufficient to confine examination to simple interpretation of these phenomena, to understand why they take place.

Rainbow

The rainbow forms during scattering of solar rays on large rain droplets. We examine the incidence of light rays on a water droplet assuming that the droplet is spherical (Fig. 5.18). Let it be that the ray is incident on the droplet at point A. In interaction with the boundary between the air and water, the light shows reflection and refraction. The light, refracted at A, is again incident on the surface of the sphere already from the inward at point B and is again reflected (inside) and refracted (outward); the reflected ray is again incident from the inside on the surface at point C where it is reflected and refracted, and so on. According to the law of the position of the incident, reflected and refracted rays in the same plane, with anu number of reflections, the pattern remains planar as previously, i.e. the circle in Fig. 5.18 is fully adequate to the entire sphere.

Let us assumed that the beam, incident on the droplet externally at point A has the angle of incidence θ_0. According to the Snell's law, for the angle of refraction θ_1 we can write the following equation

$$\sin\theta_1 = \frac{1}{n}\sin\theta_0, \qquad (5.6.2)$$

where n is the refractive index of the material of the sphere which is regarded as real. According to definition, the angle of scattering of light θ is the deflection of light from the initial direction. This means that to determine the angles of scattering in the examined geometry, we should be interested in the variation of the angles in relation to the initial direction of incidence, indicated by the broken line in Fig. 5.18.

Because of the equality of the angles of incidence and reflection, the ray of the first reflection outward at point A gives, as indicated by Fig. 5.18, $\theta = \pi - 2\theta_0$. The angle of incidence from the inside at the point B is equal to the angle of refraction θ_1 at point A as base angles of the isosceles triangle AOB. However, then the angle of refraction at point B – the angle of exit of the ray from the sphere – should be equal to θ_0 according to the Snell's law. Using identical considerations for point C and other points of intersection of the ray with the sphere, we conclude that the angle of incidence of the ray from the inside is always equal to θ_1 and the angle of exit from the sphere is θ_0. Consequently, at any point, as at A, the deflection of the direction in relation to the

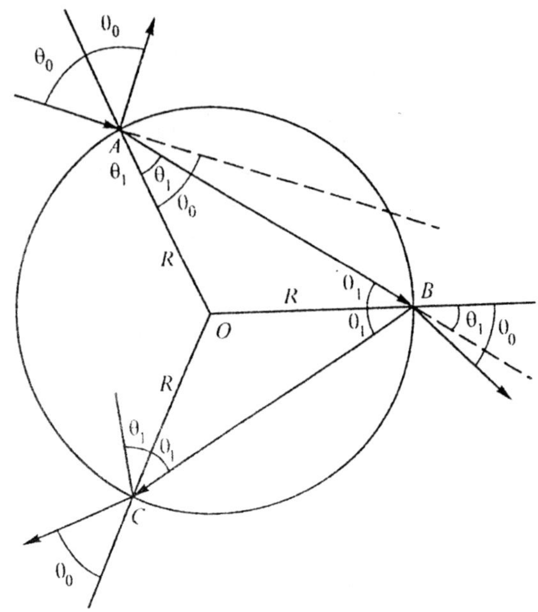

Fig. 5.18. Geometry of the path of rays in a rain droplet [6].

incident (at the same point) ray under the angle ψ in the reflection is $\pi-2\psi$, only at the point A $\psi = 0$, and at other points it was $\psi = \theta_1$. In refraction at point A the deviation of direction is $\theta_0-\theta_1$ ans the same deviation is at all other points as clearly indicated by Fig. 5.18 (the broken line at point B). The angle of scattering at the point B is the sum of deflections at refractions at the points A and B: $\theta = 2\theta_0 - 2\theta_1$. The direction of reflection at B has the deflection $\theta_0 - \theta_1 + \pi - 2\theta_1$. The angle of scattering at the point C is the sum of this deflection and $\theta_0-\theta_1$, i.e. $2\theta_0 - 4\theta_1 + \pi$, etc.

Thus, after every reflection it is necessary to add $\pi-2\theta_1$ to the direction of deflection. Therefore, after k reflections, assuming that $k = 1$ at point B, we have the direction of incidence $\theta_0 - \theta_1 + k(\pi - 2\theta_1)$. Adding to it the deflection in at refraction $\theta_0 - \theta_1$, we finally obtain the angle of scattering

$$\theta = k\pi + 2\theta_0 - 2(k + 1)\theta_1.$$

In this equation, the angle of scattering θ should be reduced to the interval $[0, \pi]$ but this is not required for our purposes. Taking into account (5.6.2), we finally obtain

$$\theta = k\pi + 2\theta_0 - 2(k+1)\arcsin\frac{1}{n}\sin\theta_0. \tag{5.6.3}$$

We have examined scattering at a single point A. In the incidence of the direct solar rays on the entire surface of the droplet (i.e., the beam of rays parallel to the ray incident on A), the angle of incidence θ_0 will evidently change from θ_0 to $\pi/2$. Thus, the equation (5.6.3) gives the dependence of the angle of scattering on the angle of incidence $\theta(\theta_0)$, where $0 < \theta_0 < \pi/2$. We examine the derivative of this function $\frac{d\theta}{d\theta_0}(\theta_0)$. This can be interpreted as the variation of the angle of scattering ($d\theta$) with the variation of the angle of incidence on $d\theta_0$. If this derivative at a specific angle of incidence is equal to 0, this means that the angle of scattering does not change with the variation of the angle of incidence by $d\theta_0$, i.e. all the incident rays are collected (focused) from some range of directions in a single angle of scattering. Correspondingly, this angle should correspond to the maximum intensity of scattered light. Differentiating (5.6.3) we obtain

$$2 - 2(k+1)\frac{\frac{1}{n}\cos\theta_0}{\sqrt{1 - \frac{1}{n^2}\sin^2\theta_0}} = 0,$$

$$\frac{\cos^2\theta_0}{n^2 - 1 + \cos^2\theta_0} = \frac{1}{(k+1)^2}.$$

Consequently

$$\cos\theta_0 = \sqrt{\frac{n^2-1}{k(k+2)}}. \tag{5.6.4}$$

The mean parameter of refraction of water in the visible range $n = 1.333$. Using this value in (5.6.4), setting $k = 1$, we obtain θ_0 from (5.6.4) and, subsequently, from (5.6.3) the corresponding angle of scattering θ. The identical procedure is applied for $k = 2$. We obtain $\theta = 138°$ for the first rainbow ($k = 1$) and $\theta = 129°$ for the second rainbow ($k = 2$). These two rainbows are detected in the atmosphere (the second less frequently than the first one), and are distributed around the antisolar point of the sky sphere under the angles of 42 and 51°, respectively. Since the antisolar point is situated below the horizon, we can observe the rainbow only if the Sun is sufficiently low. Usually, the rainbow is visible in the evening on the background of departing storm clouds. The rainbows with higher orders ($k > 2$) are not detected in the atmosphere and their intensity are very low. However, in the laboratory conditions and when using different liquids, it is possible to detect rainbows of higher orders. The record here belongs to the solution of sugar syrup when the dispersion of the syrup in the form of droplets was accompanied by the formation of rainbows up to the 17th order [6]. The colour of the rainbow (well known seven colours) is associated with the dependence of the reflective index of water on the wavelength. For example, for violet rays, it is equal to $n = 1.343$, and for the red rays $n = 1.331$. Substitution of these values into (5.6.4), (5.6.3) gives the width of the rainbow 1.7 and 3.1° for the first and second order, respectively. The band of these angles also contains the colour pattern referred to as the rainbow.

The halo

The halo – the rainbow ring around the Sun – is detected in scattering of the direct solar rays on the icy crystals in the clouds (usually in cirrus and altostratus clouds). It is interesting to note that the halo is a clear indication of the non-equivalence of scattering on an ensemble of randomly oriented icy crystals in relation to the scattering on the ensemble of icy spheres: for the spheres, the Mie theory does not reproduce the halo.

As the model of the crystal, we examine a triangular prism with the reflective index n and the angle Δ at the tip onto which the ray is incident and forms the angle θ_0 (Fig. 5.19) with the normal to side surface. The ray undergoes two refractions: with the angle θ_1 after the first one which determines the angle of incidence from the inside θ_2, and the angle θ_3 after the second which is the angle of exit of the ray from the prism. To describe deflections in the direction of the rays in refraction it is evident that we can use the same equations as in the case of the sphere and, consequently, the scattering angle is:

$$\theta = \theta_0 - \theta_1 + \theta_3 - \theta_2.$$

The angles θ_0, θ_1 and the angles θ_3 and θ_2 are linked together by the equation (5.6.2). θ_2 is expressed through θ_1. Both these angles are additional to the angles BAC and BCA of the triangle ABC. Since the sum of the angles of the triangle is equal to π, we obtain such $\pi/2 - \theta_1 + \Delta + \pi/2 - \theta_2 = \pi$, i.e. $\theta_2 = \Delta - \theta_1$. Now

$$\theta = \theta_0 + \theta_3 - \Delta, \tag{5.6.5}$$

and θ depends only on θ_0 but, in order to avoid the accumulation of the sinuses and arcsinuses, we shall not write this dependence in the explicit form.

The random orientation of the crystals is equivalent to the change of the angle of incidence θ_0 from 0 to $\pi/2$. However, in this case using the same considerations as for the sphere, we obtain that the maximum of the intensity of scattered light should correspond to the zero of the derivative function $\theta(\theta_0)$. Differentiating (5.6.5) on the basis of the rules of the complex function and equating the derivative to zero, we obtain

$$1 + \frac{d\theta_3}{d\theta_2}\frac{d\theta_2}{d\theta_1}\frac{d\theta_1}{d\theta_0} = 0,$$

$$1 - \frac{d}{d\theta_2}\arcsin(n\sin\theta_2)\frac{d}{d\theta_0}\arcsin\left(\frac{1}{n}\sin\theta_0\right) = 0,$$

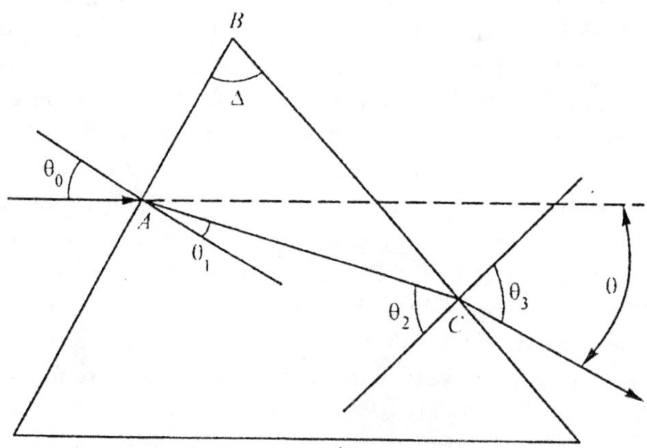

Fig. 5.19. Geometry of the rays in a triangular prism [6, 62].

$$1-\frac{\cos\theta_2\cos\theta_0}{\sqrt{1-n^2\sin^2\theta_2}\sqrt{1-\frac{1}{n^2}\sin^2\theta_0}}=0.$$

Squaring (5.6.2) give $1-\cos^2\theta_1=\frac{1}{n^2}\sin^2\theta_0$ and expressing from the same relationship of the cosine of the angle of refraction, $1-\cos^2\theta_3 = n^2\sin^2\theta_2$. Taking this into account, we finally obtain

$$\frac{\cos\theta_0\cos\theta_2}{\cos\theta_1\cos\theta_3}=1.$$

The equality is fulfilled at $\cos\theta_2=\cos\theta_1$, i.e. $\theta_2=\theta_1$ and in this case we automatically obtain the second equality $\cos\theta_3=\cos\theta_0$, i.e. $\theta_3=\theta_0$. This condition shows that the refracted ray should always travel in the prism in the direction parallel to its base. This geometry is well known in optics and it corresponds to the minimum deflection of the ray at exit from the prism [62]. From (5.6.5) we obtain $\theta = 2\theta_0 - \Delta$, the incidence angle θ_0 should be such that the triangle *ABC* is isosceles, i.e. the angle *BAC* is $1/2\,(\pi-\Delta)$, but it is also equal to $\pi/2-\theta_1$, from which $\theta_1=\Delta/2$, and the required angle of scattering is finally:

$$\theta=2\arcsin\left(n\sin\frac{\Delta}{2}\right)-\Delta. \tag{5.6.6}$$

Ice crystals have the form of a prism whose base contains the hexahedron (the hexagonal crystal structure (Fig.5.20)). Correspondingly, in incidence through the side faces $\Delta = 60°$, and through the base $\Delta = 0°$. The mean refractive index of ice in the visible range $n = 1.313$, and the equation (5.6.6) gives the halo angles of 22 and 46°, and the halo with 22° is more wide-spread in the nature. The variation of the refractive index of ice is in the range from $n = 1.318$ for violet rays to $n = 1.3084$ for red rays which makes the halo to have rainbow colours and the resultant width of the colour band is 0.7° for the 22° halo and 2.2° for the 46° halo. It should be mentioned that the width of the first halo is not large and, consequently, its rainbow colour is often difficult to see.

The halo is a relatively wide-spread optical phenomenon in the atmosphere, and it forms almost always when the sunshine sifts through cirrus and cirrostratus clouds. Another point is that the brightness of the halo is not high and unprofessional observers do

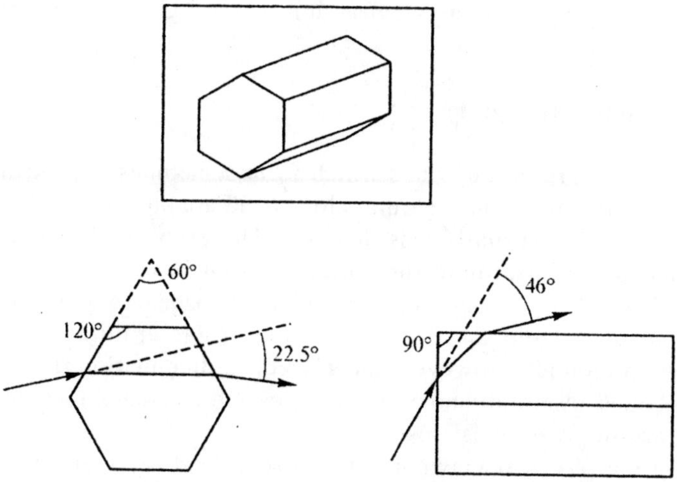

Fig. 5.20. Refraction of rays in an ice crystal [6].

not see it on the sky. The appearance of cirrus or altostratus clouds is connected with the approach of a warm atmospheric front and, therefore, the formation of a halo is a relatively reliable indication of worsening of the weather. The halo can also be observed at ice crystals in the troposphere which form at low temperatures and a high air humidity. Halos of this type are often very bright. They are usually observed in polar regions but can also form at moderate latitudes during severe frost.

False suns, solar pillars and crosses

Sometimes, bright spots, i.e. false suns, form at a hallo with the angle of 22°. They are situated on one side and above the 'true' Sun (Fig. 5.21). In many cases, they are connected with the Sun by bright lines, forming a Sun cross. They are not observed all the time and all together; in most cases, there are horizontal false suns and a vertical line, referred to as the solar pillar. These phenomena have not as yet been satisfactorily explained. It is assumed that they form when the orientation of ice crystals in the atmosphere is not completely random and, consequently, effects of the increase of the intensity of light as a result of identical refraction in the crystals with the same orientation become apparent. If the brightness of these phenomena is high, natural halo, columns, crosses, secondary false suns, tangential to the main halo of the arc, may appear around the false suns [43]. In all likelihood, they are all caused by secondary refraction of the light already refracted on ice crystals.

Solar coronas and glory

Solar coronas are colour rings which form sometimes around the Sun when the Sun shines through thin clouds and are observed under small scattering angles (usually less than 5°). The glory is identical colour rings but detected around the antisolar point in reflection from the clouds. Since this point is situated below the horizon, the glory is usually observed only from an aircraft (rainbow rings around the aircraft shadow on clouds). However, under specific illumination, the glory is also detected in mountains on the background of clouds located below or on the side of clouds (fog).

From the physical viewpoint, the coronas and glory are diffraction patterns of light scattering on water droplets, detected in the transmitted or reflected light, respectively. This phenomenon is connected with the wave nature of light and, consequently, it is 'explained' by Mie theory, which describes diffraction on a sphere. Calculations by Mie formulas give the angles of coronas and glory.

Figure 5.22 shows the points of the 'special' atmospheric optics phenomena. It should also be mentioned that in addition to the Sun, the Moon is also a source of light in the atmosphere. The intensity of moonlight is very low and not sufficient to enable observation of the phenomena in reflected light, i.e., rainbow and glory, but the phenomena formed in transmitted light, i.e., coronas, halo and false suns, can be observed.

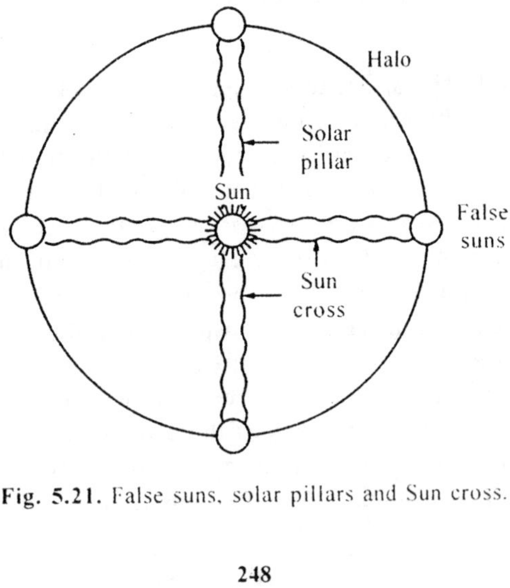

Fig. 5.21. False suns, solar pillars and Sun cross.

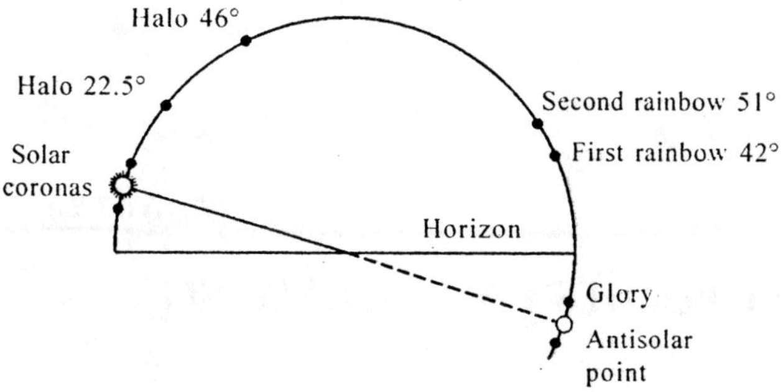

Fig. 5.22. Observation points of 'special' optical phenomena.

CHAPTER 6

OPTICAL PROPERTIES OF UNDERLYING SURFACES

6.1. Main special features of reflection of radiation

On many planets, the atmosphere is bounded by the solid or liquid medium – the surface of the planet. Naturally, this surface has a strong effect on the transfer of solar radiation in the atmospheres of the planet and also generates thermal radiation.

In atmospheric optics, the surface of the planet is referred to as the underlying surface. On the Earth, and later we shall discuss mainly this type of surface from the viewpoint of its optical properties, this surface greatly varies. In particular, it may be the land surface or water surface. It is evident that water, snow, plants, open areas of soil, etc. have different optical properties. For some practical problems (for example, cosmic photography) it is convenient to also refer to the underlying surface of the cloud.

The role of the underlying surfaces in atmospheric optics is very important: in the visible range of the spectrum, the surface reflects and absorbs solar radiation. For the water surface, part of solar radiation penetrates into water and in this case we are concerned with the transfer radiation in the atmosphere–water complex system.

In the thermal range of the spectrum (infrared and microwave ranges) the surfaces emit electromagnetic energy and also reflect incident radiation – solar and downwelling atmospheric radiation of the atmosphere. In the long wave part of the spectrum (radio waves), the radiation forms not only in the thin layer of the interface of two media, but also in the finite thickness of the underlying surface and here, as in the case of the water surface, we are again concerned with the formation and propagation of radiation in a system of at least two media.

Different types of reflection

The structure of the surface of the interface of two media – atmosphere and underlying surface – determines different types of reflection. The simplest case of reflection is observed for the ideally smooth surface and is often referred to as mirror reflection (Fig. 6.1a). This model is close to the flat surface of water (without waves) or ice (without irregularities).

However, the actual surfaces are always characterised by different deviations from the flat surface due to the presence of heterogeneities on them. Even the water surface is covered with ripples in the presence of already slight wind. All actual surfaces are characterised by irregularities: lumps of soil, sands and stones, elements of the plant cover – from grass blades to trees. If the effect of heterogeneities of the surface is not strong, there is the second type of reflection – quasi-mirror (Fig. 6.1b). This type of reflection is often detected in reflection by the water surface in the presence of waves. In this case, almost the entire reflection of radiation takes place in a relatively small solid angle around the direction of mirror reflection (in the range 15–25° along altitude and azimuthal angles).

Let us assume that a parallel beam of rays is incident on a rough surface. Therefore, reflecting from every small element of the surface (which may be regarded as flat) in accordance with the laws for the flat surface, radiation from each irregularity and surface roughness is reflected in different (in almost all) directions. This type of reflection is referred to as diffusion reflection. Strictly speaking, diffusion reflection also includes the case of quasi-mirror

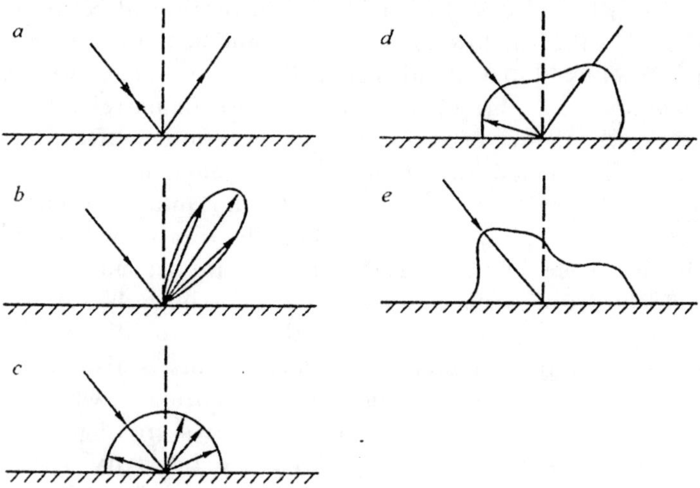

Fig.6.1. Different types of reflection from the surface: a) mirror; b) quasi-mirror; c) orthotropic; d) quasi-orthotropic; e) retroreflection.

reflection and also other types of reflection from actual surfaces which will be examined later.

A special case of diffusion reflection in which reflection takes place in the same manner in all directions is referred to as orthotropic (Lambert) reflection (Fig. 6.1c). Like a mirror reflection, in the majority of situations, this is the idealization of actual reflection even for specially prepared surfaces. However, this type of reflection is often used in the theory of radiation transfer. The real surfaces are not orthotropic reflecting surfaces and, therefore, another type of reflection – quasi-orthotropic reflection is often introduced (Fig. 6.1d).

Finally, we mention another type of diffusion reflection – retroreflection (Fig. 6.1e). This type of reflection is typical of many natural surfaces. For these surfaces, maximum reflection is found in the direction opposite to the direction of incident radiation. This effect will be explained using Fig. 6.2. In the figure, the surface roughness is indicated by 'hillocks' which cast shadows under illumination (position of the Sun, for example at point *A*). In examination from different directions, the device will see the areas of these shadows as being different. In the direction of mirror reflection (position 1) – maximum, because the entire shadow will fall into the field of view of the device, including the shadow situated on the hillock sides opposite to the light source. In the direction of position 2 the shadow area is smaller because we do not see the opposite hillock sides. Finally, in the direction of retroreflection (position 3) the shadow area is minimum because the shadow is hidden from us by the illuminated hillock sides. Thus, the examined roughness surfaces are characterised by minimum reflection in the direction of mirror reflection and by maximum reflection in the direction of reversed reflection. Measurements show that the value of the minimum is not large and at the maximum the intensity of reflection is 1.5–2 times higher than reflection into the nadir.

The presence of the previously examined shadows from the irregularities and surface roughness results in daily variation, i.e. the dependence on the altitude of the Sun, for different optical characteristics of the underlying surfaces. This is also clearly seen in Fig. 6.2. Actually, if the Sun is in the position *B*, which is lower than position *A*, the area of the cast shadows crosshatched on the figure, is greater and, consequently, the intensity of radiation reflected in the direction of the observer is smaller. As regards the orthotropic surfaces, the snow cover is the closest to the orthotropic surface with small surface roughness [13].

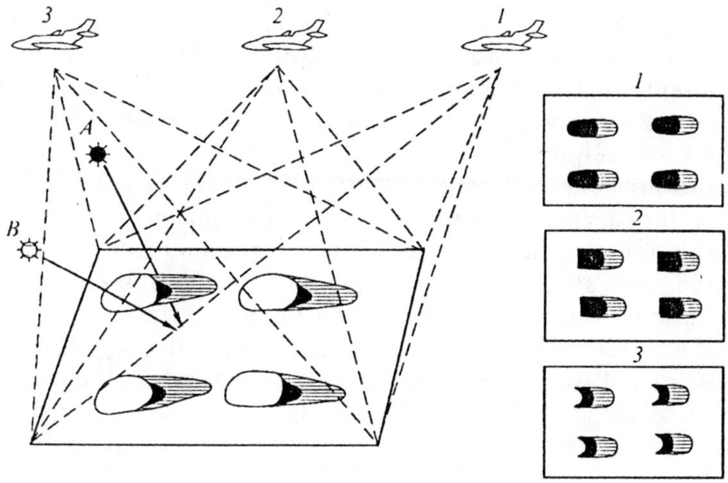

Fig.6.2. Explanation of retroreflection and daily variation of the reflection characteristics of rough surfaces; 1,2,3) different positions of the observer.

6.2. Quantitative characteristic of reflection of radiation (mirror reflection)

Although the model of mirror reflection is encountered only seldom in nature, it is regarded as important as some idealised case for which exact equations have been derived, and is used for modelling the optical characteristics of the actual surfaces.

Let us assume that we have an ideal flat interface of two media: air for which the complex refractive index (CRI) is assumed to be equal to unity, and the medium with CRI equal to $m = n - i\kappa$. The reflection plane is the plane formed by the direction of an incident electromagnetic wave and the normal to the interfacial surface of the two media. The initial vector of the strength of the electrical field E_0 in a general case of an elliptical polarised wave is expanded into two mutually perpendicular components (Chapter 3): $E_{0,\parallel}$ – parallel situated in the reflection plane, and $E_{0,\perp}$ normal to it. The vector of strength of the electrical field of the reflected wave is also expanded into the same components. In examination of the reflection from the horizontal surface investigators often use different terminology: $E_{0,\parallel} = E_{0,v}$ – the vertical component of the field, $E_{0,\perp} = E_{0,\parallel}$ – the horizontal component of the field.

The relationship between the incident, reflected and refracted components of the strength of electrical field is given by the Fresnel equations. They are derived in examining the boundary conditions in electrodynamics [6, 62]. In derivation, the existence of the reflected and refracted waves, the equality of the frequencies of the incident, reflected and refracted waves (for the stationary interface of two media), and also three well-known laws of reflection and refraction are proven:

1) the incident, reflected and refracted waves and the normal to the surface are situated in the same plane;
2) the angle of incidence θ_0 (between the incident wave and the normal) is equal to the angle of reflection (the angle between the reflected wave and the normal);
3) the Snell's law: the refraction angle θ_r between the refracted wave and the normal) is connected with the angle of incidence θ_0 by the relationship

$$\sin\theta_0 = m\sin\theta_r, \tag{6.2.1}$$

where θ_r is the angle of refraction; m is the CRI.

The Fresnel equations for the reflected wave are written as follows [6,7,39]:

$$E_{\parallel} = \frac{m\cos\theta_0 - \cos\theta_r}{m\cos\theta_0 + \cos\theta_r} E_{0,\parallel},$$
$$E_{\perp} = \frac{\cos\theta_0 - m\cos\theta_r}{\cos\theta_0 + m\cos\theta_r} E_{0,\perp}. \tag{6.2.2}$$

Equation (6.2.1.) may be written in the form

$$\cos\theta_r = \frac{1}{m}\sqrt{m^2 - 1 + \cos\theta_0}. \tag{6.2.3}$$

Substitution of (6.2.3) into (6.2.2) gives

$$E_{\parallel} = \frac{m^2\cos\theta_0 - \sqrt{m^2 - 1 + \cos^2\theta_0}}{m^2\cos\theta_0 + \sqrt{m^2 - 1 + \cos^2\theta_0}} E_{0,\parallel},$$
$$E_{\perp} = \frac{\cos\theta_0 - \sqrt{m^2 - 1 + \cos^2\theta_0}}{\cos\theta_0 + \sqrt{m^2 - 1 + \cos^2\theta_0}} E_{0,\perp}, \tag{6.2.4}$$

Equations (6.2.4) are suitable for practice because they give the explicit dependence of reflection on the angle of incidence and the CRI. It should be mentioned that the quantity $\varepsilon = m^2$ is the dielectric permittivity of the medium. In the radiowave range, the complex dielectric permittivity is presented in the form [5, 85]

$$\varepsilon = \varepsilon_1 - i\varepsilon_2 = \varepsilon_3 - i60\sigma\lambda, \tag{6.2.5}$$

where ε_1 and ε_2 are its real and imaginary ($60\sigma\lambda$) parts; σ is conductivity; λ is wavelength. Evidently, at $\lambda \to 0$ ($\nu \to \infty$)ε, tends to its real part, i.e. $\varepsilon = \varepsilon_1$. Therefore, in radiophysics it is often denoted by ε_∞.

Equation (6.2.1) may be re-written in the form

$$\varepsilon = \varepsilon_1\left(1 - i\frac{60\sigma\lambda}{\varepsilon_1}\right) = \varepsilon_1\left(1 - i\,\mathrm{tg}\delta\right) \tag{6.2.6}$$

where $\mathrm{tg}\delta$ is the tangent of the loss angle.

Equation (6.2.2) and (6.2.4) have the form of the relationship introduced in Chapter 3 (3.6.3):

$$E_{\parallel} = S_1 E_{0,\parallel},\ E_{\perp} = S_4 E_{0,\perp}. \tag{6.2.7}$$

By analogy, equations (3.6.5) can be used to determine the reflection matrix which has the form (3.6.4). Thus, the Fresnel equations determine fully the problem of the description of the reflection of radiation from the ideally flat boundary of two media, in particular, the reflection matrix [116].

Let us assume that non-polarized directional radiation with the intensity I_0 (numerically equal to the radiation flux) is incident on the interface [116]. Reflection is determined by the element D_{11} of the matrix D (chapter 3):

$$D_{11} = \frac{1}{2}(S_1 S_1^* + S_4 S_4^*). \tag{6.2.8}$$

Denoting D_{11}=$R(\theta)$ – reflection coefficient, we have

$$I(\theta) = R(\theta) I_0. \tag{6.2.9}$$

Consequently, the following equation can be written for the reflection coefficient

$$R(\theta) = \frac{1}{2}\left(\left|\frac{m^2\cos\theta_0 - \sqrt{m^2 - 1 + \cos^2\theta_0}}{m^2\cos\theta_0 + \sqrt{m^2 - 1 + \cos^2\theta_0}}\right|^2 + \right. \tag{6.2.10}$$

$$+\left|\frac{\cos\theta_0-\sqrt{m^2-1+\cos^2\theta_0}}{\cos\theta_0+\sqrt{m^2-1+\cos^2\theta_0}}\right|^2\Bigg)$$

The terms in equation (6.2.10) are the reflection coefficients for the parallel and perpendicular components of radiation.

The data on the complex dielectric permittivity and refractive index for different natural media have been published in many monographs and handbooks [2, 5, 22, 26, 78, 85, 94, 115] (see also Chapter 5). As an example, Fig. 6.3 shows the real and imaginary parts ε for fresh and sea water in the microwave range [115]. The presence of the spectral dependence of dielectric permittivity and the refractive index for different substances leads to the spectral dependence of the reflection characteristics.

We examine a case of normal incidence ($\theta_0 = 0$). Using equation (6.2.10) we can determine the following relationships:

$$R = R_h = R_v = \left|\frac{m-1}{m+1}\right|^2 = \frac{(n-1)^2+\kappa^2}{(n+1)^2+\kappa^2} = \qquad (6.2.11)$$

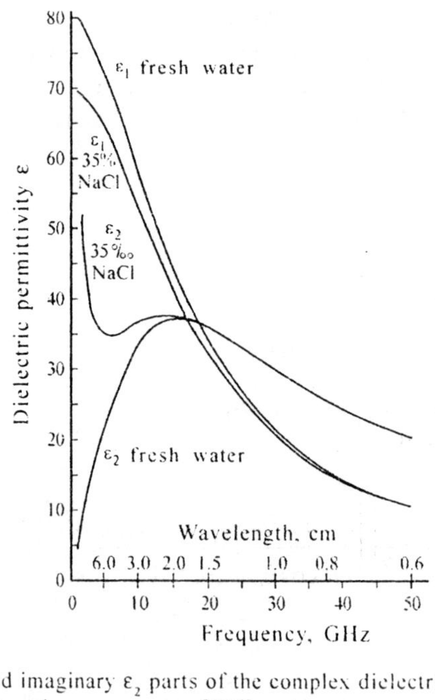

Fig.6.3. Real ε_1 and imaginary ε_2 parts of the complex dielectric permittivity for fresh and sea water in the microwave range [116].

$$= 1 - \frac{4n}{(n+1)^2 + \kappa^2}.$$

If there is no absorption in the medium ($\kappa = 0$), the reflection coefficient is determined by the real part of the refractive index

$$R = \left(\frac{n-1}{n+1}\right)^2 \qquad (6.2.12)$$

In the case $k \gg n$ (absorption band) R tends to unity. For example, if the surface is illuminated by radiation in a relatively wide spectral range, the reflection spectrum will contain maxima corresponding to the position of the absorption bands of substances included in the composition of the underlying surface. In the literature, this effect is referred to as the re-emission effect and can be used to determine the physical–chemical nature of the underlying surface on the basis of the reflection spectra.

There is a special angle θ_B – Brewster angle, at which $R_v = 0$. Only the radiation for which the strength of the electrical field is normal to the plane of incidence is reflected at this angle. Therefore, in the incidence of non-polarised solar radiation on the flat surface under the Brewster angle, the reflected light is completely polarised in the direction parallel to this plane. The Brewster angle is determined by equation $\theta_B = n$.

Figure 6.4 shows an example of the angular dependence of the reflection coefficients of a smooth sea surface as a function of the angle of incidence for two wavelengths – in the visible range ($\lambda = 0.55 \mu m$) and in the microwave range ($\lambda = 3$ cm). In the visible range, the refractive index is close to 1.2–1.3 and the reflection coefficients at nadir (vertical) incidence are very small, 0.01–0.02 up to the angles of incidence of 40–60°. At large angles of incidence ($\theta > 80°$) the reflection coefficients rapidly increase and when the angle of incidence approaches 90°, the reflection coefficients tends to unity. Figure 6.4. shows for vertical (parallel) polarisation the Brewster angle ~53° at which reflection for this component is equal to zero. In the microwave range at $\lambda = 3$ cm the complex refractive index is approximately equal to $m = 52 - 37i$ and the reflection coefficient for vertical incidence is high and equals 0.61. The Brewster angle is approximately 82°.

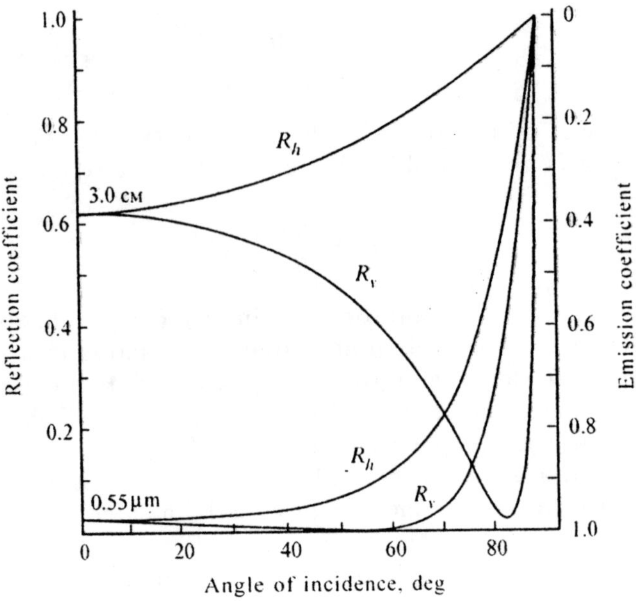

Fig.6.4. Example of the dependence of the reflection coefficients of the smooth sea surface on the angle of incidence for visible (λ = 0.55 μm) and microwave (λ = 3cm) ranges [116].

6.3. Quantitative characteristics of reflection of radiation (real surfaces)

There are various methods for describing the optical characteristics of real underlying surfaces in the theory of transfer and in atmospheric optics. As in the case of the flat surface of the interface between two media (Fresnel equations), we start with the case of illumination of the surface by radiation in one direction (with the intensity I_0 or the radiation flux F_0 with the same numerical value). The angles, determining the direction of incidence radiation, are denoted by θ_0 – the zenith angle of incidence, and φ_0 – the azimuth of incidence; the angles which determine the direction of reflection will be denoted by θ and φ. As mentioned previously, in a general case, reflection from the surface is of the diffusion type. The more general characteristic of reflection is the two-directional coefficient of reflection r which depends both on the angles of incidence (θ_0, φ_0), and on the direction of reflection (θ, φ). It is determined using the equation

$$I(\theta, \varphi) = r(\theta, \varphi, \theta_0, \varphi_0)I_0. \tag{6.3.1}$$

where $I(\theta,\varphi)$ is the intensity of radiation reflected by the surface in the direction (θ,φ). Because of the arbitrary nature of the selection of method of determination of the azimuths, the physical quantity (two-directional coefficient of reflection) cannot depend on its absolute value. Therefore, the two-directional coefficient of reflection is a function of three variables: $r(\theta, \varphi, \theta_0, \varphi_0) = r(\theta, \theta_0, \varphi - \varphi_0)$.It is usually assumed that $\varphi_0 = 0$, i.e. all azimuths are counted from the azimuth of incidence. Therefore, finally

$$I(\theta, \varphi) = r(\theta, \theta_0, \varphi)I_0. \tag{6.3.2}$$

The previously examined partial case – mirror reflection – is written in the form

$$r(\theta,\varphi,\theta_0,\varphi_0) = \begin{cases} R(\theta_0) & \text{at } \theta=\theta_0, \varphi=\varphi_0, \\ 0 & \text{at } \theta\neq\theta_0, \end{cases} \tag{6.3.3}$$

where $R(\theta_0)$ is the Fresnel coefficient of reflection (equation (6.2.10)). Equation (6.3.3) is often written in the form

$$r(\theta, \varphi, \theta_0, \varphi_0) = r(\theta_0)\delta\,(\theta_0 - \theta)\delta\,(\varphi - \varphi_0), \tag{6.3.4}$$

where $\delta(\varphi - \varphi_0)$ is Dirac's function.

It is often convenient to examine the radiation flux falling on the horizontal area, $F = F_0\cos\theta_0$ instead of the intensity of radiation or flux directed to the area normal to the direction of radiation. The characteristics corresponding to the flux is referred to as the coefficient of (spectral) brightness (CSB) of the surface and is determined by the equation [47, 64, 65]

$$I(\theta,\theta_0,\varphi) = \frac{1}{\pi}\rho(\theta,\theta_0,\varphi)F_0\cos\theta_0. \tag{6.3.5}$$

It is needed to determine the relationship between the previously introduced characteristics of reflection:

$$r(\theta,\theta_0,\varphi) = \rho(\theta,\theta_0,\varphi)\frac{\cos\theta_0}{\pi}. \tag{6.3.6}$$

In the theory of radiation transfer the characteristics of reflection (in this case not only by the surface but also by the entire atmosphere–surface system) are described by the reflection function $y(\theta,\theta_0,\varphi)$ [65], determined by the equation

$$I(\theta,\varphi)\cos\theta = \frac{1}{2\pi}y(\theta,\theta_0,\varphi)I_0\cos\theta_0. \tag{6.3.7}$$

Using definitions of different characteristics of reflection we have

$$\rho(\theta,\theta_0,\varphi) = \frac{1}{2\cos\theta} y(\theta,\theta_0,\varphi), \tag{6.3.8}$$

$$r(\theta,\theta_0,\varphi) = \frac{1}{2\pi} y(\theta,\theta_0,\varphi) \frac{\cos\theta_0}{\cos\theta}. \tag{6.3.9}$$

In many problems of the theory of radiation transfer and atmospheric optics it is interesting to examine the total energy of reflected radiation, i.e. the upwelling radiaion flux on the surface, instead of detailed (angular) characteristics of reflection

$$F^{\uparrow} = \int_0^{2\pi} d\varphi \int_0^{\pi/2} I(\theta,\varphi)\cos\theta\sin\theta\, d\theta. \tag{6.3.10}$$

The ratio of the upwelling and downwelling (falling) radiation fluxes is referred to as the albedo of the surface:

$$A = \frac{F^{\uparrow}}{F^{\downarrow}}, \tag{6.3.11}$$

where $F^{\downarrow}$ and $F^{\uparrow}$are the fluxes of the incident and reflected radiation.

The albedo of the surface, often expressed in percent, gives the fraction of the incidence energy reflected by the surface. Correspondingly, the quantity $(1-A)$ is the fraction of absorbed (and in a general case of transparent media – transmitted) energy. Case $A=1$ corresponds to the absolutely 'white' surface, $A = 0$ to the absolutely 'black' surface.

On the basis of the definition of the albedo, we obtain the following relationships between the introduced reflection characteristics:

$$A = \frac{1}{\pi} \int_0^{2\pi} d\varphi \int_0^{\pi/2} \rho(\theta,\theta_0,\varphi)\cos\theta\sin\theta\, d\theta, \tag{6.3.12}$$

$$A = \frac{1}{\cos\theta} \int_0^{2\pi} d\varphi \int_0^{\pi/2} r(\theta,\theta_0,\varphi)\cos\theta\sin\theta\, d\theta, \tag{6.3.13}$$

$$A = \frac{1}{2\pi} \int_0^{2\pi} d\varphi \int_0^{\pi/2} y(\theta, \theta_0, \varphi) \sin\theta \, d\theta. \tag{6.3.14}$$

We examine a case of orthotropic reflection for which according to definition $I(\theta,\varphi)$=const. According to equation (6.3.5), this denotes the absence of the angular dependence of CSB on angles θ and φ. Consequently, we obtain $r(\theta,\theta_0,\varphi)=A$, i.e. for the orthotropic surface, the coefficient of spectral brightness of the surface is numerically equal to its albedo.

The relationship (6.3.14) gives the physical meaning of the reflection functions $y(\theta,\theta_0,\varphi)$. Since the integral from $y(\theta,\theta_0,\varphi)$ over all directions is the albedo of the surface which because of equation (6.3.11) may be interpreted as the probability of reflection, the reflection function $y(\theta,\theta_0,\varphi)$ is the density of the probability of reflection in the direction (θ,φ). In this sense, function $y(\theta,\theta_0,\varphi)$ is some analogue of the scattering phase function (see Chapter 2). Therefore, this function or its normalized value $\frac{1}{A} y(\theta,\theta_0,\varphi)$ is sometimes referred to as the reflection phase function.

An important property of the coefficient of spectral brightness is its symmetry (the rule of reversibility or reciprocity, Helmholtz rule):

$$\rho = (\theta,\theta_0,\varphi) = \rho(\theta_0,\theta,\varphi). \tag{6.3.15}$$

This relationship may be proved by simple considerations. The entire real reflecting surface is broken up into a set of elementary areas and mirror reflection takes place from each area. Therefore, the direction of reflection (θ,φ) is determined by the orientation of the appropriate elementary area. In the coordinate system in which the elementary area is horizontal, the Fresnel equation and the law according to which the angle of reflection is equal to the angle of incidence) show evidently the equality of the reflected intensities when the direction of reflection is replaced by the direction of incidence. This equality should also be returned in the initial coordinates. However, here it must be taken into account that the intensity of radiation is determined as the energy per unit surface area. Therefore, the equality of intensity should be understood taking into account the reduction of these intensities to the unit area of the reflecting surface which, as indicated by the elementary geometry leads to the equation

$$\frac{I(\theta,\theta_0,\varphi)}{\cos\theta_0}=\frac{I(\theta_0,\theta,\varphi)}{\cos\theta} \tag{6.3.16}$$

From this relationship and taking into account equation (6.3.5) we directly obtain the rule of reciprocity (6.3.15). Taking into account the equation (6.3.15), for other detailed characteristics of reflection we have the following relationship

$$r(\theta,\theta_0,\varphi)\cos\theta = r(\theta_0,\theta,\varphi)\cos\theta_0, \tag{6.3.17}$$

$$y(\theta,\theta_0,\varphi)\cos\theta_0 = y(\theta_0,\theta,\varphi)\cos\theta, \tag{6.3.18}$$

Another characteristic of reflection from the underlying surfaces is the parameter of anisotropy of reflection determined by the following equation:

$$\gamma(\theta,\varphi,\theta_0,\varphi_0)=\frac{\pi}{A}r(\theta,\varphi,\theta_0,\varphi_0). \tag{6.3.19}$$

It is assumed that we are concerned here with a non-Lambert surface with albedo A. Consequently, the parameter $\gamma(\theta,\varphi,\theta_0,\varphi_0,)$ gives the ratio of the intensity of the radiation, reflected by the surface, to the intensity of the radiation reflected from the Lambert surface with the same albedo A. This parameter may be higher than unity for the surface reflecting in specific directions more than the Lambert surfaces, and vice versa.

It is very important to mention that all the previously examined optical characteristics of reflection are functions of the wavelength, as already demonstrated on the example of the Fresnel reflection coefficients. Therefore, when discussing the optical characteristics of some surface, it is necessary to specify the spectral range or spectral region to which this characteristic belongs.

Previously, it was assumed that directional radiation is incident on the surface. We examine a more general and more appropriate case in which the surface is illuminated by diffusion radiation consisting of the direct and scattered (by the sky) solar radiation. For this case, we also introduce the concept of the coefficient of spectral brightness and albedo but in this case the incidence flux is the total flux consisting of the direct and scattered solar radiation. Sometimes these different definitions of the CSB and albedo are used without taking these differences into account. If the incident flux is the total radiation, then CSB and the albedo depend not only on the properties of the surface but also on the properties of the atmosphere. For example, if we write in detail a radiation flux incident on the surface (according to the equation (6.3.10) for an upwelling flux), and the definitions of the CSB and the albedo give directly their dependence on the angular distribution of the downwelling radiation $I^{\downarrow}(\theta_0,\varphi_0)$ incident from the atmosphere onto the surface. Naturally, according to these considerations, the albedo, determined for the total incident radiation flux, is the function of the zenith angle of the Sun. Therefore, although this is not always the case, all measurements of the albedo should be accompanied by at least a qualitative description of the state of the atmosphere and by the definition of the zenith angle of the Sun. Thus, in examination of the surfaces of planets the surface albedo is not an unique characteristic of the optical properties of the underlying surface and it is a function of the atmospheric state and the conditions of illuminations of the underlying surface–atmosphere system. It should be mentioned that in actual experiments aimed at examining the optical properties of the underlying surfaces measurements are taken of just these quantities (related to the total flux of incident radiation). The albedo, determined by this definition, does not depend greatly on the condition of the atmosphere (the presence of clouds, aerosol) whereas the CSB may greatly differ for different conditions..

In the theory of radiation tansfer we examine the characteristics of reflection of the entire atmosphere–underlying surface system [47, 65]. These characteristics can be regarded as functions of the altitude in the atmosphere, for example

$$\bar{A} = (z) = \frac{F^{\uparrow}(z)}{F^{\downarrow}(z)}, \tag{6.3.20}$$

$$\bar{\rho}(\theta,\theta_0,\varphi,z) = \pi\frac{I(\theta,\varphi)}{F^{\downarrow}(z)}. \tag{6.3.21}$$

According to this definition, the albedo $\bar{A}(z = 0) \neq A$, i.e. the albedo of the atmosphere–underlying surface is not the albedo of the surface, although this is often disregarded.

The albedo and the CSB at the upper boundary of the atmosphere are very important in atmospheric optics. According to the definition

$$I(\theta, \varphi, z = \infty) = \frac{1}{\pi} \overline{\rho}(\theta, \theta_0, \varphi, z = \infty) F_0 \cos\theta_0, \tag{6.3.22}$$

$$\overline{A}(z = \infty) = \frac{F^{\uparrow}(z = \infty)}{F_0 \cos\theta_0}, \tag{6.3.23}$$

where $F_0 \cos\theta_0$ is the extra-atmospheric flux of solar radiation incident on a horizontal area. The quantity $\overline{A}$ ($z = \infty$) is the fraction of the energy reflected by a planet into space and is referred to as a flat albedo [47,65]:

$$\overline{A}(z = \infty) = \frac{1}{\pi} \int_0^{2\pi} d\varphi \int_0^{\pi/2} \overline{\rho}(\theta, \theta_0, \varphi, z = \infty) \cos\theta \sin\theta \, d\theta. \tag{6.3.24}$$

The flat albedo depends on the examined point of the planet and the angles of incidence of solar radiation. For the characteristic of reflection of solar radiation from the planet as a whole we should use the concept of the spherical albedo A_s [47,65]. If a planet is regarded as a sphere with radius R_0, the angle of incidence of the solar rays θ_0 for the sphere changes from 0 to 90°. Therefore, the ratio of the energy of the solar radiation reflected from the entire planet (from the area of the illuminated part $2\pi R_0^2$) to the entire energy of solar radiation incident on the planet (on the projection area πR_0^2), i.e. spherical albedo, is determined by the relationships

$$A_s = \frac{1}{\pi} \int_0^{2\pi} d\varphi \int_0^{\pi/2} \int_0^{\pi/2} \overline{\rho}(\theta, \theta_0, \varphi, z = \infty) \times \\ \times \cos\theta \cos\theta_0 \sin\theta \sin\theta_0 \, d\theta d\theta_0. \tag{6.3.25}$$

In conclusion, we present Table 6.1 which characterises different methods of description of the optical characteristics of the surface of the atmosphere–underlying surface system.

6.4. Examples of the optical characteristics of underlying surfaces

We presented previously the Fresnel coefficients of reflection for the smooth surfaces, in particular, for the water surface. We analysed in greater detail the optical characteristics of different surfaces, paying attention to their complicated dependencies on the physical–chemical properties of the surfaces, the altitude of the Sun

Table 6.1. Methods of describing the optical characteristics of the surface or of the atmosphere-underlying surface system

Incident radiation	Reflected radiation	Characteristic
Intensity (flux of directional radiation)	Intensity	Two-directional reflection factor; coefficient of (spectral) brightness, reflection function
Flux (total)	Intensity	Coefficient of (spectral) brightness
Intensity (flux of directional solar radiation)	Flux reflected at a specific point of planet	Flat albedo
Intensity (flux of directional solar radiation)	Flux reflected from entire planet	Spherical albedo
Flux (total)	Flux	Albedo

and the atmospheric state. The data for the actual surfaces are determined in theoretical and numerical modelling of these surfaces and also by measurements in field and laboratory conditions.

Theoretical models

We describe briefly a scheme of determination of the theoretical reflection characteristics for the case of a wind-driven water surface. This surface is simulated by different methods. Two approaches to describing the complex spatial–temporal pattern of waves on the water surface. The first of them consists of the representation of the relief of the sea surface by regular functions of the spatial–temporal variables [85]. Thus, assuming that the amplitude of the wave is infinitely small in comparison with its length, the profile of the wave may be approximated by a sinusoidal. The second approach, statistical, is based on describing the characteristics of the surface of the sea in the form of the probability laws of distribution of different elements of the waves–altitude, period, length, slope of inclination, etc. For example, the model proposed by Cox and Munk [85] gives the probability of distribution of the slopes of the surfaces of the waves in the form of a Gaussian distribution (assuming that the distribution is independent of the direction of wind – isotropic case):

$$P(z_x,z_y)=\frac{1}{\pi\sigma^2}\exp\left[-\frac{(z_x^2+z_y^2)}{\sigma^2}\right], \quad (6.4.1)$$

where z_x and z_y are the characteristics of the slopes in respect of two orthogonal directions; σ is the dispersion of the slopes (independent of the direction) and is determined by the velocity of the wind v:

$$\sigma^2 = 0.003+0.00512v \pm 0.004. \quad (6.4.2)$$

The approximations for $P(z_x, z_y)$ and σ^2 were determined from the experimental data and held in the range of the wind speed of 0–14 m/s. This was followed by examination of the reflection (or radiation) from an arbitrary oriented area of the wind-driven surface of water. The characteristics of reflection of the wind-driven surface of the sea are determined by averaging taking into account the probability of distribution of the slopes of the water surface. In the presence of foam on the water surface, formed at high wind speeds, the model is more complicated and the medium near the surface is regarded as heterogeneous consisting of different components – water and air.

In examination of the optical characteristics of the real land surfaces, it is necessary to construct different models of the surface layer of land. These models include the characteristics of the spectral distribution of the heterogeneities and dielectric permittivity of different components. Various models of the transfer of radiation in these media are then used. In the simplest case for infrared and microwave ranges the emissivity of the surfaces is determined by solving the integral equation of transfer of radiation without taking scattering into account. In more generalized models, the effects of scattering in the investigated inhomogeneous media are taken into account.

Classification of the spectra of reflection of natural surfaces

A large number of experimental data is available for different optical characteristics of different underlying surfaces. There are detailed data on the albedo and CSB of different surfaces because these measurements are relatively simple. As an example, Fig.6.5. gives the values of the albedo of different surfaces in relation to the altitude of the Sun. Figure 6.5 shows that the variations of the albedo are most significant at low altitudes of the sun. The figure

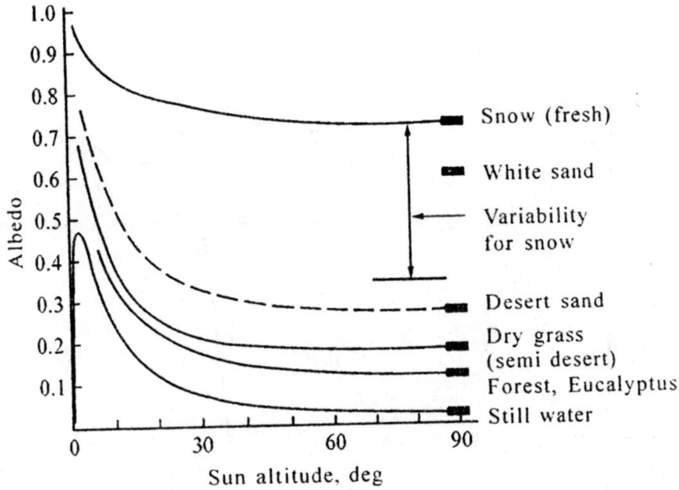

Fig.6.5. Dependence of the albedo of different surfaces on the altitudes of the Sun [111].

also shows the strong variability of the albedo of snow at high altitudes of the Sun.

The spectral special features of the albedo and CSB differ quite considerably. However, it appears that regardless of the differences in the spectral albedo of the underlying surfaces, visible and near infrared ranges of the spectrum, those can be grouped into four main classes. This classification was proposed by E.L. Krinov on the basis of the results of the first aircraft measurements of the albedo carried out in the former USSR in the forties of the previous century [36]. The characteristic spectra of the albedo of the four classes are given in Fig. 6.6. The first class includes snow and clouds. Their albedo is high and slightly increases along the spectrum from ultraviolet to the beginning of the near infrared range of the spectrum. The second class includes soil, sand and open rock. They are characterised by a smooth, almost linear increase of the albedo with increasing wavelength. The third class is the water surface. The albedo of the water is small and its spectrum is almost constant or slightly decreases with increasing wavelength. Finally, the fourth class is plants. The albedo of green plants is characterised by a complicated spectral dependence: a local maximum in the range 0.55 μm causing its green colour, followed by a decrease and rapid increase after 0.7 μm, where the albedo of the plants is close to that of snow and clouds. Yellow plants (dry

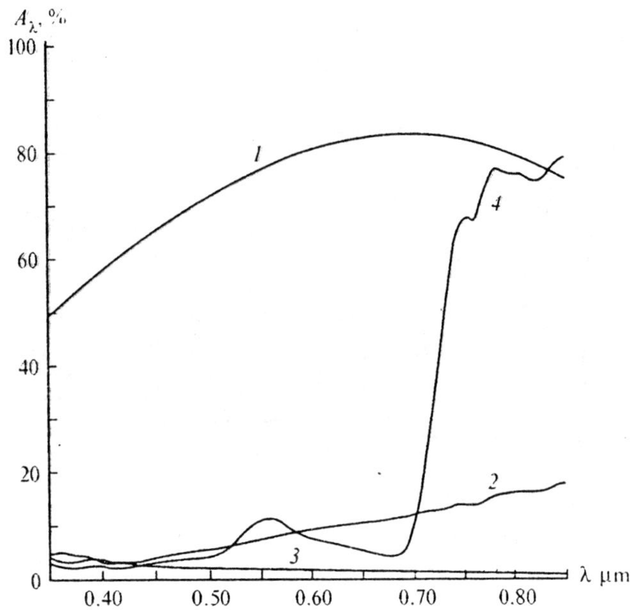

Fig.6.6. Classification proposed by E.L. Krinov for the spectral dependences of the reflecting characteristics of natural surfaces [36]. 1).Snow and cloud; 2) soil, sand, rock; 3) water surfaces; 4) vegetation (characteristic spectral albedo representatives of these classes are given).

grass in the steppes, autumn leafy forests) have a similar spectal dependence but the maximum of the albedo in the vicinity of 0.55 μm is less distinctive.

To provide more detailed characteristics of the albedo and its spectral dependences, we present Table 6.2 and 6.7.

Variability of albedo

As already mentioned, the albedo of the surfaces is dependent on their type and state, the altitude of the sun and also the atmospheric state (the presence of clouds). We examine the variability of the albedo in relation to different factors, with special reference of the albedo for the solar range of the spectrum. Almost all surfaces are characterised by one special feature – the largest changes of the albedo take place from sunrise to altitudes of the sun of 20–30°. Sharp changes of the albedo of the underlying surface are detected only in the periods of thawing and the formation of snow, i.e. at changes of the type of underlying surface. In these periods, the

Table 6.2. Values of the albedo of surfaces (in percent) for the visible ranges of the spectrum.

Surface types	Conditions	Albedo
Water surfaces	Equator	6
	Winter, 30° latitude	9
	Winter, 60° latitude	21
	Summer, 30° latitude	6
	Summer, 60° latitude	7
Snow	Fresh	75–95
	Old	40–70
Sea ice	–	30–40
Sand (dunes)	Dry	35–45
	Moist	20–30
Soil	Dark	5–15
	Grey, moist	10–20
	Dry, grey clay	25–35
	Dry light sand	25–45
Concrete	Dry	17–27
Road	Black	5–10
Desert	–	25–30
Savannah	Dry season	25–30
	Humid season	15–20
Bush	–	15–20
Meadow	Green	10–20
Forest	Leafy	10–20
	Coniferous	5–15
Tundra	–	15–20
Grain cultures	–	15–25

difference between the values of the albedo in adjacent days may reach 20–30°. The albedo of a dry snow cover with the cloudless sky varies in the range 52–99%. The albedo of a dirty, wet snow may decrease to 20–30%. With increase of the amount of clouds, the albedo of the snow surface increases. In the presence of a continuous cloud cover it may increase by 2–10%. The albedo of melting snow decreases to 30–40%.

The albedo of the grass cover varies from 12 to 28% depending

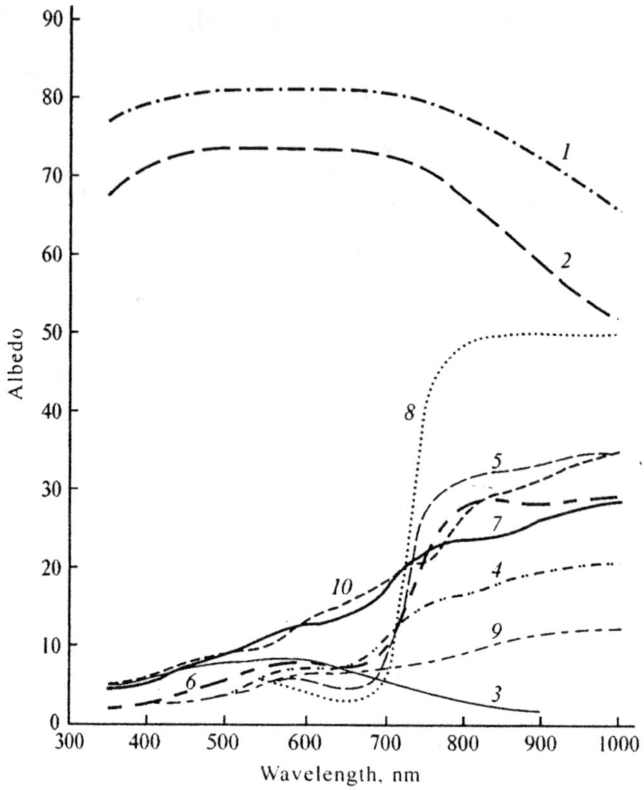

Fig.6.7. The spectral dependence of the albedo of different surfaces [111]. 1) Snow, altitude of Sun 38°; 2) wet snow, altitude of sun 27°; 3) lake water, altitude of sun 56°; 4) soil after melting of snow, altitude of sun 24° 30'; 5) wheat after ensilage, altitude of sun 54°; 6) tall green wheat, altitude of sun 56°; 7) yellow wheat, altitude of sun 46°; 8) Sudanese grass, altitude of sun 52°; 9) black soil, altitude of sun 40°; 10) stubble, altitude of sun 35°.

on the density, colour and moisture of the grass. The albedo of wet grass is 2–3% lower than that of dry grass. The reflectivity of the grass cover also depends on the altitude of the Sun and, consequently, the albedo of dry green grass in morning and evening hours is 2–9% higher than at midday. In the spring, the albedo of dry grass cover and last year's grass varies in the range 10–24%.

The reflecting properties of the surfaces, free from the vegetation cover, depend on the type of soil, structure and moisture content. The albedo of non-moistened soil is 8–26%. The highest albedo is shown by white sand, up to 40%. The albedo of wet soil is 3–8% lower than that of the albedo of the dry soil, that of the white sand by 18–20%. With a decrease of the surface roughness soil, the albedo increases. During the day the albedo of the soil changes from the maximum values at low altitudes of the Sun to minimum values at midday. The amplitude of the daily variations of the albedo of soil is 11–17%.

The albedo of the surface of water depends on a number of factors: the altitude of the sun above the horizon, the ratio between the direct and diffusion components of solar radiation, the amount of clouds, the strength of waves and the characteristics of water reservoirs – depth, transparency of water, etc. The daily variation of the albedo of the water surface is most distinctive in the absence of waves when its amplitude may reach 30% or more. In the presence of strong waves, the albedo remains almost constant throughout the day. Under a continuous cloud cover, the daily variation of the albedo is also minimum. Waves and cloud decrease the albedo of the water surface at altitudes of the sun smaller than 30°. At a large altitude of the Sun, the clouds and waves have the reversed effect which is however considerably weaker.

We present a number of specific examples of the spectral reflecting properties of different underlying surfaces [115, 116]. Figure 6.8a shows the spectral behaviour of the albedo of two agricultural cultures – wheat prior to the heading period and soya beans in comparison with the spectral characteristics of two types of bare soil – dry and wet. The figure shows clearly the variability of the optical characteristics of soils (wet soil has lower albedo values) and different types of vegetation. Figure 6.8b gives the spectral characteristics of reflection of a lucerne field in different periods of the life cycle. The curves shown in the graph correspond to different amounts of the biomass and different degrees of cover of the lucene soil – the case of exposed soil corresponds to the zero values of the biomass and the covering. The transformation of the spectral curves of reflection with increase of the biomass and the covering makes it possible to measure the reflected solar radiation for inspection of the state of cultures. It is important to note a very important special feature of the spectral behaviour of reflection from the vegetation cover, shown clearly in Fig. 6.8a and Fig. 6.8b – a large increase of the albedo in the vicinity of 0.8 μm. The green colour of vegetation is connected with this special feature. The presence of chlorophyll in the plants results in strong absorption of solar radiation for wavelengths shorter than 0.7 μm.

The variability of the spectral characteristics of the reflection of leaves of a plane tree in relation to their moisture content is shown

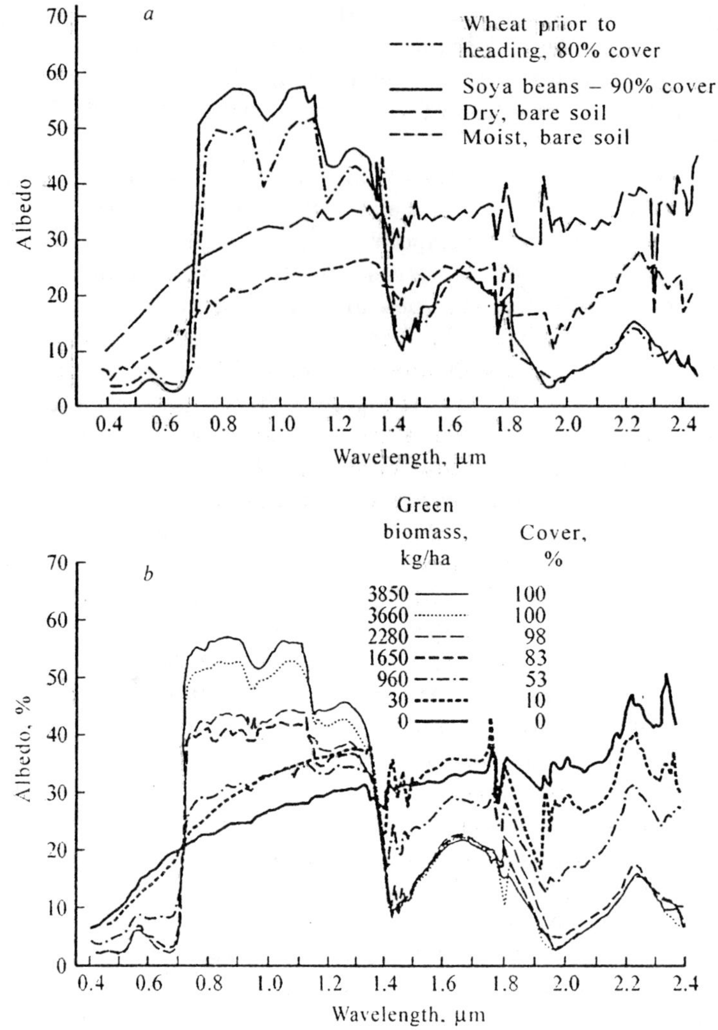

Fig.6.8. Spectral behaviour of albedo [114] of two agricultural cultures – wheat prior to the heading period, and soya beans in comparison with spectral characteristics of two types of bare soil – dry and moist (a) and spectral characteristics of reflection [115] of a lucerne field in different periods of the life cycle (b).

in Fig. 6.9. With increase of the moisture content the albedo of the leaves rapidly decreases in the near infrared range of the spectrum. It should be mentioned that the albedo values in this figure are reduced to the albedo of a specially prepared sheet of magnesium oxide.

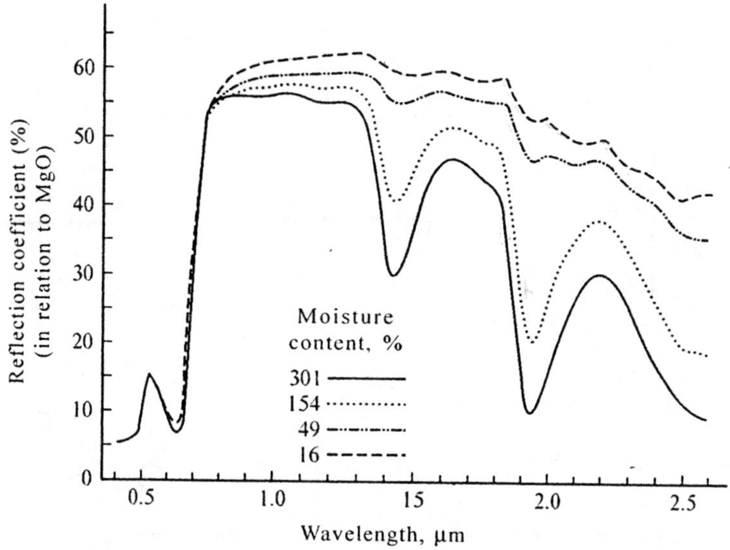

Fig.6.9. Variability of spectral characteristics of reflection of plane tree leaves in relation to the moisture content [115] (in relation to the reflection coefficient of MgO).

The two-directional reflection coefficients are characterized in Fig. 6.10 which gives the factors of the anisotropy of the reflection of the atmosphere – underlying surface system obtained as a result of satellite measurements. (As mentioned previously, the concept of the albedo and other reflection characteristics can be applied directly to both the underlying surfaces and to the entire atmosphere – surface system). The figure shows two cases of underlying surface – the ocean (Fig. 6.8a) and the snow cover (6.8b). The factors of anisotropy of reflection are given as functions of the observation zenith angle of satellite devices (in relation to the local vertical) for eight sub-intervals of the azimuthal angle in relation to the azimuths of the sun. These measurements relate to cases in which the zenith angle of the sun is in the range 25.84–36.87°. The maximum values of the factor of anisotropy of reflection above the ocean with a cloudless sky were recorded for the angles of examination in the vicinity of 30° and the range of azimuthal angles of 0–30° and correspond to the examination of solar track (quasi-mirror reflection). Another special feature of the reflection is the increase of the factor of anisotropy of reflection with increase of the angle of examination (when approaching approach to the horizon of the planet) when the azimuth of observation is in the angle range 90–180°, i.e. on approaching the

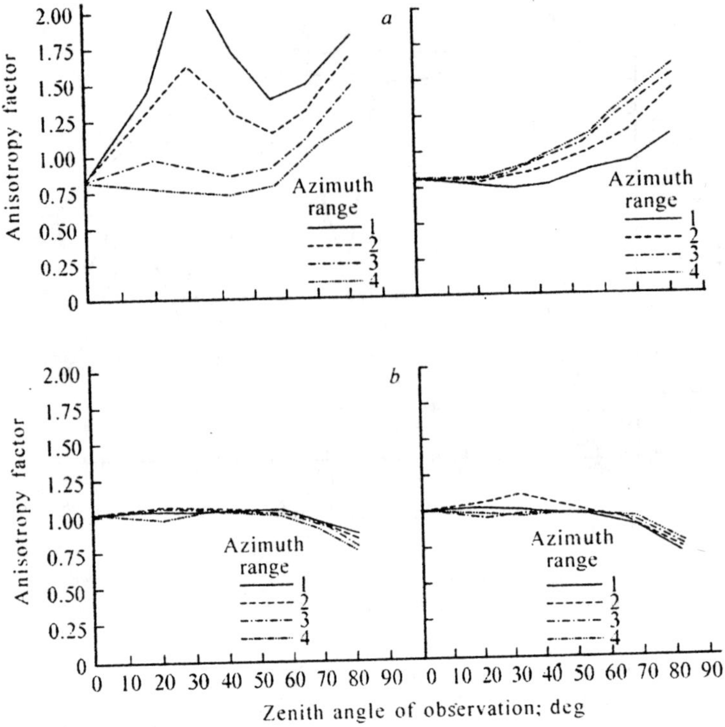

Fig.6.10. Factors of reflection anisotropy for the atmosphere–underlying surface system as a function of zenith angles of the Sun, measured in satellite measurement above the water surface (a) and snow (b) [114]. Observation azimuth ranges, deg: 1) 0–9°; 2) 9–30°; 3) 30–60°; 4) 60–90°; 5) 30–120°; 6) 120–150°; 7) 150–171°; 8) 171–180°.

opposite direction from the sun. Figure 6.10b corresponds to the indentical examination for the snow surface. In this case, the reflection characteristics are more homogeneous in comparison with the water surface.

6.5. Emitting properties of underlying surfaces

In the infrared and microwave regions of spectrum, the underlying surfaces of the planets are important sources of the generation of natural radiation. The characteristics of these surfaces as emitters are described by the coefficients of radiation or the emissivity of the surfaces [34, 45, 85, 102, 1156, 119]. The emissivity of a surface (λ, T) is the ratio of intensity of radiation of the surface with temperature T to the radiation of the absolutely black body at the same temperature

$$\varepsilon(\lambda,T)=\frac{I_s(\lambda,T)}{B(\lambda,T)}. \tag{6.5.1}$$

The radiation of a black body is isotropic which cannot be said about the radiation of real surface. Therefore, in a general case, the emissivity depends on the direction of radiation, wavelength and in a number of cases the temperature. Of course, the emissivity also depends greatly on the physical–chemical properties of the surface – its nature, form of the surface, etc.

An important relationship used widely in atmospheric optics is a relationship between the absorption, reflection and emissivity of the medium. To derive this relationship, the conversion of the radiation energy during its interaction with the medium is examined. The incident radiation may be reflected from the medium, be absorbed by it, and also part of the radiation may transfer to the medium. To characterise the processes, the reflectivity (the albedo of the medium) A_λ, the absorptivity B_λ, the transmission function P_λ are introduced. They are the ratios of the appropriate components of radiation to incident radiation. According to the law of conservation of energy, the sum of these quantities should be equal to unity

$$A_\lambda + B_\lambda + P_\lambda = 1. \tag{6.5.2}$$

It is implicitly assumed that in these processes we can ignore phenomena such as Raman scattering and fluorescence (see Chapter 5) which lead to the redistribution of radiation energy over wavelengths. Assuming that the entire radiation incident on the medium is absorbed (P_λ=0), we obtain a simpler relationship

$$A_\lambda + B_\lambda = 1. \tag{6.5.3}$$

Further, in examining the relationship between different characteristics of the medium we use the Kirchoff law defining the relationship between the emissivity and absorbing properties of the medium. Here it is confirmed that $B_\lambda = \varepsilon_\lambda$. Consequently, equation (6.4.3) gives the relationship between the emission and reflecting characteristics of medium

$$\varepsilon_\lambda=1-A_\lambda. \tag{6.5.4}$$

It must be remembered that this ratio holds for a fixed wavelength. The emissivity of the underlying surfaces strongly depends on their type, the form of the surface, the wavelength and the angle of observation. To determine ε_λ we use the experimental and numerical methods. In the latter case, equation (6.5.4) is used widely. For example, the Fresnel coefficients of reflection for the smooth surface (examined in paragraph 6.2) enable us to determine its emissivity.

Emissivity of surfaces in the infrared range of the spectrum

We present examples of the emissivity of different underlying surfaces in the infrared range of the spectrum. The most detailed data on these characteristics are available for 'transparency windows' of the atmosphere which is used in satellite meteorology to determine the temperature of the underlying surfaces (Chapter 10). Table 6.3. gives the emissivity ε in the spectral range 8–14 μm [45].

Recently, a large number of theoretical and experimental investigations have been carried out for determining the spectral dependence of the emissivity of different surfaces in the infrared region of the spectrum. As an example, Fig.6.11 gives the spectral dependences of the emissivity of the water surface, measured (at different wind speeds in the vicinity of the water surface) and calculated using the model identical with that described previously in subsection 'Theoretical models' [108]. The figure indicates that the experimental and calculated results are in relatively good agreement.

At present, there are special databanks for the spectral dependences of the emissivity properties of different surfaces, obtained by experimental methods. These data are used for modelling radiation fields in the atmosphere–underlying surface system and also for remote measurement of the characteristics of the underlying surfaces, especially their temperature.

Table 6.3. Emissivity of different surfaces in the infrared range

Surface	ε	Surface	ε
Granite	0.898	Clay	0.963–0.968
Bazalt	0.934	Asphalt	0.956
Dolomite	0.958	Grass, dense cover	0.976
Sandstone	0.935–0.985	Snow	0.99
Gravel	0.943	Water	0.98–0.993
Sand, quartz, dry	0.914	Water with thin oil film	0.954–0.972
Sand, wet	0.934	Concrete	0.942–0.966
Black sol	0.965	Water with machine oil film	0.960
Loam	0.98		

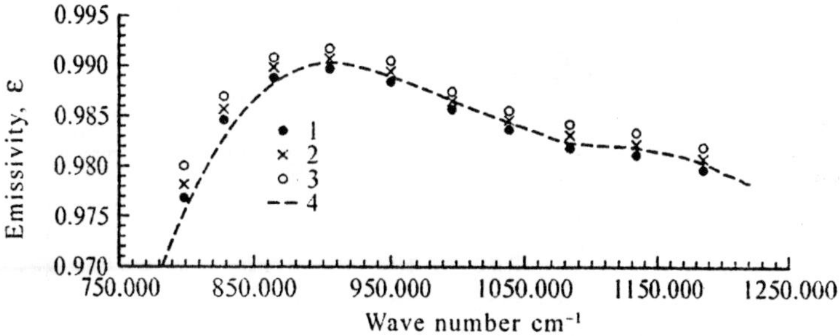

Fig.6.11. Measured (1,2,3) and calculated (4) spectral dependences of the emissivity of the water surface in the infrared range of the spectrum [108].

Emissivity of surfaces in the microwave range of spectrum

The main special feature of the emissivity of the natural surfaces in the microwave range of the spectrum is their considerable variability in comparison with the infrared range of the spectrum. For example, the emissivity of the water surface is in the range 0.2–0.95 depending on its temperature, salt content, the velocity of driving wind, the condition of the surface and the observation angle. It also depends on the polarization characteristics of radiation (see, for example, Fig. 6.6. showing the coefficient of reflection of the smooth sea surface for a wavelength of 3 cm). On the other hand, a number of dry land surfaces have the emissivity close to unity (Table 6.4) [45, 119].

To describe the emissivity of the surface with waves, various researchers developed a number of semi-empirical models taking into account both the process of formation of waves and the possible presence of foam formation on the surface arising in disruption of the waves [85, 119]. According to the data by various authors, the process of breakage of the waves starts at wind speeds of 3–7 m/s. The determination of the relationship between the emissivity of the sea surface and, for example the velocity of the driving wind enables remote methods of determination of the velocity to be developed.

Table 6.4. Emissivity of different surfaces at λ = 3.2 cm

Surface	Description	Air temperature, °C	Emission coefficient
Soil (sample)	Layer thickness 20 cm,		
	Moisture content 6.4%	21	0.947
	Moisture content 19.5%	20	0.919
Soil	Moisture content on surface 21%, at a depth of 20 cm 17.5%	23	0.923
"	Moisture content on surface 34.6% at a depth of 20 cm 17.5%	22.5	0.668
Frozen soil (sample)	Temperature 0.4 °C Moisture content 13.9%	1.2	0.923
Defrosted soil	Temperature 0.4 °C		
	Moisture content 14%	0	0.892
Snow cover	Snow on soil, layer thickness 20–30 cm density 0.408 g/cm^3	1.2	0.956
Frozen soil	Thickness of frozen layer 11–12 cm, moisture content at surface 45.6%	0.4	0.941
Peat	Moisture content 114%	18	0.943
	Moisture content 159%	15	0.918
Clay	Moisture content 9%	15	0.902
Plywood	Sheet layer thickness 8 mm	21	0.829
"	Sheet covered with aluminium paint	23	0.728
Concrete covered with snow	30 cm thick plate snow thickness 8–10 cm	0	0.906
Wet concrete		0	0.669
Asphalt	Motorway	15	0.974

CHAPTER 7

FUNDAMENTALS OF THE THEORY OF TRANSFER OF ATMOSPHERIC RADIATION

Atmospheric radiation

In Chapter 3 we examined different types of atmospheric radiation. We introduced important concepts of thermal (equilibrium) radiation and described its main laws. Formally speaking, all other types of generation of radiation of the atmosphere may be related to another large class – non-equilibrium radiation of the atmosphere.

At present, it is accepted to subdivide the non-equilibrium radiation into non-equilibrium infrared radiation and glow of the atmosphere. This division is based on the spectral principle – non-equilibrium radiation in the ultraviolet, visible and near infrared ranges of the spectrum is related to the glow of the atmosphere, and non-equilibrium radiation in the middle and far infrared range – to non-equilibrium infrared radiation. Another difference is that the main mechanisms of formation of the glow of the atmosphere and non-equilibrium infrared radiation differ. The glow of the atmosphere is caused in most cases by the processes of excitation of electronic states of the molecules and atoms as a result of the adsorption of high energy radiation of the Sun in the ultraviolet and visible ranges and the energy of fluxes of different particles. Non-equilibrium infrared radiation forms because of the disruption of local thermodynamic equilibrium (LTE) in the upper layers of the atmosphere. In this case, the formation of non-equilibrium radiation is caused by the relatively small number of collisions of the molecules (transferring the system in the lower layers of the atmosphere to the equilibrium state), the non-isothermal nature of the atmosphere and the losses of the energy of the system as a result of the radiation of the atmosphere outgoing into the space. However, the absorption of the solar radiation arriving in the atmosphere is also an important factor in this case.

In this chapter, special attention is given to the fundamentals of the theory of transfer of the thermal radiation of the atmosphere [20, 34, 37, 91, 103]. In the final paragraphs of the chapter we examine briefly the non-equilibrium infrared radiation and glow of the atmosphere.

7.1. Transfer of thermal radiation

In chapter 3 we derived equation (3.4.28) for the intensity of monochromatic natural thermal radiation $I_\nu^{\uparrow}(z)$ at arbitrary altitude z. Using this relationship, we can write the following expression for the intensity of rising radiation at altitude z:

$$I_\nu^{\uparrow}(z,\theta) = I_{\nu,0}\exp\left(-\sec\theta\int_0^z k_\nu(z')dz'\right) + \\ +\sec\theta\int_0^z k_\nu(z')B_\nu[T(z')]\exp\left(-\sec\theta\int_{z'}^z k_\nu(z'')dz''\right)dz'. \tag{7.1.1}$$

Here $I_{\nu,0}$ is the radiation of the underlying surface which must be determined separately. This expression corresponds to upward radiation (i.e. the radiation into the upper atmosphere) for the model of the plane-parallel horizontally homogeneous atmosphere. Similarly, we can write an expression for the intensity of monochromatic downward thermal radiation for the same model of the atmosphere:

$$I_\nu^{\downarrow}(z,\theta) = I_{\nu,\infty}\exp\left(-\sec\theta\int_z^\infty k_\nu(z')dz'\right) + \\ +\sec\theta\int_z^\infty k_\nu(z')B_\nu[T(z')]\exp\left(-\sec\theta\int_z^{z'} k_\nu(z'')dz''\right)dz', \tag{7.1.2}$$

where $I_{\nu,\infty}$ is the intensity of radiation at the upper boundary of the atmosphere. Usually, for the infrared range of the spectrum this radiation is assumed to be equal to zero. For the microwave range of the spectrum in equation (7.1.2) we must take into account the relict microwave radiation arriving from space. Its brightness temperature is at present assumed to be equal to 2.7 K.

In the equations (7.1.1) and (7.1.2) there is the transmittance function of the atmospheric layer (z_1, z_2):

$$P_\nu(\theta, z_1, z_2) = \exp\left(-\sec\theta \int_{z_1}^{z_2} k_\nu(z')dz'\right). \qquad (7.1.3)$$

Using these transmittance functions, the expressions (7.1.1) and (7.1.2) can be written in a more compact form:

$$I_\nu^{\uparrow}(z,\theta) = I_{\nu,0} P_\nu(0,z) + \int_0^z B_\nu[T(z')] \frac{dP_\nu(z',z)}{dz'} dz', \qquad (7.1.4)$$

$$I_\nu^{\downarrow}(z,\theta) = I_{\nu,\infty} P_\nu(z,\infty) - \int_z^\infty B_\nu[T(z')] \frac{dP_\nu(z,z')}{dz'} dz'. \qquad (7.1.5)$$

Further, in this chapter we examine in greater detail the methods of determining the transmittance functions of the atmosphere.

Radiation of the underlying surface in expression (7.1.1) and (7.1.4) is often represented by the emissivity of the surface ε (see for more details chapter 6), i.e.

$$I_{\nu,0} = \varepsilon_\nu B_\nu(T_0), \qquad (7.1.6)$$

where T_0 is the surface temperature. However, the difference of the emissivity of the surface from unity assumes the existence of reflection of downward thermal radiation of the atmosphere from the surface. The intensity of reflection may be written in different form depending on the given reflection model (see chapter 6). In the simplest case of mirror reflection it can be presented in the form

$$I_\nu(\theta) = (1 - \varepsilon_\nu)\, I_\nu^{\downarrow}(0), \qquad (7.1.7)$$

where $I_\nu^{\downarrow}(0)$ is the intensity of downward thermal radiation. Thus,

$$I_{\nu,0} = \varepsilon_\nu B_\nu(T_0) + (1 - \varepsilon_\nu)\, I_\nu^{\downarrow}(0).$$

In various practical problems of atmospheric physics we have not been interested in many cases in the transfer of monochromatic radiation (with the exception of, for example, theoretical study of special features of radiation transfer or in the propagation or 'almost' monochromatic radiation of lasers). In fact, calculating, for example, the values of the radiation influxes of heat for determining the variation of the atmospheric temperature as a result of radiation heat exchange, it is necessary to integrate the appropriate influxes of radiation in respect of frequency (or wavelength). In the same manner, analysing the results of measurements of specific characteristics of the radiation field, we examine the radiation in finite spectral ranges. To obtain the intensity of thermal radiation in finite spectral ranges it is necessary to integrate the monochromatic intensities in respect of frequency (or wavelength)

$$I_{\Delta\nu} = \int\limits_{\Delta\nu} I_\nu d\nu, \tag{7.1.8}$$

where $I_{\Delta\nu}$ is the intensity in the spectral range $\Delta\nu$. Substituting into equation (7.1.8), for example, equation (7.1.4) and ignoring the dependence on θ, we obtain the intensity of upward radiation in finite spectral range $\Delta\nu$:

$$I_{\Delta\nu}(z) = \int\limits_{\Delta\nu} I_{\nu,0} P_\nu(0,z) d\nu + \int\limits_{\Delta\nu}\int\limits_{0}^{\bar{z}} B_\nu[T(z')] \frac{dP_\nu(z',z)}{dz'} dz' d\nu. \tag{7.1.9}$$

Equation (7.1.9) or identical expressions for downward radiation can be used to calculate the intensity of thermal radiation in finite spectral ranges but it can also be greatly simplified. For this purpose, it is necessary to take into account the special features of the spectral behaviour of the Planck function and the monochromatic transmittance functions present in all members of the equation (7.1.9). As shown in Chapter 4 (Fig.4.6), the monochromatic transmittance functions are very rapidly changing functions of frequency (or wavelength). At the same time, the Planck function changes quite slowly with frequency. Therefore, if we examine relatively narrow spectral ranges (~50–100 cm^{-1}), in which the variation of the Planck function can be ignored, we can write an approximate equation for the intensity of thermal radiation in finite spectral ranges. For upward radiation, for example

$$I^{\uparrow}_{\Delta\nu}(z) = I_{\bar{\nu},0} \int\limits_{\Delta\nu} P_\nu(0,z) d\nu + \int\limits_{0}^{\bar{z}} B_{\bar{\nu}}[T(z')] \int\limits_{\Delta\nu} \frac{dP_\nu(z',z)}{dz'} dz' d\nu, \tag{7.1.10}$$

where $B_{\bar{\nu}}[T(z')]$ and $I_{\nu,0}$ are the Planck function and the contribution of the underlying surface at some mean frequency $\bar{\nu}$ of the examined spectral range. The integrals in respect of frequency present in equation (7.1.10) determine the transmittance functions and the derivatives for the finite spectral ranges. Therefore, expression (7.1.10) can be written in a new form

$$I^{\uparrow}_{\Delta\nu}(z) = I_{\bar{\nu},0} P_{\nu}(0,z)\Delta\nu +$$

$$+\Delta\nu \int_{0}^{z} B_{\bar{\nu}}[T(z')] \int_{\Delta\nu} \frac{dP_{\Delta\nu}(z',z)}{dz'} dz'. \qquad (7.1.11)$$

In particular it should be mentioned that the assumption on the weak dependence of the Planck function of frequency enables us to write the equation for the intensity of thermal radiation in finite spectral ranges in the same form as for monochromatic intensity. The difference between the equations (7.1.4) and (7.1.11) is that equation (7.1.11) contains the functions of transmittance for finite spectral ranges in contrast to the monochromatic functions of transmittance in equation (7.1.4). This approximation reduces the problem of integration of monochromatic radiation (equation (7.1.8)) to the problem of obtaining the transmittance functions for finite spectral ranges. Naturally, similar equations can also be obtained for downward thermal radiation.

7.2. Transmittance functions of atmospheric gases

The transmittance functions of the atmosphere are of fundamental importance in atmospheric optics. This is due to the fact that they are used for solving various direct problems of atmospheric optics - calculations of intensities, fluxes and influxes of radiation. They are also used for interpreting the measured results, for example, when using different remote methods of measurements of parameters of the atmosphere and the surface.

Determination of the transmittance function

We now return to describing the process of attenuation of radiation in the atmosphere of the planets. In accordance with the Bouguer's law (equation (3.4.5)) the monochromatic intensity of radiation, transmitted through the homogeneous layer of the atmosphere, containing amount of the absorbing substance $u = \rho l$, is equal to

$$I_\nu(u) = (I_{0,\nu} \exp(-k_\nu u), \qquad (7.2.1)$$

where in the present case k_ν is the mass molecular absorption coefficient. If the weakening coefficient is determined by molecular absorption in the multicomponent gas atmosphere, then in accordance with the equations (4.4.1) and (4.4.3) given previously, the relationship (7.2.1) may be written in the new form

$$I_\nu(u) = I_{0,\nu} \exp\left(-\sum_j \sum_i S_{ij}(T) f_{ij}(p,T,\nu-\nu_{0,ij}) u_j\right), \qquad (7.2.2)$$

where U_j is the content of the absorbing j-th gas. The relationship

$$P_\nu(u) = \frac{I_\nu(u)}{I_{0,\nu}} = \exp\left(-\sum_j \sum_i S_{ij}(T) f_{ij}(p,T,\nu-\nu_{0,ij}) u_j\right) \qquad (7.2.3)$$

is the monochromatic transmittance function for molecular absorption. The transmittance function is expressed either in percent or in fractions of unity. The case $P_\nu=1$ (at 100%) corresponds to the absorption, in the case $P_\nu = 0$ to the complete absorption of radiation on the examined path. Equation (7.2.3) describes absorption on a homogeneous path containing j absorbers (different gases), i.e. on the path characterised by constant pressure p and temperature T. (Strictly speaking, at constant partial pressures of the absorbing gases p_j and constant pressure of secondary gases). Horizontal paths along the surface of the Earth 5–10 km long are a good approximation of homogeneous paths. In a general case when radiation propagates on vertical (or inclined) paths, we are concerned with the propagation of radiation in media with pressure and temperature changing from point to point. In this case, equation (7.2.3) is generalized for the monochromatic transmittance function of a inhomogeneous medium as follows:

$$P_\nu(l) = \exp\left(-\sum_j \sum_i \int_0^l S_{ij}(T)(l)) \times f_{ij}(p(l),T(l),\nu-\nu_{0,ij}) du_j(l)\right). \qquad (7.2.4)$$

These monochromatic transmittance functions are also found in equation (7.1.4). The differential of the content of the absorbing gases du may be presented in different forms depending on the characteristics of the gas content in the atmosphere used in this case. For molecular absorption, it is expressed through the density of the absorbing gas ρ_j:

$$du_j(l) = p_j(l)dl. \tag{7.2.5}$$

As mentioned previously, for practical problems it is interesting to use different quantities (for example, radiation intensity) calculated for finite spectral ranges. In this case, as followed from equation (7.1.9), we are concerned with the transmittance functions (over the derivatives) in finite spectral ranges. For example, let us assume that our device records the solar radiation passing in the atmosphere along the path l. Consequently, the signal recorded by the device according to (7.2.1) and (7.2.4) can be written in the form

$$I_{\Delta\nu}(\bar{\nu},l) = \int_{\Delta\nu} \varphi(\bar{\nu}-\nu')I_0(\nu')\times$$

$$\exp\left(-\sum_j\sum_i\int_0^l k_{ij}(\nu',p(l'),T(l'))\rho_j(l')dl'\right)d\nu', \tag{7.2.6}$$

where $\Delta\nu$ is the frequency interval determined by the spectral resolution of the device; $\bar{\nu}$ is the interval to wich the measured intensity is assigned; $I_0(\nu')$ is the intensity of solar radiation at the upper boundary of the atmosphere; $\varphi(\bar{\nu}-\nu')$ is the slit function of the device characterizing the sensitivity of the device to radiation at different frequencies inside the interval $\Delta\nu$. (Strictly speaking, the device integrates radiation not only in respect of frequency but also the solid angle and measurement time but to simplify the considerations this is not taken into account in equation (7.2.6)). As in (7.2.3) and (7.2.4), we can introduce the concept of the transmittance functions for finite spectral ranges. For example, for a inhomogeneous medium, we have

$$P_{\Delta\nu}(\bar{\nu},l) = \frac{I_{\Delta\nu}(\nu',l)}{I_{\Delta\nu}(\bar{\nu},o)} =$$

$$= \frac{\int_{\Delta\nu}\varphi(\bar{\nu}-\nu')I_0(\nu')\exp\left(-\sum_j\sum_i\int_0^l k_{ij}(\nu' p(l'),T(l'))\rho_i(l')dl'\right)d\nu}{\int_{\Delta\nu}\varphi(\bar{\nu}-\nu')I_0(\nu')d\nu'}. \tag{7.2.7}$$

If it is assumed that in the examined spectral interval $\Delta\nu$ the radiation incident on the layer is not selective, i.e. $I_0(\nu)$ = const, taking into account the standard normalisation of the slit function of the device

$$\int_{\Delta\nu} \varphi(\bar{\nu}-\nu')d\nu' = \Delta\nu \tag{7.2.8}$$

we obtain from (7.2.7) a simple expression for the transmittance function often used in atmospheric optics:

$$P_{\Delta\nu}(\bar{\nu},l) = \frac{1}{\Delta\nu}\int_{\Delta\nu} \varphi(\bar{\nu}-\nu')\times$$

$$\times\exp\left(-\sum_j\sum_i\int_0^l k_{ij}(\nu',p(l'),T(l'))\rho_j(l')dl'\right)d\nu'. \tag{7.2.9}$$

Naturally, in calculations of radiation energy, for example, radiation fluxes in the atmosphere, the slit function is not present or, formally, it is equal to unity. On the other hand, in analysing the results of measurements of different characteristics of the radiation fields, it is important to take into account the slit function.

7.3. Methods of determination of transmittance functions

Because of the extensive use of transmittance functions in the atmospheric physics, special attention is being paid to their determination. Various experimental and calculation methods of determination have been developed. The most important are [34]:

- direct calculation method;
- method of the models of absorption bands;
- method of integration in respect of the absorption coefficient;
- experimental methods (laboratory and field size).

Direct method of calculating transmittance functions

If information on the parameters of the fine structure (position, intensity, line half width, etc) for different absorption bands is available, the transmittance functions can be calculated from the equations presented previously. To determine the transmittance functions in finite spectral intervals, this method requires integration in respect of the frequency range of the monochromatic transmittance functions. This approach is referred to as the line-by-line method of calculating transmittance functions.

The advantages of the approach are:

1. The possibility of obtaining the transmittance functions for homogeneous and inhomogeneous media for different geometries of the propagation of radiation for any mixtures of gases, arbitrary slit functions of spectral devices, etc. In this sense, the direct method is universal. Of course, in this case it is necessary to have information not only on all parameters of the fine structure, describing molecular absorption, but also on their dependences on the parameters of the physical state of the medium – temperature, pressure, etc.

2. The highest potential accuracy because the direct method does not require any simplifications (for example, approximated consideration of the inhomogeneity of the atmosphere, etc). The actual accuracy of the method is determined by the accuracy of the initial information on the parameters of the fine structure and the accuracy of defining these functions dependencies on the parameters of the physical state of the medium.

The main shortcoming of the direct method is the need to carry out a large volume (even on the scale of currently available computers) of calculations. In this case, the considerable computing time is required in the calculation of the monochromatic coefficients of absorption and numerical integration of the monochromatic transmittance in respect of the frequency. If it is taken into account that the operation of integration should be carried out with strict approach for every new model of the atmosphere, the path of propagation of radiation, the slit function of the device, etc, then it is understandable why in addition to the direct method many problems of atmospheric physics are also solved using other methods of determination of the transmittance functions.

The method of the models of absorption bands

The concept is based on the replacement of the actual spectral structure of the absorption bands by a specific analytical or statistical model of the mutual position of the absorption lines, distribution of their intensities, etc [20, 91, 102]. The introduction of these approximations makes it possible to integrate the monochromatic functions of transmittance in the analytical manner in respect of frequency in many cases. It is important that the analytical expressions for the transmittance functions in the modelling approach depend on a small number of parameters, sometimes 2–3. Because of the considerable saving of computer resources, the operating speed of this method is several times orders of magnitude higher than that of the direct method of calculations. Naturally, the parameterization of the absorptions spectra leads to additional errors in the resultant values of the transmittances functions. It is therefore necessary to ensure optimum selection of the model of absorption band for a specific spectral range or spectral interval and to analyse the accuracy of the approach.

When using the absorption models, attention is often given to the current data on the parameters of the fine structure of the specific investigated model for determination of the model parameters. In a number of cases, the expressions for the transmittance functions, obtained within a specific modelling approach, are used as approximations for the transmittance functions, obtained in experiments or in direct calculations. A large number of different models of absorption bands have been proposed: the model of isolated lines, the Elsasser model (regular model), various random models, quasi-random Plass model and many others. A detailed review and analysis of different models of the bands were published in, for example, monographs [20, 37, 91]. We present only the most frequently used models of absorption bands.

Model of the isolated line

The simplest model of the absorption band is the model of the isolated spectral line. If it assumed that the main factor of broadening the line is collisions and the appropriate line shape is of the Lorentz type, then we have

$$P_{\Delta\nu} = \frac{1}{\Delta\nu}\int_{\Delta\nu} \exp\left(-\frac{1}{\pi}\frac{S\alpha_L u}{(\nu-\nu_0)^2+\alpha_L^2}\right)d\nu. \tag{7.3.1}$$

As indicated by the name of the method, it is assumed that only one spectral line is found in the interval $\Delta\nu$.

Equation (7.3.1) is converted. We introduce new variables:

$$x = \frac{\nu-\nu_0}{\Delta\nu},\ y = \frac{\alpha_L}{\Delta\nu},\ z = \frac{Su}{2\pi\alpha_L}. \tag{7.3.2}$$

It is assumed that the examined line is isolated, i.e. does not overlap with any other lines so that we can extend the range of integration from $-\infty$ to $+\infty$. We introduce the absorption function

$$A_{\Delta\nu} = 1 - P_{\Delta\nu} = \frac{1}{\Delta\nu}\int_{\Delta\nu}\left(1-\exp\left(-\frac{1}{\pi}\frac{S\alpha_L u}{(\nu-\nu_0)^2+\alpha_L^2}\right)\right)d\nu.$$

Consequently for the absorption function in the interval $\Delta\nu$ we can write

$$A_{\Delta\nu} = \int_{-\infty}^{+\infty}\left(1-\exp\left(-\frac{2zy^2}{x^2+y^2}\right)\right)dx. \qquad (7.3.3)$$

The integral (7.3.3) for the absorption functions $A_{\Delta\nu}$ is expressed through the special functions – cylindrical functions of the first kind referred to also as Bessel functions:

$$A_{\Delta\nu} = 2\pi yze^{-z}(J_0(iz) - iJ_1(iz)) = 2\pi yL(z), \qquad (7.3.4)$$

where $J_0(iz)$ and $J_1(iz)$ are the zero and first Bessel function of the purely imaginary argument.

The function

$$L(z) = ze^{-z}(J_0(iz) - iJ_1(iz)) \qquad (7.3.5)$$

is often referred to as the Ladenburg–Reiche function. Examining the asymptotic behaviour of the function $L(z)$, we can obtain important approximations for the absorption in the isolated line. At low values of z, the term e^{-z} and $J_0(iz)$ in (7.3.5) tend to unity, and $J_1(iz)$ to zero. Consequently, $L(z) \sim z$ and we have

$$A_{\Delta\nu} = 2\pi yz = \frac{Su}{\Delta\nu}. \qquad (7.3.6)$$

The low values of z correspond to the low values of the product Su, i.e. the case of weak absorption. This case is also referred to as the region of linear absorption as a result of the linear dependence of $A_{\Delta\nu}$ on u. It should be mentioned that in this limiting case absorption does not depend on pressure. The law of linear absorption can also be determined directly from equation (7.3.1.). It is written in the following form

$$P_{\Delta\nu} = \frac{1}{\Delta\nu}\int_{\Delta\nu}\exp(-k(\nu)u)d\nu.$$

It is assumed that $k(\nu)u$ is very small even in the centre of the absorption line ($\nu = \nu_0$). Consequently, expanding the exponent into a series ($e^x = 1 + x + \ldots$) and retaining only the first two members of the expansion, we have

$$P_{\Delta\nu} = \frac{1}{\Delta\nu}\int_{\Delta\nu}(1-k)(\nu)u)d\nu = 1 - \frac{1}{\Delta\nu}\int_{\Delta\nu}k(\nu)u\,d\nu = 1 - \frac{u}{\Delta\nu}\int_{\Delta\nu}k(\nu)d\nu.$$

Passing as previously to the infinite limits of integration in respect of frequency and taking into account normalization (4.4.4) we get

$$A_{\Delta\nu} = \frac{Su}{\Delta\nu}$$

It should be mentioned that this conclusion shows that the law of linear absorption at low values of Su holds for any and not only for the Lorentz contour of the absorption line.

At high values of z we use the well-known asymptotic equation for the Bessel function

$$J_m(x) = \sqrt{\frac{2}{\pi x}}\sin\left(x - \frac{m\pi}{2} + \frac{\pi}{4}\right) \quad \text{at } x \to \infty.$$

Consequently

$$L(z) = ze^{-z}\left|J_0(iz) - iJ_1(iz)\right| =$$

$$= ze^{-z}\left|\left(\sqrt{\frac{2}{\pi iz}}\left(\sin(iz)\frac{\sqrt{2}}{2} + \cos(iz)\frac{\sqrt{2}}{2} - i\sin(iz)\frac{\sqrt{2}}{2} + i\cos(iz)\frac{\sqrt{2}}{2}\right)\right|\right. =$$

$$= \sqrt{\frac{2z}{\pi}}e^{-z}\frac{\sqrt{2}}{2}\left|\sqrt{-i}\left((1-i)\frac{e^{i(iz)} - e^{-i(iz)}}{2i} + (1+i)\frac{e^{i(iz)} - e^{-i(iz)}}{2}\right)\right|$$

Ignoring terms e^{-z} below the modulus, we obtain

$$L(z) = \sqrt{\frac{2z}{\pi}}e^{-z}e^{z}\left|\sqrt{-i}\frac{i-1+i-1}{2i}\right| =$$

$$= \sqrt{\frac{2z}{\pi}}\frac{\sqrt{2}}{2}\left|\sqrt{-i}(1+i)\right| = \sqrt{\frac{2z}{\pi}}\sqrt{\frac{2}{2}}\left|\left(\frac{\sqrt{2}}{2} - i\frac{\sqrt{2}}{2}\right)(1+i)\right| = \sqrt{\frac{2z}{\pi}}.$$

Consequently,

$$A_{\Delta\nu} = 2\pi y\sqrt{\frac{2z}{\pi}} = \frac{2\sqrt{Su\alpha_L}}{\Delta\nu}. \tag{7.3.7}$$

The equation (7.3.7) corresponds to the case of strong absorption or the square root law. The latter name is associated with the fact that the dependence of absorption on the amount of absorbing matter and pressure (it should be mentioned that α_L is proportional to p) is presented as $\sqrt{up}$. It is also important to mention that in the case of strong absorption in the centre of the line (when $P_{\nu 0} = 0$) the variation of absorption $A_{\Delta\nu}(u)$ takes place as a result of the variation of the extent of absorption only in the 'wings' of the isolated spectral line. This holds not only for the Lorentz but also Voigt line because the wings of the Voigt line are determined mainly by the Lorentz line shape.

The examined model of the isolated line is used for upper layers of the atmosphere where the spectral lines are very narrow and may be regarded as isolated.

Elsasser model

Analysis of the absorption spectra of different linear molecules shows the presence of a regular structure in the arrangement of the lines corresponding to the same vibration transition, and the intensities of these lines change only very slowly. Therefore, the regular model of the absorption band, or the Elsasser model, has been proposed, Fig.7.1. For a regular model consisting of an infinite sequence of identical spectral Lorentz lines, situated at distance $\Delta\nu$ from each other, the absorption coefficient can be written in the form:

$$k(\nu) = \sum_{n=-\infty}^{n=+\infty} \frac{1}{\pi} \frac{S\alpha_L}{(\nu - n\Delta\nu)^2 + \alpha_L^2}. \tag{7.3.8}$$

Passing in (7.3.8) to the absorption function using the variable x, y, z introduced for the model of the isolated line, we obtain

$$A_{\Delta\nu} = \int_{-1/2}^{+1/2} \left[1 - \exp\left(-\sum_{n=-\infty}^{n=+\infty} \frac{2zy^2}{(x-n)^2 + y^2}\right)\right] dx,$$

where the integration limits correspond to the range $[\nu_0 - \Delta\nu/2, \nu_0 + \Delta\nu/2]$. For the sum in the exponent we have the analytical

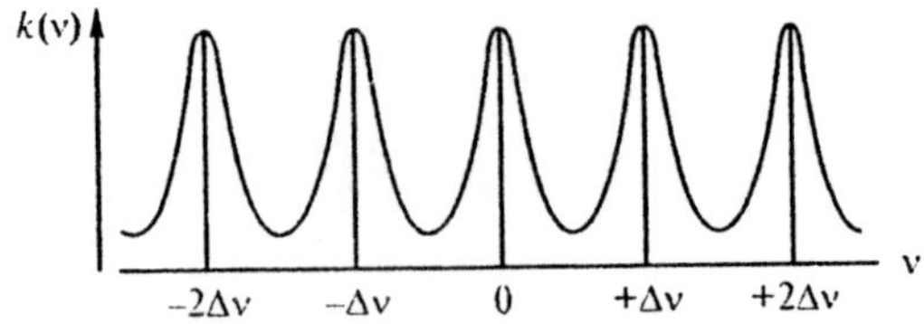

Fig.7.1. Elsasser model [20].

expression (obtained from the theory of Loran series). Consequently we obtain

$$A_{\Delta\nu} = 1 - E(y, z),$$

where $E(y,z)$ is the Elsasser function

$$E(y,z) = \int_{-1/2}^{+1/2} \exp\left(-\frac{2\pi yz \operatorname{sh} 2\pi y}{\operatorname{ch} 2\pi y - \cos 2\pi x} \right) dx. \qquad (7.3.9)$$

It should be mentioned that the hyperbolic sinuses and cosines are determined as follows

$$\operatorname{sh} x = \frac{1}{2}(e^{x} - e^{-x}), \ \operatorname{ch} x = \frac{1}{2}(e^{x} + e^{-x}). \qquad (7.3.10)$$

Let it be that $y \to \infty$, and, therefore, according to (7.3.10), $\frac{\operatorname{sh} 2\pi y}{\operatorname{ch} 2\pi y} \to 1$

and from (7.3.9) we obtain

$$A_{\Delta\nu} = 1 - \exp(-2\pi yz) = 1 - \exp\left(-\frac{Su}{\Delta\nu} \right). \qquad (7.3.11)$$

The examined case corresponds to the situation when $\alpha_L >> \Delta\nu$, i.e. the lines are very close to each other and greatly overlap. Here, as indicated by (7.3.11), the transmittance function depends exponentially on u, as in the case of monochromatic radiation, and there is no spectral dependence of transmittance. It should also be mentioned that the transmittance function does not depend on pressure. The examined situation is actually observed when the half width of the lines is considerably greater than the distance between them, for example, at high pressures (it should be mentioned that α_L is proportional to p). These special features of absorption in this case are characteristic of all models of the absorption bands. In particular, they are found (with different degrees of approximation) in electronic absorption bands, and also in infrared spectra of heavy molecules consisting of very closely distributed absorption lines.

If parameter y is small ($y \to 0$) then, as shown directly from (7.3.9), (7.3.10), the Elsasser function changes to the previously examined model of the isolated line. The condition $y \ll 1$ denotes that the distance between the lines is considerably greater than their width. This case corresponds to low pressures. However, it should be remembered that at low pressures the contribution of Doppler broadening may become considerable in comparison with Lorentz broadening and the relationship (7.3.8) for the Voigt line shape will no longer be valid.

The approximation of strong lines is realized when complete absorption is found in the centre of the Lorentz lines. In this case, the transmittance function changes as a result of the change of the absorption in the wings of the lines and in expression (7.3.8) we can ignore the value y^2 in the denominator. Consequently, for the infinite sum the analytical expression is known and after a number of mathematical transformations using the special functions, the Elsasser function is expressed by the well-known probability integral

$$E(y,z) = 1 - \mathrm{erf}(\pi y(\sqrt{2z}),\ A_{\Delta\nu} = \mathrm{erf}(\pi y\sqrt{2z}), \qquad (7.3.12)$$

where

$$\mathrm{erf}(x) = \frac{1}{\sqrt{\pi}}\int_0^x \exp(-\xi^2)d\xi.$$

Substituting into (7.3.12) the parameters y and z from (7.3.2), we have

$$A_{\Delta\nu} = \mathrm{erf}\left(\frac{\sqrt{\pi}}{\Delta\nu}\sqrt{Su\alpha_L}\right). \qquad (7.3.13)$$

It may be seen that in the approximation of the strong line, the absorption function depends on the product $Su\alpha_L$. Since the Lorentz half width is proportional to pressure, relationship (7.3.13) shows that absorption dependence on the product Sup. Taking into account that in (7.3.13) the product is under the root, by analogy with (7.3.7) this circumstance is also sometimes referred to as the 'law of the square root'.

Random models

Investigations of the absorption spectra of different atmospheric gases (H_2O, O_3 etc) showed the 'random' distribution of individual absorption lines. Therefore, various models of the absorption bands with specific random properties were developed. The best known and widely used is the Goody random model proposed by the American scientist R.M. Goody in 1952 [20]. It should be mentioned that this model was previously developed by Mayer, a participant of the well-known American program for the development of atomic weapons at Alamos, in the solution of the problem of propagation of radiation in the atmosphere after a nuclear explosion. In fact, the work of Mayer became known considerably later. Therefore, the examined model should be referred to as the Mayer–Goody statistical model.

Let it be that $\Delta\nu$ is a spectral integral consisting of n lines situated at mean distance δ from each other, i.e. $\Delta\nu = n\delta$. Let it be that $p(S_i)$ is the probability density of the i-th line having the intensity S. Therefore

$$\int_0^\infty p(S_1)dS = 1 \tag{7.3.14}$$

(the probability of S being in the range from 0 to ∞ is equal to one). It is assumed that any line may be found with equal probability in any area within the limits of the range $\Delta\nu$ and have any intensity. Therefore, according to the probability theory, the transmittance function in this spectral interval $\Delta\nu$ is found by averaging in respect of all possible intensities and positions of the lines:

$$P_{\Delta\nu} = \frac{1}{(\Delta\nu)^n} \int_{\Delta\nu} d\nu_1 \ldots \int_{\Delta\nu} d\nu_n \int_0^\infty p(S_1) e^{-k_1 u} dS_1 \ldots \tag{7.3.15}$$

$$\ldots \int_0^\infty p(S_n) e^{-k_n u} dS_n,$$

where k_n is the coefficient of absorption in the n-th line. According to the previously made assumption on the equal probability of the position and intensity of the lines all integrals in (7.3.15) are identical. Consequently, using the relationship (7.3.14) we can write

$$P_{\Delta\nu}' = \left(\frac{1}{\Delta\nu}\int_{\Delta\nu} d\nu \int_0^\infty p(S)e^{-ku}dS)\right)^n = \left(1 - \frac{1}{\Delta\nu}\int_{\Delta\nu} d\nu \int_0^\infty p(S)(1-e^{ku})dS\right)^n \cdot (7.3.16)$$

Taking into account that $\Delta\nu = n\delta$, it may easily be shown that the expression (7.3.16) at high n tends to an exponential function. In fact, according to one of the definitions of the exponent $\lim_{x\to\infty}(1+x/n)^n = e^x$, and consequently $(1-x/n)^n \to e^{-x}$. However, formally at $n\to\infty$, the range $\Delta\nu\to\infty$ and therefore, the integration range in respect of frequency may be set from $-\infty$ to $+\infty$. Taking this into account, we obtain

$$P_{\Delta\nu} = \exp\left[-\frac{1}{\delta}\int_{-\infty}^{+\infty}\int_0^\infty p(S)(1-\exp[-k(S,\nu)u])dS\,d\nu\right]. \quad (7.3.17)$$

Different variants of the statistical model can be determined using different expressions for the probability density $p(S)$.

The Goody model uses the simple exponential distribution

$$p(S) = \frac{1}{S_0}\exp(-S/S_0), \quad (7.3.18)$$

where S_0 is the mean intensity of the line. Substituting (7.3.18) into (7.3.17), taking into account $k(S,\nu) = Sf(\nu-\nu_0)$, and after integration in respect of S, we obtain

$$P_{\Delta\nu} = \exp\left[-\frac{1}{\delta}\int_{-\infty}^{+\infty}\int_0^\infty \frac{1}{S_0}\exp[-S/S_0](1-\exp[-Suf(\nu-\nu_0)])dS\,d\nu\right] =$$

$$= \exp\left[-\frac{1}{\delta}\int_{-\infty}^{+\infty}\frac{S_0uf(\nu-\nu_0)}{1+S_0uf(\nu-\nu_0)}d\nu\right].$$

Further, using the Lorentz line shape $f(\nu-\nu_0)$, we finally obtain

$$P_{\Delta\nu} = \exp\left[-\frac{S_0u}{\delta}\left(1+\frac{S_0u}{\pi\alpha_L}\right)^{-1/2}\right]. \quad (7.3.19)$$

If it is assumed that the Malkmus distribution

$p(S)=\frac{1}{S}\exp(-S/S_0)$, is fulfilled for the intensities of the lines, the expression for the transmittance function has the form (Malkmus statistical model) [20, 91]:

$$P_{\Delta\nu}=\exp\left(-\frac{\pi\alpha_L}{2\delta}\left[\left(1-\frac{4S_0u}{\pi\alpha_L}\right)^{1/2}-1\right]\right). \qquad (7.3.20)$$

The resultant transmittance functions are functions of two parameters: S_0/δ and $\pi\alpha_L/\delta$. For the given absorption band or spectral interval these two parameters can be determined by approximating the results of laboratory measurements of the transmittance functions in relation to the content of the absorbing gas and pressure. In addition to this they can also be obtained by approximating the parameters of the fine structure of the spectral lines of specific absorption bands. In the case of weak absorption, when $S_0\ u\ll1$, from (7.3.19) we again have

$$P_{\Delta\nu}=1-\frac{S_0u}{\delta},\ A_{\Delta\nu}=\frac{S_0u}{\delta},$$

i.e. the linear dependence of absorption on the gas content u. In the strong absorption regime $\frac{S_0u}{\pi\alpha_L}<<1$ equation (7.3.19) changes to

$$P_{\Delta\nu}=\exp\left[-\frac{\sqrt{\pi S_0u\alpha_L}}{\delta}\right],$$

i.e., as previously the 'square root law' is fulfilled. Comparing these relationships with the previously described identical limiting cases for other models, it may be concluded that the laws of linear absorption (weak absorption) and the law of the square root (strong absorption) are universal. Since calculations are simple in this case, because of the relatively high accuracy of the statistical model, the possibility of 'calibrating' the model on the basis of laboratory data or preliminary exact direct calculation of the transmittance functions and, finally, the possibility of determining the parameters of functions on the basis of the parameters of the fine structure, the random model is used widely in different calculations in atmospheric optics.

Although any model of the absorption band describes approximately the real spectral structure of absorption of the atmospheric gas, and the modelling approach is used widely in atmospheric optics, especially in infrared range of thge spectrum, due to its high computing efficiency. This is associated with the fact that regardless of the given simplifications, the models make it possible to describe the main properties of the absorption spectrum and the relationships of molecular absorption by the atmospheric gases. Certainly, there are certain principal restrictions in using the models of the absorption bands. For example, the accuracy is often greatly reduced for relatively narrow spectral ranges. In addition to this, the majority of models have been developed for the conditions of absorption of radiation in a homogeneous medium.

Method of integration of transmittance functions in respect of the absorption coefficient (*k*-method)

The direct method of calculating the transmittance functions in the case of selected gas absorption when the spectral lines of absorption, especially at low pressures are very narrow and the calculations of the radiation characteristics atmosphere should be carried out for wide spectral ranges, is extremely cumbersome even when using advanced computers. There is a completely different approach to determining the transmittance functions, proposed as early as in the thirties of the previous century [20, 32, 102]. The method is based on replacement of integration in respect of frequency by integration in respect of the absorption coefficient. In the last couple of decades this method (*k*-method) was greatly improved and has been used quite widely in different calculations of the radiation characteristics of the atmosphere [20, 37, 91, 103].

In integration by the direct method of the expression for the transmittance function in respect of frequency

$$P_{\Delta\nu}(u) = \frac{1}{\Delta\nu}\int_{\Delta\nu} \exp[-k(\nu)u]\,d\nu \tag{7.3.21}$$

it is necessary to calculate the monochromatic transmittance function for a large number of points. This case, at many points situated in different sub-intervals of the spectral range $\Delta\nu$, similar values of the absorption coefficient are obtained. To illustrate the *k*-method, we examine Fig. 7.2. This figure shows that the values of the absorption coefficient in an arbitrary range $[k_i+\Delta k]$ are found in the entire series of spectral ranges, d_{i1}, d_{i2}, etc. Examining the entire range of the absorption coefficient from k_{max} to k_{min}, realized in the examined spectral range $\Delta\nu$ (at low pressures and temperature), we can construct (using the limiting transition) the density of distribution ('probability density') of the values of the coefficient of absorption in the examined spectral range:

$$f(k)dk = \lim_{\Delta k_i \to 0} \frac{1}{\Delta\nu} \sum_{i=1}^{N} \left(\frac{\Delta\nu_j}{\Delta k_i} \right) w(k_i, k_i + \Delta k_i), \qquad (7.3.22)$$

where the function $w(k_i, k_i + \Delta k_i)$ is equal to zero everywhere with the exception of the range $(k_i, k_i + \Delta k_i)$, where it is equal to unity. Actually, the function $f(k)dk$ shows the relative part of the spectral range occupied by the sub-intervals Δk_i with the absorption coefficient $(k_i, k_i + \Delta k_i)$. The next step in the examined k-method is the transition from integration in the transmittance function in respect of frequency to integration in respect of the absorption coefficient. This can be carried out on the basis of a sequence of relationships:

$$P_{\Delta\nu}(u) = \frac{1}{\Delta\nu} \int \exp(-k_\nu u) d\nu = \frac{1}{\Delta\nu} \lim_{\Delta\nu_j \to 0} \sum_{j=1}^{N} \exp(-k_\nu u) \Delta\nu_j =$$

$$= \frac{1}{\Delta\nu} \lim_{\Delta k_i \to 0} \sum_{j=1}^{N'} \left(\frac{\Delta\nu_j}{\Delta k_i} \right) w(k_i, k_i + \Delta k_i) \exp(-k_i u) \Delta k_i = \qquad (7.3.23)$$

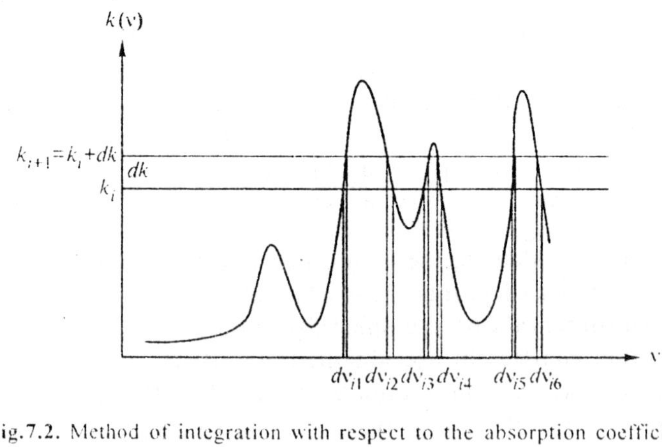

Fig.7.2. Method of integration with respect to the absorption coefficient.

$$= \int_0^\infty f(k)\exp[-ku]dk,$$

where we have used the transition $k_{min} \to 0$ and $k_{max} \to \infty$. The definition of the probability density $f(k)$ shows that

$$\int_0^\infty f(k)dk = 1. \quad (7.3.24)$$

The probability distribution of the absorption coefficient in the examined spectral range can also be characterised by the distribution function (sometimes referred to as the integral distribution function)

$$g(k) = \int_0^k f(k)dk. \quad (7.3.25)$$

In this case it holds that: $g(0)=0$, $g(k\to\infty)\to 1$, $dg(k)=f(k)dk$. The transmittance function can now be written in a different form:

$$P_{\Delta\nu}(u) = \int_0^1 \exp[-k(g)u]dg. \quad (7.3.26)$$

It should be stressed that the expression (7.3.21), (7.3.23) and (7.3.26) for the transmittance functions are equivalent. However, if in equation (7.3.21) integration is carried out in respect of frequency, then in the expression (7.3,23) it is carried out in respect of the absorption coefficient and in the expression (7.3.26) in respect of the integral distribution function g.

We examine Fig.7.3a, which shows the behaviour of the absorption coefficient in the spectral range 980–1100 cm^{-1} in the absorption band of ozone at 9.6 μm (pressure 30 mbar, gas temperature T = 220 K) on the logarithmic scale [103]. It may be seen that the values of the absorption coefficient in this range (for the given pressure and temperature) are in the range from 10^{-3} to 10^2 cm^{-1}atm^{-1}. In this case, the changes of the absorption coefficient with frequency are very 'fast'. They will be followed by similarly fast changes of the monochromatic absorption functions in the equation (7.3.21). Figures 7.3b and 7.3c show the functions $f(k)$ and $g(k)$ for the examined spectral range. Comparison of the functions $k(\nu)$, $f(k)$ and $g(k)$ shows clearly the advantages of writing the transmittance function in the form of (7.3.23) or (7.3.26). If the function $k(\nu)$ changes very rapidly and, therefore, in integration of the monochromatic transmittance we are forced to use many points in respect of frequency, the function $f(k)$ and in particular $g(k)$ is far smoother. It may formally be concluded that the transition to integration in respect of the absorption coefficient or the distribution function g leads to the integration of the only equivalent absorption 'line' $k(g)$ (see Fig.7.3d).

When using the equations (7.3.23) and (7.3.26), since functions $f(k)$ and $g(k)$ are smoother (especially $g(k)$), in numerical calculations of the transmittance functions we can confine ourselves to quadratures with a small number of nodes (or sub-intervals of the absorption coefficients). This is also carried out in the examined k-method:

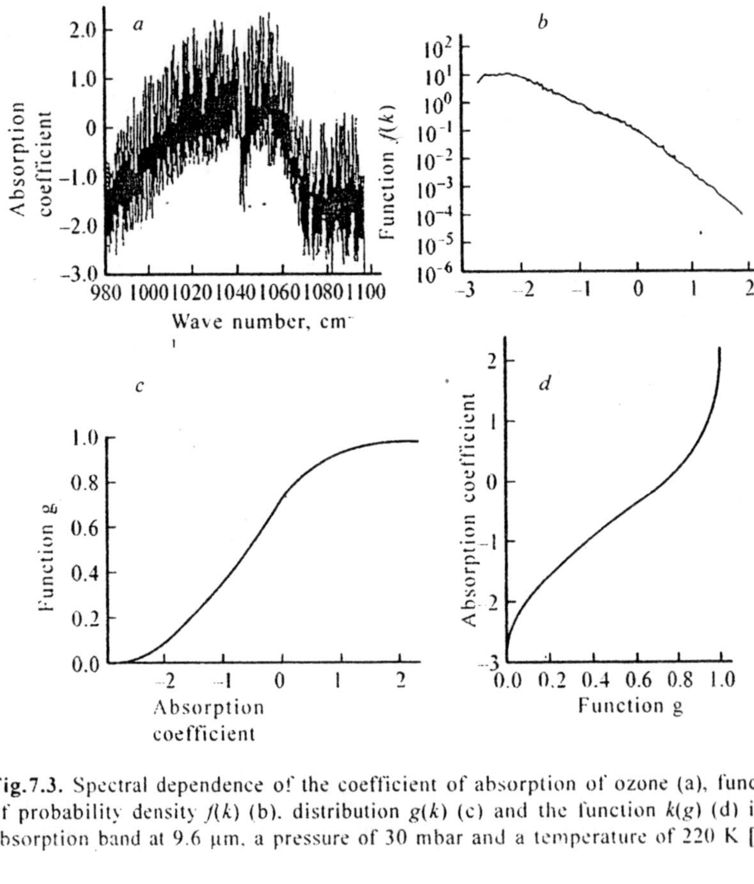

Fig.7.3. Spectral dependence of the coefficient of absorption of ozone (a), functions of probability density $f(k)$ (b), distribution $g(k)$ (c) and the function $k(g)$ (d) in the absorption band at 9.6 μm, a pressure of 30 mbar and a temperature of 220 K [103].

$$P_{\Delta\nu}(u) \cong \sum_{k=1}^{L} f(k_k) e^{-k_k u} \Delta k_k \approx \sum_{k=1}^{L} e^{-k_k(g)u} \Delta g_k. \qquad (7.3.27)$$

In this case, the number of terms in equation (7.3.27) for calculating the transmittance functions is relatively small (~5–10) in comparison with hundreds and thousands of points in respect of frequency when using the equation for the transmittance function (7.3.21). The method of calculating the transmittance functions using equation (7.3.27) is often referred to as the approximation of the exponential expansion of the transmittance functions. To use the method, it is necessary to know the values of k_k and weight g_k (quadrature coefficients) for specific spectral ranges (and physical conditions). There are a number of approaches to solving this problem [91]:

1. Detailed calculations of $k(\nu)$ can be carried out for the examined spectral range and sampling functions $f(k)$ and $g(k)$ can be constructed using the 'grading' of the absorption coefficient. Consequently, the appropriate k_k and g_k can be obtained dividing in some manner the entire range $g(k)$ into sub-intervals. An example of such a region into 10 sub-intervals is shown in Fig.7.4. The figure shows an example corresponding to the random Malkmus model for three values of parameter $y = \pi\alpha_L/\delta$ (i.e. at three pressures). Attention should be given to the fact that the selected ten sub-intervals in respect of the variable g are identical for all three pressures.

2. We can calculate the transmittance function $P_{\Delta\nu}(u)$ by, for example, the direct method for a set of values of the absorbing mass u_i, and approximate $P_{\Delta\nu}(u_i)$ by equation (7.3.26). Using some numerical method, we solve the problem of minimisation of the function

$$R(k_1, k_2, \ldots k_L, g_1, g_2, \ldots g_L) = \sum_i [P(u_i) - g_k \exp(-k_k u_i)]^2 \to \min, \qquad (7.3.28)$$

i.e. minimisation of the mean square errors of approximation of the transmittance functions, calculated by the direct method for different values of the absorbing masses (in the general case also pressure) using equation (7.3.26).

3. Equation (7.3.23) can be regarded as the Laplace transformation of the function of $f(k)$:

$$P(u) = \mathbf{L}[f(k)], \qquad (7.3.29)$$

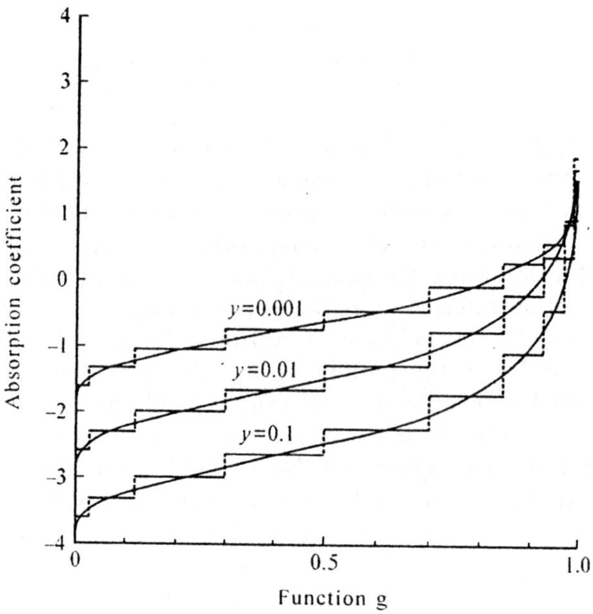

Fig.7.4. Example of dividing function $g(k)$ into sub-intervals [91] at different values x of parameter g.

where **L** is the Laplace operator. Since function $P(u)$ is continuous, there should be an inverse Laplace transform, i.e.

$$f(k)=\mathbf{L}^{-1}[P(u)], \tag{7.3.30}$$

using which we obtain $f(k)$ for a number of analytical representations of the transmittance functions, determined using the method of the models of absorption bands.

Analytical expressions for $f(k)$ are also available for a number of other models – Elsasser, Schnaydt models, etc. In the absence of an analytical representation for the transmittance functions or their Laplace transformations, we can use the current numerical methods of computing the inverse Laplace transforms.

There are the following advantages and shortcomings of the method of integration of the transmittance functions in respect of the absorption coefficient [91]:

1. The k-method makes it possible to take rigorously into account the selectivity of radiation (absorption) in scattering problems). The models of the absorption bands take into account quite strictly the selectivity of absorption in the calculations of

thermal emission (because of the weak spectral dependence of the Planck function) in finite spectral intervals. However, they are not suitable for scattering problems because the function of the source in these problems depends strongly on the wavelength. This special feature of the *k*-method is also one of the reasons for the relatively extensive use of the method in recent years.

2. The *k*-method is efficient for relatively wide spectral intervals, whereas the modelling approach has certain restrictions in this sense or requirements on the homogeneity of the statistical distribution (structure) of the absorption bands. Thus, the entire CO_2 band at 15 μm is 'statistically inhomogeneous' – it includes sets of lines in *P* and *R* branches with a specific structure and the *Q* branches with a completely different structure.

3. An important advantage of the *k*-method is the high accuracy of approximation of the transmittance functions using a series with a relatively small number of exponential terms.

4. On the other hand, it should be mentioned that the modelling approach makes it possible to describe explicitly the dependence of the transmittance functions on pressure and temperature. In the majority of cases the *k*-method evens out these dependences. This causes that, generally speaking, the functions $f(k)$ and $g(k)$ should be calculated for every pair of p and T, i.e. for different layers of the atmosphere. However, in the case in which in the modelling approach there is the inverse Laplace transform for analytical expressions for the transmittance function, the dependence of $f(k)$ and $g(k)$ on p and T in the explicit form is maintained.

5. The *k*-method and the models of the absorption bands have been developed for homogeneous media. The calculation of the different radiation characteristics is associated in the majority of cases with the propagation of radiation in a inhomogeneous medium. Both methods use in this case the approximate reduction of inhomogeneous media to homogeneous ones. If for the transmittance function in the conventional form (integration in respect of frequency) the transition from inhomogeneous to homogeneous paths may be carried out using physically reasonable approximations, then in the *k*-method this transition is difficult. This problem will be examined more particularly in paragraph 7.4.

Experimental methods

The previously examined methods of determining the transmittance function should be classified as calculation methods. Another group of methods is based on the application of experimental data. These are mainly laboratory methods in which attention is given to the absorption of radiation in special optical cells in strictly controlled physical conditions of the state of the gas mixture. The problem of detailed examination of the transmittance functions in this approach is very complicated, expensive and cumbersome. Infact, in the ideal situation we should examine the dependence of the absorption of radiation as a function of the following variables:

- the amount and type of absorbing gas;
- spectral interval $\Delta\nu$;
- pressures of absorbing and broadening gas;
- temperature

These investigations should be carried out for all atmospheric gases and all spectral intervals $\Delta\nu$ in the relevant (possibly very wide) range of the spectrum.

Regardless of the complicated nature of the experiments and difficulties in the detailed examination of the transmittance functions, the laboratory method was extensively used until recently. At present, it is used mainly to obtain the parameters of the fine structure of the individual spectral lines – their positions, intensities, half width, etc. The results of laboratory investigations of the transmittance functions were often approximated using some analytical expressions as functions of the amount of the absorbing gas, pressure and temperature. For example, the following representation was widely used:

$$P_{\Delta\nu} = \exp(-\beta_{\nu} u^{m} p^{n} T^{l}), \tag{7.3.32}$$

where β_{ν} is the generalized absorption coefficient; m, n and l are the empirical parameters approximately taking into account the dependence of the transmittance function on the content of the absorbing gas, pressure and temperature. In some cases these approximating equations were selected on the basis of the results of application of the modelling approach. In this case, it should be remembered that the range of applicability of approximation (7.3.32) is restricted by at least the ranges of measurement of u, p and T in the laboratory investigations. The use of equations (7.3.32) for other physical conditions may result in large errors in calculations of the transmittance functions. In addition to this, it should be remembered that the experimental transmittance functions are determined using specific spectral devices and, consequently, these transmittance functions correspond to specific slit functions of the device (see, for example, equation (7.2.6)).

Field investigations of the transmittance functions are conducted in a real atmosphere on different paths (horizontal and slant) using artificial sources of radiation or solar radiation. The main advantage of this approach is that the radiation transfer is examined in a real atmosphere (not in artificial mixtures), when using the Sun as a source of radiation for an inhomogeneous atmosphere. Shortcomings are caused by the fact that the control of the state of the atmosphere is a complicated task and requires considerable additional expenditure. Field measurements of the transmittance function are often used in analysis of the accuracy of various methods of calculating the transmittance functions. In addition to this, they are useful for analysis of the contributions of different extinction mechanisms to total attenuation in the real atmosphere, for example continual molecular or aerosol extinction.

7.4. Approximate methods of radiation transfer theory

Approximate methods of taking into account the inhomogeneity of the atmosphere

The majority of methods of determining the transmittance functions provide information on the characteristics of absorption of homogeneous media, i.e. the media with constant pressure and temperature. In a real atmosphere, radiation transfer takes place in the majority of cases along paths on which the pressure, temperature and content of the absorbing gas are not constant.

Pressure and temperature influence the half-width and intensities of the spectral lines and, consequently, the absorption coefficients. Thus, the transfer of radiation in a real atmosphere takes place in a medium with changing line shapes and values of the coefficient of absorption of the spectral lines. This special feature of transfer is reflected in equation (7.2.4) by the integral in respect of the spatial variable in the index of the exponent. As already mentioned, the spectral integration of monochromatic transmittance functions is the final operation in determining the transmittance functions in finite spectral intervals $\Delta\nu$. Consequently, the integration, strictly speaking, should be carried out for every new model of the atmosphere.

Naturally, it is desirable, on the one hand, to increase the efficiency of calculations of different radiation characteristics and, on the other hand, use the transmittance functions of homogeneous media which result from laboratory measurements and models of the absorption bands. Therefore, special approximate methods of taking into account the inhomogeneity of the atmosphere have been developed [20, 32, 34, 37, 91, 102]. The aim of these procedures is to show the rule which can be used in replacing an inhomogeneous atmospheric layer, i.e a layer with variable p and T, by a homogeneous layer at constant pressure and temperature. Formally, this means that constant pressure $\tilde{p}$, temperature $\tilde{T}$ and gas content $\tilde{u}$ can be determined from the obvious condition which for a single gas can be written in the form

$$\int_{\Delta\nu} \exp\left(-\int_{l_1}^{l_2} k(\nu, p(l), T(l))\rho(l)dl \right) d\nu = \int_{\Delta\nu} \exp(-k(\nu, \tilde{p}, \tilde{T})\tilde{u})d\nu, \quad (7.4.1)$$

where $k(\nu, p(l), T(l))$ is the mass absorption coefficient.

The most widely used approximate methods of taking the inhomogeneity of the atmosphere into account will be discussed.

1. The effective mass method

We examine the left-hand part of equality (7.4.1) for the Lorentz line shape of the absorption line.

$$P_{\Delta\nu}(l_1, l_2) = \quad (7.4.2.)$$
$$= \frac{1}{\Delta\nu} \int_{\Delta\nu} \exp\left(-\int_{l_1}^{l_2} \frac{S(T(l))}{\pi} \frac{\alpha_L(p(l), T(l))\rho(l)}{(\nu - \nu_0)^2 + \alpha_L^2(p(l), T(l))} dl \right) d\nu.$$

In the case of strong absorption when the transfer process takes place only in the wings of the line, we can set $|\nu - \nu_0| >> \alpha_L(p(l), T(l))$ and ignore the term α_L^2 in the denominator. Further, we use the dependence of the Lorentz half width on pressure and temperature (4.5.24):

$$\alpha_L(p, T) = \alpha_L(p_0, T_0)\left(\frac{p}{p_0}\right)\left(\frac{T_0}{T}\right)^m. \quad (7.4.3)$$

Taking into account (7.4.3), the transmittance function of the inhomogeneous medium (7.4.2) is written in the form

$$P_{\Delta\nu}(l_1, l_2) = \frac{1}{\Delta\nu} \int_{\Delta\nu} \exp\left(-\frac{\alpha_L(p_0, T_0)}{\pi(\nu - \nu_0)^2} \times \right.$$

$$\times \int_{l_1}^{l_2} S(T(l)) \left(\frac{T_0}{T(l)} \right)^m \left(\frac{p(l)}{p_0} \right) \rho(l) dl \Bigg) d\nu. \tag{7.4.4}$$

Expression (7.4.4) can be regarded as the transmittance function of the homogeneous medium at pressure p_0 and temperature T_0 with the effective (reduced) amount of the absorbing gas $\tilde{u}$:

$$\tilde{u} S(\tilde{T}) = \int_{l_1}^{l_2} S(T(l)) \left(\frac{T_0}{T(l)} \right)^m \left(\frac{p(l)}{p_0} \right) \rho(l) dl. \tag{7.4.5}$$

This becomes evident if we expand the right hand part of the equality (7.4.1) by analogy with (7.4.4).

Equation (7.4.5) or, more accurately, its generalized form

$$\tilde{u} = \int_{l_1}^{l_2} \left(\frac{T_0}{T(l)} \right)^m \left(\frac{p(l)}{p_0} \right)^n \rho(l) dl. \tag{7.4.6}$$

corresponds to the approximate method of taking into account the inhomogeneity of the atmosphere which in the literature is referred to as the method of effective (reduced) absorbing mass. The transition from the equation (7.4.5) derived here to equation (7.4.6) is determined by the fact that it is convenient to approximate the temperature dependences of the intensity of the line and of its half width by a single procedure in the form of the weight multiplier $(T_0/T)^{m_1}$. Here it should be mentioned that exponent m_1 is not equal to exponent m introduced previously for approximating the dependence $a_L(T)$ in equation (7.4.3)). Finally, the appearance of the exponent n in the equation (7.4.6) for describing the dependence of absorption on pressure is associated with the fact that expression (7.4.4) was derived for strong absorption. For another limiting case – weak absorption – there is no dependence on pressure (see previous paragraph) and consequently $n = 0$. For some 'compensation' of this contradiction it has been proposed to use the expression for determining the effective absorbing mass in the form (7.4.6). Parameters m_1 and n are often found for specific spectral intervals $\Delta\nu$ from the analysis of the laboratory measurements of the dependences $P_{\Delta\nu}(u, p, T)$ or direct calculations of the transmittance functions. In a number of applications where the high accuracy of calculating the radiation characteristics of the atmosphere is not required we can ignore the temperature dependence of the transmittance function, i.e. in the equation (7.4.6) we can set $m_1 = 0$.

Because of its simple nature, the effective mass method is used widely in solving various problems of atmospheric physics. However, the accuracy of this approximate method it often low and the errors of calculation of the transmittance functions of the inhomogeneous atmosphere may reach 10–20%. This is unacceptable, for example, when solving inverse problems of atmospheric optics.

2. Curtis–Godson method. The reduced mass method is sometimes referred to as the one-parameter method because only one parameter $\tilde{u}$ should be calculated when using this method. The relatively low accuracy of this method has stimulated the development of other, multiparametric methods of taking into account the inhomogeneity of the atmosphere. The best known and most widely used method in this group is the Curtis–Godson two-parameter method. In this method, the inhomogeneous atmospheric layer is replaced with a homogeneous layer with reduced pressure $\tilde{p}$ and the content of the absorbing substance $\tilde{u}$:

$$\tilde{p} = \frac{\int_{l_1}^{l_2} p(l)\rho(l) f(T(l))dl}{\int_{l_1}^{l_2} \rho(l) f(T)(l))dl}, \quad \tilde{u} = \int_{l_1}^{l_2} \psi(T(l))\rho(l)dl.$$

In the simplest variant of the method the temperature dependence of the transmittance function is not taken into account and $f(T) = \psi(T) = 1$. Theoretical substantiation of the method has shown that: 1. The method is suitable for limiting cases of strong and weak absorption, 2. The temperature dependences of the transmittance function should be taken into account as follows:

$$f(T) = \frac{\sum_i S_i(T)}{\sum_i S_i(T_0)}, \quad \psi(T) = \left(\frac{\sum_i \sqrt{S_i(T)\alpha_{L,i}(T)}}{\sum_i \sqrt{S_i(T_0)\alpha_{L,i}(T_0)}} \right)^2.$$

The Curtis–Godson method is far more accurate than the method of reduced mass although it requires calculation of two parameters $\tilde{u}$ and $\tilde{p}$ and availability of the transmittance function as a function of the two variables $P_{\Delta\nu}(\tilde{u}, \tilde{p})$.

3. In the k-method, the inhomogeneity of the atmosphere is taken into account using the following procedure. As shown previously, the k-method of determining the transmittance functions was developed for inhomogeneous media. Therefore, it is necessary to examine the problem of using these transmittance functions for inhomogeneous media. In particular, it is quite evident that in this case we can use the method of the effective (equivalent) mass. Actually, using the relationships (7.4.4.), the inhomogeneous layer of the atmosphere is reduced to a homogeneous one for which it is necessary to find the appropriate coefficients of absorption k_k and quadrature coefficients g_k in expansion (7.3.26). The only difference is that when using this method for a inhomogeneous atmosphere we are concerned not with the actual content of the absorbing gas but with its reduced value. It may be shown that the k-method is suitable for inhomogeneous media also in cases of weak absorption, the models of the absorption bands proposed by Elsasser and Schnaydt. A special approach referred to in the literature as the correlated k-method (kk-method), was proposed for a more general case. We shall illustrate this method on the basis of the conclusion published by Liou [103].

According to the definition, the optical thickness for the homogeneous medium is given by the equation

$$\tau_\nu = \int_0^{u_0} k(\nu)du = \sum_i k_i(u)\Delta_i, \qquad (7.4.7)$$

where summation is carried out in respect of i homogeneous layers; u_0 is the total content of the absorbing gas in the examined layer of the atmosphere. The transmittance function of the inhomogeneous medium is presented in the form

$$P_{\Delta\nu}(u) = \frac{1}{\Delta\nu}\int_{\Delta\nu} \exp(-\tau_\nu)d\nu = \int_0^{\infty} \exp(-ku)f(k)dk, \qquad (7.4.8)$$

Where $k = \frac{\tau_\nu}{u} = \sum_i k_i a_i = \sum_i \frac{k_i \Delta u_i}{u}$, $a_i = \frac{\Delta u_i}{u}$; $f(k)$ is the distribution density of the absorption coefficient k which depends in the inhomogeneous medium on the distribution of pressure and temperature along the path of propagation of radiation.

The integral distribution function for the absorption coefficient is determined by the equation

$$g(k) = \int_0^k f(k)dk. \qquad (7.4.9)$$

For the given pressure and temperature the integral distribution function is:

$$g_i(k_i) = \int_0^{k_i} f(k_i)dk_i. \qquad (7.4.10)$$

and the transmittance function may be presented in the form

$$P_{\Delta\nu}(u) = \int_0^1 \exp\left(-u\sum_i k_i(g_i)a_i\right)dg_i. \qquad (7.4.11)$$

In equation (7.4.11) the integral distribution function g_i, determined by equation (7.4.10) depends on pressure and temperature along the inhomogeneous path of propagation of radiation. The main assumption of the *kk*-method is that the integral function is independent of p and T, i.e.

$$P_{\Delta\nu}(u) = \int_0^1 \exp\left(-u\sum_i k_i(g)a_i\right)dg. \qquad (7.4.12)$$

Thus, the *kk*-method requires knowledge of the dependence of the absorption coefficient on function g for different layers of the atmosphere (different pressures and temperatures, realized along the path of propagation of radiation). Finally, dividing integration range in respect of g from 0 to 1 into identical sub-intervals Δg_j (Fig.7.4) we can write

$$P_{\Delta\nu}(u) = \sum_j \exp\left(-u\sum_i k_{ij}(g)\Delta u_i\right)\Delta g_j. \qquad (7.4.13)$$

The large number of verifications of the accuracy of the *kk*-method on the basis of comparison with the results of direct reference calculations of the transmittance functions, and also radiation fluxes and influxes showed its high accuracy and economic efficiency. As an example, Fig.7.5 shows the rate of radiation changes of the temperature of the atmosphere in the absorption bands of ozone 9.6 μm (Fig. 7.5a, spectral range 975–1175 cm^{-1}) and carbon dioxide (7.5b, spectral range 500–1000 cm^{-1}), determined by the direct ('accurate') method and the *kk*-method [91]. The graphs indicate good agreement between the two types of calculations.

For this reason, the method has recently been used widely in solving different problems of atmospheric optics both in the thermal spectral range and in analysis of the transfer of solar radiations [91]. The *k*-method is especially efficient for use in the problems of calculating the scattering of solar radiation (Chapter 8). In this case, it is necessary to solve the integro-differential equation of radiation transfer for monochromatic quantities. This is associated

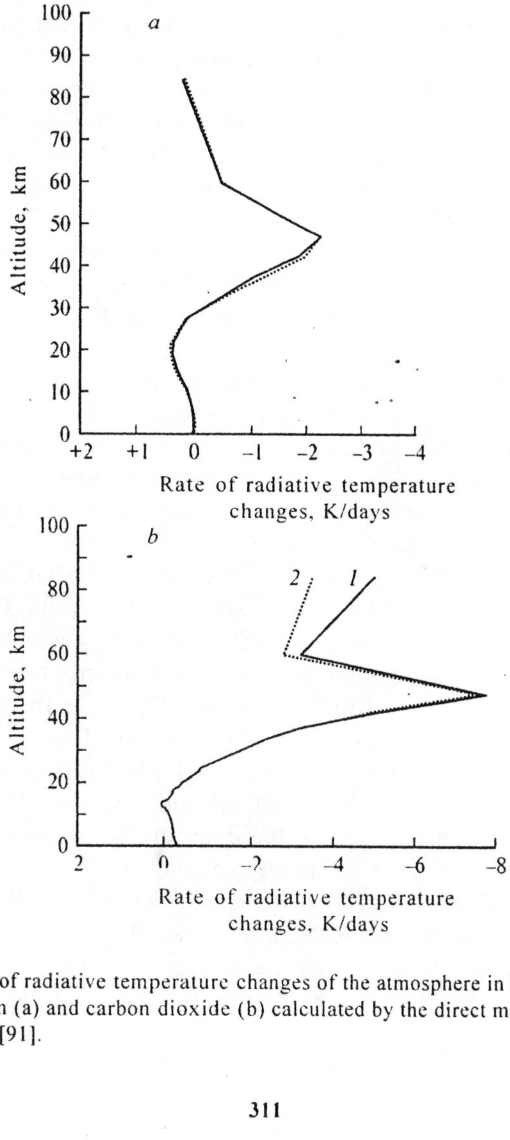

Fig. 7.5. Rates of radiative temperature changes of the atmosphere in absorption bands of ozone 9.6 μm (a) and carbon dioxide (b) calculated by the direct method (1) and the *kk*-method (2) [91].

with the fact that in the problems of scattering the functions of sources (in contrast to the case of thermal radiation where the function of sources is the Planck function) is extremely selective, like the monochromatic transmittance functions. Therefore, the transition operation carried out for the intensity of thermal radiation in finite spectral intervals (transition from equation (7.1.10) tends to (7.1.11) is not correct for scattering problems. The calculations of the intensity of scattered solar radiation in a general case with taking into account multiple scattering are relatively cumbersome and, therefore, the *k*-method which reduces the integration of the monochromatic quantities in respect of many thousands of spectral points to calculating 5–10 radiation components for 5–10 values of the absorption coefficient is extremely efficient in calculations. However, it must be remembered that the practical realisation of the *k*-method requires extensive preliminary calculations. Previously we showed how to obtain the distribution density functions of the $f(k)$ or the integral functions of distribution $g(k)$ required for realization of the *k*-method. The overall diagram of application of the *k*-method in the calculations of the radiation characteristics of the atmosphere, used at present is shown in Fig.7.6. As indicated by the schema, the approach is based on the use of currently available data on the quantitative characteristics or molecular absorption in the form of the parameters of the fine structure of the absorption bands of atmospheric gases (for example, data bank HITRAN) and the coefficients of continual absorption. These initial data are used to calculate the monochromatic coefficients of absorption for the examined spectral range at different pressures and temperature. The calculated absorption coefficients are used to obtain, by some method, the distribution density functions $f(k)$ or integral distribution functions $g(k)$. This is followed by calculation of different radiation characteristics – transmittance functions, intensities, fluxes and influxes of radiation – using the direct method (reference calculations) and the *k*-method. Comparison of the results of calculations using two methods makes it possible to control the accuracy of the *k*-method and optimise the calculation procedure. The schema shows the computing advantages of the *k*-method. For the examined specific example, the direct method of calculating the radiàtion characteristics of the atmosphere requires calculating 10^6 monochromatic quantities. The *k*-method uses only 16 absorption coefficients and appropriate weights in the quadrature of the type (7.3.26) for every absorption band of the atmospheric gases.

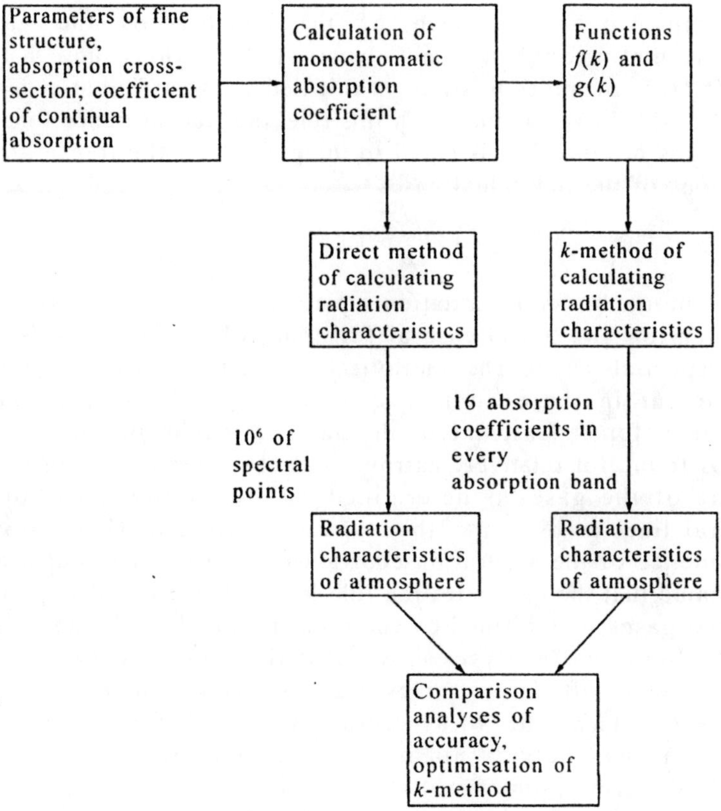

Fig.7.6. Application of the *k*-method in calculations of radiation characteristics of the atmosphere [106].

Transmittance of gas mixtures

The overlapping of the absorption bands of different atmospheric gases creates the problem of calculations of the transmittance functions of a mixture of gases (see equation (4.4.2)). A typical example of overlapping of the absorption bands is the spectral region of the 'wing' of the 15 μm CO_2 band, which also contains a large number of absorption lines of water vapour. It is not difficult to take into account overlapping in the direct calculations of the transmittance function. However, the laboratory measurements and the method of the models of absorption bands give the transmittance functions for individual atmospheric gases. Therefore, the following problem arises: having the transmittance functions, for example, of two gases, CO_2 and H_2O, how to obtain the transmittance function of the mixture of the gases. Experimental investigations and numerical calculations show that the simple 'rule of multiplication' is fulfilled with sufficient accuracy [20, 37, 91]. In accordance with this rule, the transmittance function of the gas mixture P_{1+2} is equal to the product of the transmittance functions of the individual gases

$$P_{1+2}=P_1 \cdot P_2. \qquad (7.4.14)$$

For monochromatic radiation equation (7.4.14) is accurate. We can mention one trivial case of fulfilling relationship (7.4.14) for finite spectral ranges. The multiplication rule holds if the absorption coefficient in the examined spectral range is independent of frequency for at least one of the gases. A situation close to this case is found for relatively narrow spectral ranges where absorption of one of the gases is determined mainly by the wings of far spectral lines or by some other type of continuum whose spectral dependence of the absorption coefficient is weak. It has also been confirmed that the multiplication rule is fulfilled if the absorption of the two gases is described by the random model of the absorption band. Numerical analysis shows that the multiplication rule is fulfilled with sufficient accuracy, also for spectral ranges containing a large ($N \gg 1$) number of absorption lines of different gases. For these reasons, the multiplication rule is used widely in solving greatly differing problems of atmospheric optics.

We shall describe the method of calculating the transmittance of a mixture of gases on the example of the k-method [103]. In this method, as in the traditional approach to calculating the radiation characteristics of the atmosphere, it is necessary to take into account the overlapping of the absorption bands of different atmospheric gases. In this case, using the multiplication rule (7.4.14), it is necessary to approximate the transmittance function for two gases in the form of the double sum

$$P_{\Delta\nu}(u_a, u_b) = \sum_{i=1}^{N} \sum_{j=1}^{M} \exp[-(k_i u_a + k_j u_b)] \Delta g_i \, \Delta g_j. \qquad (7.4.15)$$

This approach requires increasing greatly the computing time, and, consequently, calculation procedures have been proposed which greatly accelerate the calculation of the transmittance functions of gaseous mixtures. One of them is based on the concept of introduction of one effective absorbing gas. In this case, equation

(7.4.15) is replaced by the following equation

$$P_{\Delta\nu}(u_a,u_b)=\sum_{i=1}^{N}\exp[-k_i(u_0+fu_b)\Delta g_i]=\sum_{i=1}^{N}\exp[-k_iu_c]\Delta g_i, \qquad (7.4.16)$$

here $u_c=(u_a+fu_b)$ is the amount of the effective absorbing gas, and parameter f which determines the relative contribution to the absorption of both gases, may be given in the form $f = S_b/S_a$, where S_a and S_b are the intensities of the examined overlapping bands. It should be mentioned that the same approach in calculating the transmittance function of gas mixtures – the introduction of the effective absorbing gas – is also used in the traditional method of calculating the transmittance functions, for example, in the calculation of integral transmittance functions. Parameter f is often regarded as a fitting parameter and is determined ensuring good agreement between the results of calculations using equation (7.4.16) and reference calculations.

7.5. Thermal radiation fluxes

The expressions for the intensities (7.1.1), (7.1.2) may be simplified if we use the concept of optical thickness $\tau(z)$ introduced in Chapter 3 [37, 102, 103]:

$$\tau(z)=\int_z^{\infty}\alpha(z')dz', \qquad (7.5.1)$$

where α is the volume extinction coefficient which coincides for the examined case with the volume coefficient of molecular absorption k. Differentiating (7.5.1) we obtain $d\tau(z)/dz=-\alpha(z)$ or $-\alpha(z)dz=d\tau(z)$. Since the volume extinction coefficient $\alpha(z)$ is non-negative, equation (7.5.1) shows that $\tau(z)$ monotonically decreases with altitude from the value $\tau_0=\int_0^{\infty}\alpha(z)dz$ at the lower boundary of the atmosphere (the underlying surface at z=0) to the value 0 at the upper boundary of the atmosphere. However, consequently, there is also the reversed function $z(\tau)$. Formally, we substitute the function $z(\tau)$ everywhere instead of the dependence on z. Further, introducing the notation $\cos\theta = 1/\sec\theta = \eta$ and taking into account $\sin\theta d\theta = -d(\cos\theta)$, we obtain

$$I_\lambda^{\downarrow}(\tau,\eta) = I_\lambda^{\downarrow}(\infty)\exp(-\tau/\eta + \int_0^{\tau} B_\lambda[T(\tau')]\exp[-(\tau-\tau')/\eta]\frac{d\tau'}{\eta}, \quad (7.5.2)$$

$$I_\lambda^{\uparrow}(\tau,\eta) = I_\lambda^{\uparrow}(\tau_0,-\eta)\exp[-(\tau_0-\tau)/\eta)] + \int_{\tau}^{\tau_0} B_\lambda[T(\tau')]\exp[-(\tau'-\tau)/\eta]\frac{d\tau'}{\eta},$$

To produce the downward and upward fluxes, it is necessary to integrate (7.5.2) in respect of the hemispheres (upper and lower) which, according to (3.2.9), (3.2.10) gives

$$F_\lambda^{\downarrow}(\tau) = \int_0^{2\pi}\int_0^{\pi/2} I^{\downarrow}(\tau,\eta)\cos\theta\sin\theta\, d\theta; \; F_\lambda^{\uparrow}(\tau) = \int_0^{2\pi}\int_0^{\pi/2} I^{\downarrow}(\tau,-\eta)\cos\theta\sin\theta\, d\theta.$$

Taking into account that $\sin\theta d\theta = d\eta$ and neglecting, as mentioned previously, the term with $I_\lambda^{\downarrow}(\infty)$ for the infrared range of the spectrum, and also the reflection of the incidence radiation from the surface, from equation (7.5.2) we obtain

$$F_\lambda^{\downarrow}(\tau) = 2\pi\int_0^1\int_0^{\tau} B_\lambda(T(\tau'))\exp(-(\tau-\tau')/\eta)d\eta d\tau',$$

$$F_\lambda^{\uparrow}(\tau) = 2\pi\varepsilon_\lambda B_\lambda(T_0)\int_0^1 \exp(-(\tau_0-\tau)/\eta)\eta d\eta + \quad (7.5.3)$$

$$+2\pi\int_0^1\int_{\tau}^{\tau_0} B_\lambda(T(\tau'))\exp(-(\tau'-\tau)/\eta)d\eta d\tau'.$$

Integration in respect of η in (7.5.3) may be carried out analytically. We use one among special functions – the integral exponential n-power function which is determined as

$$E_n(x) = \int_1^{\infty}\frac{\exp(-xy)}{y^n}dy = \int_0^1 \frac{\exp(-x/y)}{y^{2-n}}dy$$

It should be mentioned that using for $E_n(x)$ the equation of integration by parts, we obtain a recurrent equation

$$nE_{n+1}(x) = e^{-x} - E_n(x).$$

Thus, all integral exponential functions are expressed through $E_1(x)$, and the algorithms of calculating this function are well-known. Using the integral exponential functions, the expressions for the thermal radiation fluxes are finally written in the form

$$F_\lambda^{\downarrow}(\tau) = 2\pi \int_0^{\tau} B_\lambda(T(\tau'))E_2(\tau - \tau')d\tau',$$
$$F_\lambda^{\uparrow}(\tau) = 2\pi\varepsilon_\lambda B_\lambda(T_0)E_3(\tau_0 - \tau) + 2\pi \int_{\tau}^{\tau_0} B_\lambda(T(\tau'))E_2(\tau' - \tau)d\tau'. \quad (7.5.4)$$

Comparing (7.5.4) with (7.1.1) and (7.1.2), we note a similarity in the expressions for the fluxes and intensities, only in the case of the fluxes the role of the exponents (transmittance functions) is played by the integral exponents $E_2(x)$ and $E_3(x)$, which are referred to as the transmittance functions (and their derivatives) for radiation fluxes (diffusive transmittance function and its derivative). We determine a relationship between the transmittance functions for the radiation fluxes and the transmittance functions for radiation intensity. This is important because we have examined previously in detail different methods of determining the transmittance functions for radiation intensity. The transmittance function for the radiation intensity is written in the form

$$P(\tau) = \exp\left(-\frac{\tau}{\eta}\right). \quad (7.5.5)$$

Consequently, the transmittance function for the radiation flux and its derivative can be written as follows:

$$P^F(\tau) = 2\int_0^1 \exp\left(-\frac{\tau}{\eta}\right)\eta d\eta = 2E_3(\tau), \quad (7.5.6)$$

$$\frac{dP^F(\tau)}{d\tau} = 2\int_0^1 \exp\left(-\frac{\tau}{\eta}\right)d\eta = 2E_2(\tau). \quad (7.5.7)$$

Taking into account the isotropic nature of the Planck function (i.e. the function is independent of variable η), the expressions for the monochromatic fluxes of thermal radiation may be written in the same form as for the intensity of radiation (equations (7.1.4) and

(7.1.5)), but transmittance functions for radiation fluxes should be used.

In the rigorous approach to calculating the radiation fluxes for finite spectral ranges $\Delta\nu$, we can use equations identical to (7.1.8) by integrating the monochromatic fluxes in respect of frequency within the limits of the examined interval. However, we can utilise again the large difference in the spectral dependence of the Planck function and monochromatic transmittance functions. In this case, the expressions for the thermal radiation fluxes for finite spectral ranges are identical to the expression for radiation intensity but these expressions contain diffusive transmittance functions.

Approximate calculation of the diffusive nature of radiation

In a general case when calculating thermal radiation fluxes it is necessary to carry out a large number of integration operations – in respect of frequency, solid angle, spatial variable in the transmittance function. For each new model of the atmosphere all these integrations should be carried out separately. Because of the multiplicity of calculations of the radiation fluxes in different numerical models of the atmosphere, this strict approach is in most cases unacceptable from the computing viewpoint. Therefore, to simplify calculations, equations were derived for the intensity and radiation fluxes in finite spectral ranges using the transmittance functions in finite spectral ranges. To simplify the calculations of radiations fluxes, the diffusion transmittance functions are replaced by the transmittance functions for the radiation intensity at some zenith angle of propagation of radiation. In this case, we can write a simple relationship between the diffusion transmittance function and the transmittance function for intensity (for example, as functions of the absorbing substance u):

$$P_{\Delta\nu}^{F}(u) = P_{\Delta\nu}(\beta u), \tag{7.5.8}$$

where β is the diffuseness factor. The results of a large number of calculations and analysis of the behaviour of the diffusion transmittance function show that although the diffuseness factor depends on the conditions of the atmosphere in the examined spectral range, we can say with satisfactory accuracy β=1.66.

Methods of calculating radiation fluxes

Analysis of the material presented above shows that the radiation fluxes may be calculated by different methods. We shall mention the main methods used at present. Criteria for the classification of individual approaches are represented by the methods of spectral integration of radiation quantities.

1. Reference calculations. In this case, we use exact expressions for the monochromatic intensity of thermal radiation, the direct method of calculating monochromatic transmittance function for a inhomogeneous atmosphere, integration of the intensities in respect of frequency and zenith angle. This approach requires long computing times and is usually used for developing banks of reference values of radiation fluxes for a small number of models of the atmosphere. In particular, this bank is useful for verifying other approximate methods of calculating radiation fluxes.

2. Narrow-band models. In this method, the entire spectral range of thermal radiation (different for different planets) is divided into a large number of sub-intervals with the width of ~50–100 cm^{-1}. The integral radiation fluxes are obtained by summation of radiation in these sub-intervals. This is carried out using various approximations for the transmittance functions in finite spectral intervals (band models, empirical equations, exponential expansions, etc). Different models of this type are characterized by spectral division into sub-intervals and also by the application of other different approximate methods of the theory of transfer of thermal radiation – methods of taking into account heterogeneities of the atmosphere, taking into account the diffuseness of radiation, etc. It should be mentioned that in most cases spectral division is carried out taking into account the location of absorption bands of atmospheric gases.

3. Wide-band models. This approach uses a relatively small number of spectral sub-intervals corresponding to the location of the absorption bands of the main atmospheric absorbing gases (for the Earth – H_2O, CO_2 and O_3). For the transmittance (or absorption) functions in the absorption bands it is necessary to use modelling representations or empirical expressions. As an example, we describe an equation for the absorption function for models of this type [8, 37]:

$$A(u) = 2A_0 \ln\left[1 + \frac{w}{\left(4 + u\left(1 + \frac{1}{\beta}\right)\right)^{1/2}}\right], \tag{7.5.9}$$

where $w = \frac{Su}{A_0}$; $\beta = \frac{4\alpha_L}{d}$; A_0 is the effective width of the absorption band (cm^{-1}); S is the intensity of the band ($cm^{-2} \cdot atm^{-1}$). For the 15 μm absorption band of carbon dioxide and the Lorentz contour of the spectral lines, these parameters are as follows:

$$A_0 = 21.3 \cdot (T/273)^{0.5}\ cm^{-1}\ ,\ S(300\ K) = 194\ cm^{-2} \cdot atm^{-1},$$

$$d = 1.56\ cm^{-1},\ \alpha_L = 0.064\ (273/T)0.5 \cdot (p/1013).$$

Integral transmittance functions, integral emissivity

The optical thickness of the atmosphere is expressed by the mass coefficient of molecular absorption [37, 103]

$$\tau_\nu(z) = \int_z^\infty k_\nu(z')\rho(z')dz', \qquad (7.5.10)$$

where $\rho(z')$ is the density of the absorbing gas at altitude z'; $du = \rho(z')dz'$. Correspondingly, for the integral (total) content of the absorbing gas in the entire thickness of the atmosphere u_0 we can write:

$$u_0 = \int_z^\infty \rho(z')dz'. \qquad (7.5.11)$$

Passing in the equations for the thermal radiation fluxes to variable u and integrating the monochromatic radiation fluxes over the entire frequency range we obtain:

$$F_\nu^\uparrow(u) = \int \pi B_\nu(T_0)P_\nu^F(u)d\nu + \int_0^\infty \int_0^u \pi B_\nu[T(u')]\frac{dP_\nu^F(u-u')}{du'}du'd\nu, \qquad (7.5.12)$$

$$F_\nu^\downarrow(u) = \int_0^\infty \int_{u_0}^u \pi B_\nu[T(u')]\frac{dP_\nu^F(u'-u)}{du'}du'd\nu. \qquad (7.5.13)$$

To simplify consideration, in equations (7.5.12) and (7.5.13) we assume that there is no radiation incident on the upper boundary from the space and that the emissivity of the underlying surface is equal to unity.

From the Stefan–Boltzmann law it follows that

$$\int_0^\infty \pi B_\nu(T)d\nu = \sigma_B T^4. \tag{7.5.14}$$

Using (7.5.14) the expressions for the integral fluxes of the upward and downward thermal radiation (equations (7.5.12) and (7.5.13)) can be presented in the form

$$F^\uparrow(u) = \sigma_B T_0^4 \tilde{P}^F(u,T_0) + \int_0^u \sigma_B T^4(u') \frac{d\tilde{P}^F(u-u',T)}{du'} du', \tag{7.5.15}$$

$$F^\downarrow(u) = \int_0^u \sigma_B T^4(u') \frac{d\tilde{P}^F(u'-u,T)}{du'} du'. \tag{7.5.16}$$

In equations (7.5.15 and (7.5.16) we introduce the quantity $\tilde{P}^F(u,T)$ – the integral (flux) transmittance function. It is the function of temperature and the content of the absorbing substance and is determined by the equation

$$\tilde{P}^F(u,T) = \frac{1}{\sigma_B T^4} \int_0^\infty \pi B_\nu(T) P_\nu^F(u) d\nu. \tag{7.5.17}$$

We can also introduce the integral (flux) emissivity

$$\varepsilon^F(u,T) = 1 - \tilde{P}^F(u,T) = \frac{1}{\sigma_B T^4} \int_0^\infty \pi B_\nu(T)[1 - P_\nu^F(u)] d\nu. \tag{7.5.18}$$

The integral transmittance functions or integral emissivitycan be determined by special experiments or calculations. For example, if we have theoretical or experimental transmittance functions for finite spectral intervals $\Delta\nu$ for the layer of the atmosphere with the amount of the absorbing substance u at temperature T, the integral emissivityis determined from the equation

$$\varepsilon^F(u,T) = \frac{1}{\sigma_B T^4} \sum_{i=1}^N \pi B_{\bar{\nu}_i}(T)[1 - P_{\bar{\nu}_i}^F(u)]\Delta\nu_i, \tag{7.5.19}$$

where the infrared spectrum is divided into N spectral ranges $\Delta\nu_i$ (i=1,2,..., N).

The flux integral radiating capacities for H_2O, CO_2 and O_3 were determined empirically by Elsasser and Culbertson, A.M. Brounshtein, and by calculations by E.M. Feigel'son et al., [20, 32, 33, 37, 91, 102, 103]. Figure 7.7 shows the dependence of the integral emissivity for water vapour and carbon dioxide on the content of the absorbing gas at different temperatures. It may be seen, for example, that at large values of u the integral emissivities of carbon dioxide depend very strongly on temperature. In order to take into account the effect of overlapping of different absorption bands of atmospheric gases, it is recommended to use the following approximate method. Let it be that u_1 and u_2 are the contents of the water vapour and carbon dioxide. In overlapping of the absorption bands of H_2O and CO_2 we can write the following equation for the monochromatic diffusion transmittance functions

$$P_\nu^F(u_{H_2O}, u_{CO_2}) = P_\nu^F(u_{H_2O}) P_\nu^F(u_{CO_2}). \tag{7.5.20}$$

Therefore, expression (7.5.18) for the integral emissivityof the gas mixture may be described approximately by the following equation:

$$\varepsilon^F(u_{H_2O}, u_{CO_2}, T) = \varepsilon^F(u_{H_2O}, T)\, \varepsilon^F(u_{CO_2}, T) - \Delta\varepsilon^F(u_{H_2O}, u_{CO_2}, T), \tag{7.5.21}$$

where the last term is the correction for the overlapping of the absorption bands of water vapour and carbon dioxide:

$$\Delta\varepsilon^F(u_{H_2O}, u_{CO_2}, T) =$$

$$\frac{1}{\sigma_B T^4} \int_0^\infty \pi B_\nu(T)[1 - P_\nu^F(u_{H_2O})][1 - P_\nu^F(u_{CO_2})]d\nu. \tag{7.5.22}$$

The application of the integral transmittance functions or the integral emissivities enables the fluxes of thermal radiation to be written in a compact form. In the availability of these characteristics of transformation of thermal radiation, the calculations of the integral radiation fluxes are quite simple. However, in many current problems the accuracy of this approach is unsatisfactory. In addition to this, the presence of many absorbing gases in the Earth's atmosphere greatly complicates the method of the integral transmittance function.

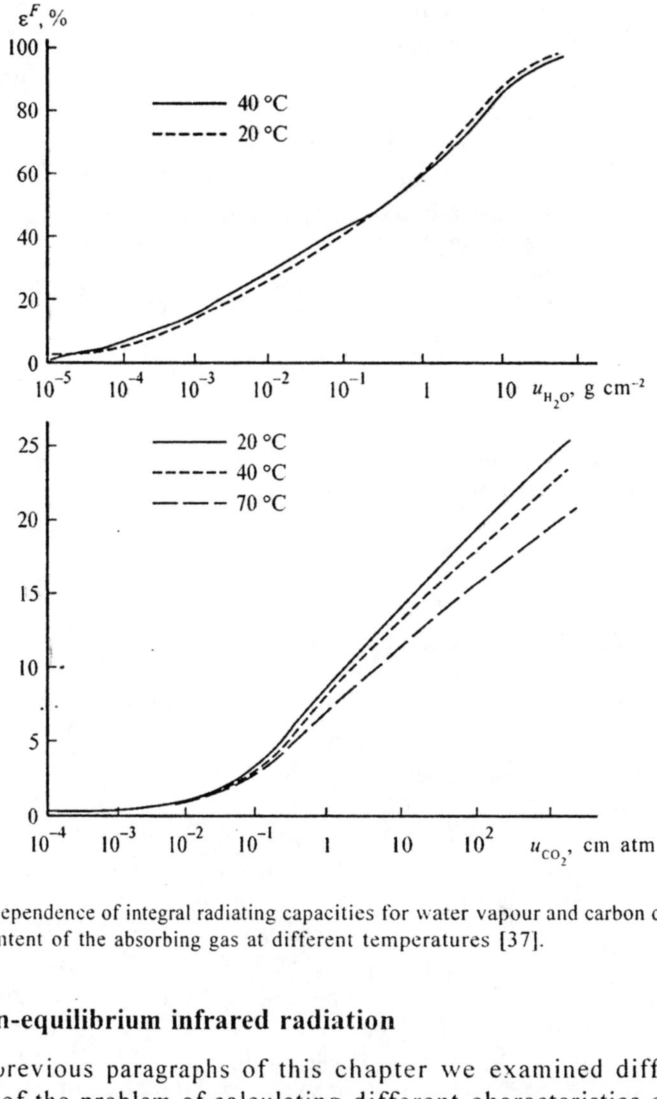

Fig.7.7. Dependence of integral radiating capacities for water vapour and carbon dioxide on the content of the absorbing gas at different temperatures [37].

7.6. Non-equilibrium infrared radiation

In the previous paragraphs of this chapter we examined different aspects of the problem of calculating different characteristics of the thermal radiation of the atmosphere. Previously, it was shown that the condition for the applicability of the Planck function of the absolutely black radiation as the function of sources is the fulfilment of the local thermodynamic equilibrium (LTE). In other words, in the case of LTE the Kirchhoff Law, connecting the atmospheric radiation and absorption coefficients of the atmosphere through the Planck function (equation (3.4.27)), is fulfilled. The LTE is usually established in the lower layers of the atmospheres of the planets. In this case, the high frequency of molecular collisions causes that the distribution of excited molecules is governed by the Boltzmann law at the kinetic temperature of the medium. Since the number of collisions of the molecules is proportional to the concentration of the air molecules, in the upper layers of the atmosphere where the number of these collisions is small, the populations of the excited states cannot be any longer governed by the Boltzmann law and this indicates the breakdown of LTE.

We introduce the collisional lifetime of an excited molecule

$$\tau_c = \frac{1}{\nu}, \qquad (7.6.1.)$$

where ν is the frequency of collisions leading to deactivation of the molecules, the radiative lifetime of excited molecules

$$\tau_r = \frac{1}{A}, \qquad (7.6.2)$$

where A is the Einstein coefficient for spontaneous emission. Simple quantitative estimates show that if the collision lifetime is comparable or longer than the radiative lifetime of the excited states, the atmosphere shows deviations from the LTE.

The radiative lifetime of vibrational states which are of principle important in the transfer of infrared radiation is of the order 1–10^{-1} s ($8 \cdot 10^{-1}$ s for the 15 μm CO_2 band, $8 \cdot 10^{-2}$ s for the ozone band at 9.6 μm, $5 \cdot 10^{-2}$ s for the water vapour band at 6.3 μm). The collisional lifetime of the molecules on the level of the Earth is considerably shorter and, for example, for the carbon dioxide band at 15 μm it is approximately $2.5 \cdot 10^{-5}$s in the temperature range 150–200 K. Since this lifetime changes in inverse proportion to the atmospheric pressure, the radiative and collisional lifetimes become approximately equal at an altitude of 70 km. This altitude is also a rough estimate of the altitude level of LTE breakdown for the vibrational states taking part in the formation of the 15 μm band of CO_2.

Identical analysis of the collisional and radiative lifetimes for the rotational levels of molecule energies shows that the height levels of disruptions of the rotational LTE (i.e. the disruption of the Boltzmann law for the populations of the rotational levels) is found considerably higher in the atmosphere of the Earth (above 100 km). For this reason, the majority of investigations of non-equilibrium infrared radiation of the atmosphere were carried out for breakdowns of only vibrational LTE (molecules of CO_2, H_2O, O_3 and so on).

The problems of transfer of non-equilibrium infrared radiation are solved on two stages:

– initially we calculate the population of the vibration states of the molecule on the basis of solutions of the system of kinetic equations, including different processes of excitation and de-excitation of vibrational states of the investigated molecule;

– this is followed by calculating the intensity and other characteristics of the radiation field of non-equilibrium infrared radiation.

The kinetic equation for the population of the vibrational state, described by the set of the quantum numbers l in a general form may be written as follows:

$$n_l \sum_{l'} (R_{ll'} + C_{ll'}) = \sum_{l'} n_{l'} (R_{l'l} + C_{l'l}) + Y_l, \tag{7.6.3}$$

where n_l is the concentration of molecules in state l; $R_{ll'}$ and $C_{ll'}$ are the rates of the radiative and collisional processes leading to transition from state l to state l'; Y_l is a term describing the rate of excitation of state l as a result of other processes, for example, the formation of excited molecules as a result of chemical reactions.

We describe the results of calculations of non-equilibrium populations of vibrational states for a number of atmospheric molecules using vibrational temperatures T_v. Vibrational temperatures are introduced on the basis of the Boltzmann law (4.4.12). At non-equilibrium population of the levels T_v replaces the kinetic temperature. Figure 7.8 gives the vibrational temperatures for different vibrational states of the main isotope CO_2 ($^{12}C^{16}O_2$) for night conditions [113]. It may be seen that the deviations of the vibrational temperatures of different vibrational states from kinetic temperatures (LTE breakdowns) start at heights of approximately 60–70 km). Figure 7.9. gives the vibrational temperatures for ozone molecules [107]. In this case, LTE breakdowns in daytime already start at a height of approximately 30 km.

Up to now, calculations of this type have been carried out for the main absorption bands of small gas components of the atmosphere of the Earth and other planets. It should be stressed that both the intensities and fluxes of infrared radiation and the radiant influxes of atmospheric radiation in non-equilibrium conditions may greatly differ from the corresponding values obtained assuming vibrational LTE. It should be stressed that the height levels of vibrational LTE breakdown depend strongly on the isotopic variety of the molecule, the energy of the vibrational state and for some vibrational states also on the zenith angle of the sun. For example, vibrational states of the O_3 molecule greatly differ for daytime and nighttime conditions. This is characterised in Table 7.1. which gives the estimates of the height levels of LTE breakdowns for a number of atmospheric gases and their bands.

As shown be calculations and measurements of atmospheric radiation, the atmosphere of the Earth is characterised by deviations from LTE not only for vibrational states but also for rotational and spin states of some molecules (for example, CO, OH, NO). Usually, these deviations become meaningful (with the exception of, for example, NO molecule) at large heights than the deviations for vibrational states.

7.7. Glow of the atmosphere

As mentioned previously, the glow of the atmosphere includes non-equilibrium atmospheric radiation in the ultra violet, visible and near infrared ranges of the spectrum [3, 8, 48, 81, 82]. These glows are associated with the transitions of molecules and atoms between electronic and vibrational states and are usually detected in the same spectral range in which the transfer solar radiation takes place. Because of the high probabilities of electronic transitions, the radiative lifetime of these states is very short (an exception are the so-called 'forbidden' electron transitions) equal to approximately 10^{-8} s. This means that for these states the LTE equilibrium may break down already at small heights. Another point is that at small heights it is not possible to record these glows in most cases because different sources of excited electronic states (solar and corpuscular radiation, photochemical and chemical reactions) are detected mostly in the upper layer of the atmosphere.

Types of atmospheric glow

There are a number of classifications of the glow (emission) of the atmosphere [8, 82]. The glow of the atmosphere does not include both the atmospheric radiation and often the aurora polaris . Strictly speaking, aurora polaris is also atmospheric glow. However, it is caused by other sources of excitation of molecules and atoms in the upper atmosphere – high-energy electrons and ions arriving in the

Table 7.1. Height levels of disruption of vibration LTE

Gas	Band, μm	Vibrational state	Approximate height of breakdown, km
CO_2	15	01^10	70
	4.3	00^01	60 (night)
			50 (day)
	2.0	12^01	10 (day)
H_2O	6.3	010	65 (night)
		020	40 (day)
	2.7	001	25 (day)
O_3	9.6	010	70 (night)
		111	30 (day)
		003	25 (day)
CH_4	7.6	Vibrational state ν_4	60
	6.5	As above ν_2	60
N_2O	17	01^10	70 (night)
CO	4.7	First excited state	40
NO	5.3	As above	55 (night)
			15 (day)

atmosphere from the outside. In this case, it is also taken into account that aurora polaris is of sporadic nature (in contrast to constant glow of the atmosphere) and is detected mostly in polar and sub polar regions. This separation of aurora polaris is not always justified because there are also atmospheric glows which become stronger during the settling of electrons in the period of aurora polaris.

The glows of the atmosphere form at discrete atomic or molecular transitions and those are of line nature (radiation bands) because of the presence of vibrational and rotational sub-levels of electronic states of the molecules. An exception is the relatively weak continual radiation with the maximum in the green part of the visible range of the spectrum. It is assumed that this radiation is associated with the process of radioactive association, namely

$$NO + O \rightarrow NO_2 + h\nu.$$

Historically, atmospheric glow was discovered as the component of glow of the night sky which also includes the diffusion galactic light – the radiation of stars, scattered on cosmic dust, and zodiacal light – solar radiation scattered by interplanetary particles. The glow of the atmosphere in the moonless period equals 40–50% of the total illumination of the night sky. The main atmospheric components

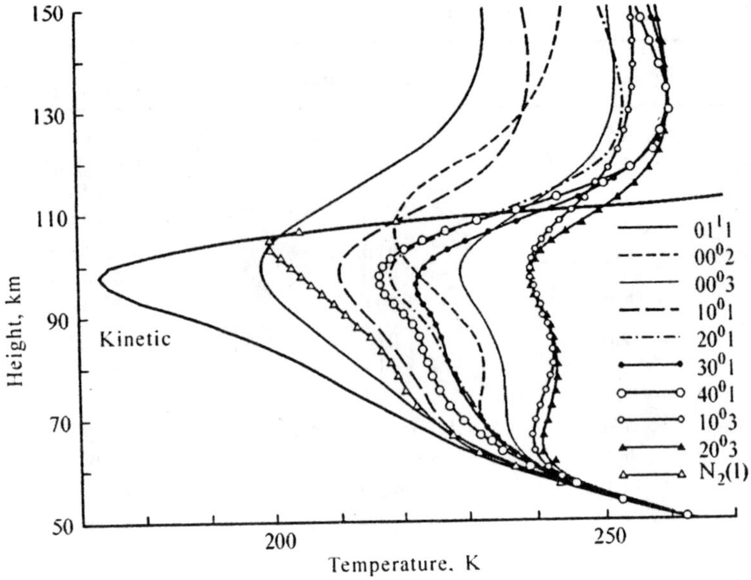

Fig.7.8. Vibrational temperatures for different vibrational states of the main isotopes CO_2 ($^{12}C^{16}O_2$)for night conditions [113].

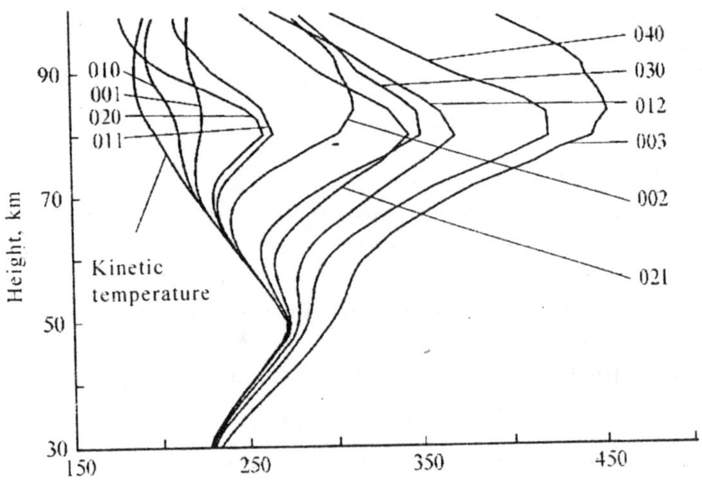

Fig.7.9. Vibrational temperatures for different vibrational states of an ozone molecule for daytime conditions [107].

responsible for the glow of the night sky are molecular and atomic oxygen, nitrogen, hydroxyl and hydrogen and also the atoms of sodium, iron, and so on. Later, the brightest glow was also recorded in twilight – from the regions of the atmosphere illuminated by direct solar radiation. In some cases, atmospheric glow can also be detected during the day but this is difficult to carry out taking into account the significance of the components of solar radiation scattered on the molecules of air and aerosols.

From the viewpoint of atmospheric chemistry and physics, the atmospheric glow is often divided into a number of classes [82]:

a) radiation formed from direct scattering of solar radiation;
b) radiation associated with the ionisation of atoms and molecules and their neutralization;
c) radiation determined by photochemical processes of neutral components of the atmosphere.

It should be mentioned that some types of glow may belong to more than one of these categories. The atmospheric glow is often characterised by a special unit – Rayleigh. One Rayleigh is equal to 10^6 photons/cm^2·s.

The glow of the day or twilight sky (i.e. the part of the atmosphere illuminated by the Sun during sunrise or sunset) is subdivided into two types:

a) resonance scattering;
b) fluorescence scattering.

These processes were briefly examined in Chapter 5. In resonance scattering, the presence of the emission lines in the spectrum of the Sun coinciding with the absorption lines of the atmospheric components results in the absorption of solar radiation and subsequent scintillation of photons with the same wavelength. This process may be represented by the following scheme:

$$X + h\nu \rightarrow X^*,$$

$$X^* \rightarrow X + h\nu.$$

The process of fluorescence scattering is associated with the re-emission of photons (after their excitation) in the more long-wave part of the spectrum than the wavelength of the absorbed solar photon. This process may be described as follows:

$$X + h\nu \rightarrow X^*,$$

$$X^{*} \to X + \sum_{i} h\nu_{i},$$

where $\sum_{i} h\nu_{i}$ is the set of the photons in transitions of the atom or molecule on the intermediate levels of the energy.

The effect of hard ultraviolet and X-ray radiation of the Sun (0.1–175 nm) may cause photoionisation of the atmospheric components. The ionization thresholds (ionization potentials) for the atmospheric components are in the range 50.4 nm (He)–134 nm (NO). The ionization process may be described by the following relationship:

$$X + h\nu \to X^{+} + e.$$

The reversed process – radiative recombination – is the reason for the formation of excited atoms and molecules and the formation of different glows

$$X^{+} + e \to X^{*} + h\nu$$

and further

$$X^{*} \to X + \sum_{i} h\nu_{i}$$

As an example, we can consider the reactions responsible for the emission of atomic oxygen:

$$O^{+}\left({}^{4}S\right) + e \to O^{*} + h\nu,$$

$$O^{*} \to O\left({}^{3}P\right) + \sum_{i} h\nu_{i},$$

where the notations of the electronic states of the atoms are given in the brackets.

The other process – dissociative recombination – usually generates a large excess of energy E:

$$XY^{+} + e \to X + Y + E.$$

This type of reaction leads to the products of dissociation that is in the excited state. Their transition to the ground state (or intermediate state) leads to glow. The resultant glow may be caused by the reaction of ion–ion recombination also causing excitation of the atmospheric components:

$$X^+ + Y^- \rightarrow X^* + Y + E.$$

An important reason for the formation of atmospheric glow are chemical reactions resulting in the formation of specific excited components of the atmosphere. In the literature, this type of glow is often referred to as chemiluminescence. It may be described as follows:

$$A + B \rightarrow C^* + D,$$

$$C^* \rightarrow C + h\nu.$$

Glow of the night sky

This type of glow for the atmosphere of the Earth has been studied in most detail initially in ground-based observations and subsequently using space systems. In the far ultraviolet range of the spectrum there are glows caused by the presence of excited neutral and ionised atoms of O, H and He. The ultraviolet and visible ranges of the spectrum, approximately from 250 nm and up to 480 nm, contain a system of Hertzberg bands of molecular oxygen (Table 4.3). The most intensive glow is detected in the band system Hertzberg I formed in transition of a molecule from the excited state $O_2(A^3\Sigma_u^+)$ to the ground state $O_2(X^3\Sigma_g^-)$. The formation of excited oxygen molecules is caused by the reaction of radiative recombination in which the oxygen atoms, situated in the ground state, take part:

$$O(^3P)+O(^3P)\rightarrow O_2+(A^3\Sigma_u^+)$$

The transition of an excited oxygen molecule to the ground electronic state generates Hertzberg I bands. The integral radiation in the bands of molecular oxygen (spectral range 250–390 nm) is 300–1000 rayleigh. In individual bands it may reach 10 rayleigh. In approximately the same spectral range (190–270 nm) there are glows of NO – bands γ and δ. These bands are also visible in the glow of the daytime sky and in twilight as a result of fluorescence scattering.

The glow of atomic oxygen is manifested in individual lines. The green (557.7 nm) and red (630 nm) lines are most well known. For many years, the reason for the formation of glow in the green line O was explained by Chapman's mechanism:

$$O + O + O \rightarrow O_2 + O^*(^1S).$$

At present, the excitation of the oxygen atom is explained by the two-stage Bart mechanism:

$$O + O + M \rightarrow O^{*}_{2} + M,$$

$$O^{*}_{2} + O(^{3}P) \rightarrow O_{2} + O^{*}(^{1}S).$$

The additional but weak emission of the green line is connected with the process of ionospheric re-combination:

$$O^{*}_{2} + e \rightarrow O^{*}(^{1}D,\ ^{1}S) + O.$$

In this reaction, the yield of the excited oxygen atom in the state ^{1}D is close to unity, and in the state ^{1}S it is around 10%. This mechanism determines the glow of the red line of atomic oxygen.

The glow of the night sky contains hydroxyl bands – Meynell bands. These bands form in vibrational–rotational transitions of the OH molecules that are in the ground electronic state. The strongest band OH is found in the range 1.5 μm. The radiation in this band reaches 10^{5} Rayleigh. The vibrational–excited molecules of OH form in the reaction

$$H + O_{3} \rightarrow OH^{*}\ (v \leq 9) + O_{2}.$$

The energy of 3.34 eV, formed in this reaction, is sufficient for the excitation of vibrational states to the ninth state, which is also shown in the brackets at the hydroxyl molecule. The glow of the hydroxyl is found in a wide spectral range starting from approximately 510 nm and involves the near infrared region.

In addition to the above bands, molecular oxygen is characterised by radiation bands in the near infrared range of the spectrum – the so-called atmospheric band (centred in the vicinity of 760 nm) and infrared atmospheric bands (in the vicinity of 1.27 μm) caused by the transitions of molecules between different vibrational states. The total radiation in the atmospheric band is approximately 30 kilo-rayleigh and in infrared atmospheric band 75 kilo-rayleigh. The relative distribution of the intensities of glow in individual rotational lines of these bands are governed by the Boltzmann distribution of the populations of rotational states at the kinetic temperature of the atmospheric layer where this glow form (in other words, rotational LTE exists). This special feature of the glow of molecular oxygen enables it to be used for the determination of kinetic temperature in the layers of formation of these emissions.

The glow of the night sky shows and intensive line of the sodium atom (589.4 nm) in which radiation shows considerable seasonal variations. The intensity of the line greatly increases during twilight because of the process of resonant scattering.

Glows of daytime and twilight sky

Previously we discussed the resonance and fluorescence glows (under the effect of solar radiation) of the atoms of sodium and molecules of NO. The resonance scattering in γ-band of NO reaches 1000 rayleigh.

Another important glow in the daytime conditions is the glow of hydrogen. Layman-α emission was detected in the glow of the night sky in rocket experiments and initially it was assumed that it is the glow of atomic hydrogen in the interplanetary medium. Analysis carried out later has shown that this emission forms as a result of the resonant scattering of hydrogen atoms at very large heights in the atmosphere of the Earth. In the outer layers of the atmosphere, hydrogen becomes the main component of the atmosphere of the Earth because of diffusion separation, and taking into account high temperatures and the small mass, the height of the homogeneous atmosphere $RT/\mu g$ reaches the value of the order of the radius of the planet.

In space investigations, resonant scattering is observed at the high-intensity coronal glow. This planetary corona is detected for all planets – from Mercury to Uranus. Layman-α emission (1250 Å) is the strongest radiation of hydrogen and reaches 8 kilo-raleigh on the daytime side of the planet. Other glows of hydrogen are resonant scattering Layman-β (1025 Å) and Balmer fluorescence in the range 6563 Å.

The spectra of the glow of the daytime sky contain resonant lines of helium – at 584 Å (neutral helium) and at 304 Å (ionised helium). The glow of helium in these lines is detected in space experiments on Jupiter and Venus. In the first case, this glow reached 1–5 kilo-rayleigh in the second case 600 rayleigh for the 584 Å line.

The glow spectra of the daytime sky of the Earth's atmosphere show the emission of a resonant triplet of atomic oxygen at 1302, 1304 and 1306 Å. This glow was detected mainly at an altitude of ~190 km and reaches ~7.5 kilo-rayleigh.

The glow of molecular oxygen, determined by resonance fluorescence, is detected in the atmospheric and infrared atmospheric bands. However, there are also other reasons for these glows in daytime. For example, the excited molecules of oxygen (state $b_1\Sigma_g^+$) originate in the following reaction:

$$O(^1D)+O_2(^3\Sigma_g^-)\rightarrow O(^3P)+O_2(b_1\Sigma_g^+).$$

The excited oxygen atoms form in this case as a result of dissociation of the oxygen molecules under the effect of radiation in the Schumann–Runge continuum range. The most important source of the excited molecules of $O_2(a^1\Delta_g)$ is the photodissociation reaction:

$$O_3 + h\nu \rightarrow O_2(a^1\Delta_g) + O.$$

The atmosphere of the Earth in daytime also shows emission of the neutral and ionised molecular nitrogen, atoms of Na, ionised Ca, Li and K. The atmospheres of Mars and Venus show different glows of the CO molecules and CO_2^+ ion, and the oxygen atom.

Measurements of different atmospheric glows of the planets represent an important source of information on the temperature and gas composition of the upper layers of their atmospheres.

CHAPTER 8

MAIN CONCEPTS OF THE THEORY OF SOLAR RADIATION TRANSFER

8.1. Multiple scattering of radiation

In Chapter 3, we presented the integro-differential equation of transfer for scattered solar radiation (3.4.35)

$$\frac{dI}{dz}\cos\theta = -\alpha(z)I + \frac{1}{4\pi}\sigma(z)\int_{4\pi} x(z,\mathbf{r})I(\mathbf{r})d\Omega, \qquad (8.1.1)$$

where I is the intensity of radiation incident at the zenith angle θ at altitude z; (z) is the volume extinction coefficient; $\sigma(z)$ is the volume coefficient of scattering; $x(z, \mathbf{r})$ is the scattering phase function; all quantities are assumed to be monochromatic at a specific wavelength. We have also stressed in Chapter 3 that equation (8.1.1) should be regarded as formal until the scattering geometry is not defined.

We examine the model of a plane-parallel atmosphere (Fig. 8.1). Axis Z – the altitude axis – is directed as usual upwards normally to the surface. Azimuth φ is counted in the plane normal to the axis Z from some specific direction; in most cases, it is selected to ensure that the azimuth of the solar rays incident on the upper boundary of the atmosphere is zero, i.e. the azimuth is counted 'from the Sun'. Now, for the intensity of scattered radiation, incoming on the level z from the direction (θ,φ) we obtain the transfer equation with the already determined geometry

$$\frac{dI(z,\theta,\varphi)}{dz}\cos\theta = -\alpha(z)I(z,\theta,\varphi) +$$

$$+\frac{1}{4\pi}\sigma(z)\int_0^{2\pi} d\varphi' \int_0^{\pi} x(z,(\theta,\varphi),(\theta',\varphi'))I(z,\theta',\varphi')\sin\theta' d\theta'. \qquad (8.1.2)$$

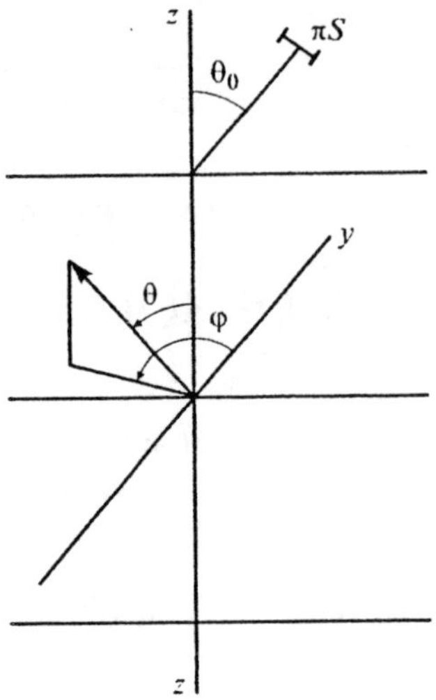

Fig. 8.1. Plane-parallel atmosphere.

Transition to cosines of angles and optical thickness

To simplify equation (8.1.2), we transform it as follows. cos θ is denoted by η (and by analogy, cos θ' by η'). If the angle θ changes from 0 to π, then cos θ, i.e. η, changes monotonically from -1 to 1. Because of this monotonicity, in the equations for the intensity and phase function we transfer from the dependence on θ to the dependence on η (formal substitution θ = arccos η), i.e. examine them not as the function of angle but as the function of variable η. Both parts are divided by $\alpha(z)$. The differential $\alpha(z)dz$ appears in the denominator of the left part. Introducing optical thickness τ (Chapter 3), we have $\alpha(z)dz = d\tau(z)$. Again we use the inverse function $z(\tau)$ and use it everywhere instead of height z. Consequently, all the functions in equation (8.1.2) become functions of optical thickness and τ – the vertical co-ordinate of the problem.

Albedo of single scattering

After dividing by $\alpha(z)$, the following relationship appears in front of the integral

$$\Lambda(\tau) = \frac{\sigma(\tau)}{\alpha(\tau)} = \frac{\sigma(\tau)}{\sigma(\tau) + k(\tau)}, \quad (8.1.3)$$

where $k(\tau)$ is the volume coefficient of adsorption. The value Λ is referred to as the albedo of single scattering or, in other words, the survival probability of a quantum. The meaning of the first term follows from (8.1.3): if there is no absorption, then $\Lambda = 1$; if there is no scattering then $\Lambda = 0$; i.e. Λ expresses the fraction of scattering in the total extinction as the albedo of the surface expresses the fraction of reflection. The meaning of the second term – if the adsorption of the quantum of light is regarded as an act of its annihilation, then Λ expresses the probability of not adsorbing, i.e. the survival probability.

Angle between two directions

Now we can write equation (8.1.2) in new notations, but we firstly introduce one standard simplification. In (8.1.1) and (8.1.2) we have written the scattering phase function in the general form. However, usually the phase function depends only on the scattering angle such as, for example, the molecular phase function and aerosol phase function for spherical particles (see Chapter 5). We shall discuss only the phase function $x(\omega)$, where ω is the cosine of the scattering angle (it should be mentioned that previously we have justified the transition from angles to their cosines). Now for substitution into equation (8.1.2) it is necessary to determine this cosine ω, i.e. the cosine of the angle between the directions (θ, φ) and (θ', φ'). The cosine of the angle between the unit vectors is equal to their scalar product. The projections of the vector with the direction (θ, φ) on the axes Z, X and Y are evidently equal to $\cos\theta$, $\sin\theta\cos\varphi$, $\sin\theta\sin\varphi$. For the direction (θ', φ') we obtain identical projections. Now

$$\omega = \cos\theta\cos\theta' + \sin\theta\cos\varphi\sin\theta'\cos\varphi' + \sin\theta\sin\varphi$$

$$\times\sin\theta'\sin\varphi' = \cos\theta\cos\theta' + \sin\theta\sin\theta'\cos(\varphi-\varphi').$$

Going over to the variables η and η', we obtain

$$\omega = \eta\eta' + \sqrt{(1-\eta^2)(1-(n')^2)}\cos(\varphi-\varphi'). \qquad (8.1.4)$$

Tranfer equation and boundary conditions

Now finally we can write the tranfer equation in new notations

$$\eta\frac{I(\tau,\eta,\varphi)}{d\tau} = -I(\tau,\eta,\varphi) + \frac{\Lambda(\tau)}{4\pi}\int_0^{2\pi} d\varphi' \int_{-1}^{1} x(\tau,\omega) I(\tau,\eta',\varphi') d\eta', \qquad (8.1.5)$$

where ω is determined by equation (8.1.4). However, the transfer equation (8.1.5) in itself is not sufficient for describing the scattering of light. The boundary conditions should be added to it. The upper boundary of the atmosphere receives solar radiation at the zenith angle $\theta_0 = \arccos \eta_0$. The intensity of solar radiation in equation (8.1.5) is represented as the sum of the intensities of direct (i.e. not scattered in the atmosphere) and scattered radiation $I(\tau, \eta, \varphi) = I_1(\tau, \eta, \varphi) + I_2(\tau, \eta, \varphi)$. Let it be that the flux incident on the area normal to the solar rays at the upper boundary of the atmosphere is πS (Fig. 8.1) (i.e. πS is the spectral solar constant). Therefore (see Chapter 3) $I_1 = \pi S\delta(\eta - \eta_0)\delta(\varphi - 0)$, where $\delta(x)$ is the δ-function. Substitution of the sum into the transfer equation (8.1.5) gives:

$$\eta\frac{I_1(\tau,\eta,\varphi)}{d\tau}+\eta\frac{I_{21}(\tau,\eta,\varphi)}{d\tau}=-I_1(\tau,\eta,\varphi)-I_{21}(\tau,\eta,\varphi)+$$

$$+\frac{\Lambda(\tau)}{4\pi}\int_0^{2\pi}d\varphi'\int_{-1}^{1}x(\tau,\omega)(I_2(\tau,\eta,\varphi)+I_1(\tau,\eta,\varphi))d\eta'. \quad (8.1.6)$$

It should be mentioned that for direct radiation $I_1(\tau,\eta,\varphi)$ we can write the transfer equation – Bouguer law (3.4.5):

$$\eta\frac{I_1(\tau,\eta,\varphi)}{d\tau}=-I_1(\tau,\eta,\varphi), \quad (8.1.7)$$

from which

$$I_1(\tau,\eta,\varphi) = \pi S \exp(-\tau/\eta_0)\delta(\eta - \eta_0)\delta(\varphi - \varphi_0). \quad (8.1.8)$$

In the transfer theory it is conventional not to include direct solar radiation (8.1.7) in the transfer equation because calculation of this radiation from equation (8.1.8) is very simple. Therefore, deducting from equation (8.1.6) equation (8.1.7), substituting (8.1.8) into this equation and taking into account the main property of the δ-function $\int f(x)\delta(x-a) = f(a)$ we obtain

$$\eta\frac{I_1(\tau,\eta,\eta_0,\varphi)}{d\tau}=-I(\tau,\eta,\eta_0,\varphi)+$$

$$+\frac{\Lambda(\tau)}{4\pi}\int_0^{2\pi}d\varphi'\int_{-1}^{1}x(\tau,\omega)I(\tau,\eta',\eta_0,\varphi')d\eta'+ \quad (8.1.9)$$

$$+\frac{\Lambda(\tau)}{4}Sx(\tau,\omega_0(\exp(-\tau/\eta_0),$$

where

$$\omega=\eta\eta'+\sqrt{(1+\eta^2)(1-(\eta')^2)}\cos(\varphi-\varphi'),$$

$$\omega_0=\eta\eta_0+\sqrt{(1-\eta^2)(1-\eta_0^2)}\cos(\varphi). \qquad (8.1.10)$$

At the lower boundary of the atmosphere there is an underlying surface which reflects the light. Taking into account the results in Chapter 6, it would be easy to write the boundary conditions for this surface. However, a different procedure is used in transfer theory: here it is conventional to consider separately the processes of scattering in the atmosphere and reflection from the surface. Therefore, the reflection from the surface is not taken into account or, which is the same, the surface is assumed to be absolutely black. The point is that after calculating the intensity of scattered radiation we have relatively simple means of taking into account the contribution provided by the reflection from the surface; this will be discussed in the next paragraph. Thus, intensity $I(\tau,\eta,\eta_0,\varphi)$ in (8.1.9) is the intensity of exclusively scattered radiation without taking direct and reflected radiation into account. We can now write the boundary conditions for it: the absence of the scattered radiation coming from outside on both the upper and lower boundaries of the atmosphere:

$$I(0,\eta,\eta_0,\varphi)=0 \quad \text{if } \eta>0$$
$$I(\tau_0,\eta,\eta_0,\varphi)=0 \quad \text{if } \eta<0. \qquad (8.1.11)$$

The relationships (8.1.10)–(8.1.11) are the final form of the transfer equation of scattered solar radiation in the plane-parallel atmosphere [65].

Function of sources. Single and multiple scattering

The contribution to scattering to the intensity (8.1.9) is controlled by the two last terms in the right-hand part. As shown in Chapter 3, this contribution is identical to the presence of additional light sources in the medium. We introduce the function of sources $\tilde{B}(\tau,\eta,\eta_0,\varphi)$ determining it as:

$$\tilde{B}(\tau,\eta,\eta_0,\varphi) = \frac{\Lambda(\tau)}{4\pi}\int_0^{2\pi} d\varphi' \int_{-1}^{1} x(\tau,\omega) I(\tau,\eta',\eta_0,\varphi') d\eta' +$$

$$+\frac{\Lambda(\tau)}{\pi} S x(\tau,\omega_0)\exp(-\tau/\eta_0). \qquad (8.1.12)$$

The integrated term in (8.1.12) is associated with the scattering of direct solar radiation. This scattering, i.e., the scattering of direct radiation, is single scattering. The integral term in (8.1.12) is associated with the scattering of the already scattered radiation $I(\tau,\eta',\eta_0,\varphi')$ and this scattering is multiple scattering.* The physical meaning of these terms is relatively simple: The term outside the integral is the direct solar radiation transmitted to the level τ, taking into account the fraction of scattering in the general extinction $\Lambda(\tau)$ (this leads to the term 'the albedo of single scattering'!), and the phase function – 'the strength' of scattering at specific angle ω_0. Integral term – similar consideration of the contribution of the scattered radiation travel to the level τ from all possible directions.

After introducing the function $\tilde{B}(\tau,\eta,\eta_0,\varphi)$ the transfer equation has the form

$$\eta\frac{I(\tau,\eta,\eta_0,\varphi)}{d\tau} = -I(\tau,\eta,\eta_0,\varphi) + \tilde{B}(\tau,\eta,\eta_0,\varphi). \qquad (8.1.13)$$

However, (8.1.13) is the linear differential equation for the intensity $I(\tau,\eta,\eta_0,\varphi)$ of the type $dy/dx = a(x)\,y(x) + b(x)$, whose general solution (this has already been discussed in Chapter 3) is

$$y(x) = y(x_0)\exp\left(\int_{x_0}^{x} a(x')dx'\right) + \int_{x_0}^{x} b(x')\exp\left(\int_{x'}^{x} a(x'')dx''\right).$$

*It should be mentioned that the concept 'single scattering' and 'multiple scattering' are used, in addition to transfer theory, in the optics of aerosols in a completely different sense. This often causes confusion and misunderstanding. In the optics of aerosols 'single scattering' denotes the scattering on a particle described regardless of the presence of other particles, for example, as in Mie theory. 'Multiple scattering' is the combined diffraction of electromagnetic waves on several particles. In transfer theory 'multiple scattering' denotes consideration of the radiation scattered several times during its passing through the atmosphere. It may be seen that the concepts are completely different. It should be mentioned that usually the volume coefficients of aerosol scattering are calculated in the approximation of single scattering in the sense of aerosol optics but used for calculating multiple-scattered radiation in transfer theory.

Here $a(x)=-\frac{1}{\eta}$, $b(x)=\frac{1}{\eta}\tilde{B}(\tau,\eta,\eta_0,\varphi)$, the integration variable – the optical thickness τ, the initial value x_0 is 0 for the direction from the upper boundary ($\eta > 0$) and τ_0 for the direction from the lower boundary ($\eta < 0$), the values of $y(x_0)$, according to (8.1.11) are equal to zero. Thus, we have

$$I(\tau,\eta,\eta_0,\varphi)=\frac{S}{4\eta}\int_0^{\tau}\tilde{B}(\tau',\eta,\eta_0,\varphi)\times$$

$$\times\exp\left(-\frac{\tau-\tau'}{\eta}\right)d\tau', \qquad \text{if } \eta>0, \tag{8.1.14}$$

$$I(\tau,\eta,\eta_0,\varphi)=-\frac{S}{4\eta}\int_{\tau}^{\tau_0}\tilde{B}(\tau',\eta,\eta_0,\varphi)\times$$

$$\times\exp\left(-\frac{\tau-\tau'}{\eta}\right)d\tau', \qquad \text{if } \eta<0.$$

Finally, equations (8.1.14) are not solutions of the transfer equation because in a general case the function of sources $\tilde{B}(\tau,\eta,\eta_0,\varphi)$ according to (8.1.12) depends on intensity. However, they do provide an explicit expression for the required intensity through the function of sources and this is very important. For example, from (8.1.14) we can immediately write a general solution of the transfer equation in the approximation of single scattering, i.e., when the function of sources takes into account the term outside the integral $\tilde{B}(\tau,\eta,\eta_0,\varphi)=\frac{\Lambda(\tau)}{4}Sx(\tau,\omega_0)\exp(-\tau/\eta_0)$:

$$I(\tau,\eta,\eta_0,\varphi)=-\frac{S}{4\eta}\int_0^{\tau}\Lambda(\tau')x(\tau',\omega_0)\times$$

$$\times\exp\left(-\frac{\tau'}{\eta_0}-\frac{\tau-\tau'}{\eta}\right)d\tau', \qquad \text{if } \eta>0, \tag{8.1.15}$$

$$I(\tau,\eta,\eta_0,\varphi)=-\frac{S}{4\eta}\int_{\tau}^{\tau_0}\Lambda(\tau')x(\tau',\omega_0)\times$$

$$\times\exp\left(-\frac{\tau'}{\eta_0}-\frac{\tau-\tau}{\eta}\right)d\tau', \qquad \text{if}\,\eta<0.$$

The approximation of single scattering (8.1.15) is often used in problems where the high accuracy of calculating scattered radiation is not required.

Integral equation for the functions of sources

Let us substitute the equations for intensity (8.1.14) in the definition of the function of sources (8.1.12). We obtain:

$$\tilde{B}(\tau,\eta,\eta_0,\varphi)=\frac{\Lambda(\tau)}{4\pi}\int_0^{2\pi}d\varphi'\times$$

$$\times\left(\int_0^1 x(\tau,\omega)\frac{d\eta'}{\eta'}\int_0^{\tau}\tilde{B}(\tau',\eta',\eta_0,\varphi')\exp\left(-\frac{\tau-\tau'}{\eta'}\right)d\tau'-\right. \qquad (8.1.16)$$

$$\left.-\int_{-1}^{0} x(\tau,\omega)\frac{d\tau'}{\eta'}\int_{\tau}^{\tau_0}\tilde{B}(\tau',\eta',\eta_0,\varphi')\exp\left(-\frac{\tau-\tau'}{\eta'}\right)d\tau'\right)+$$

$$+\frac{\Lambda(\tau)}{4}Sx(\tau,\omega_0)\exp\left(-\frac{\tau}{\eta_0}\right).$$

(8.1.16) contains only the function $\tilde{B}\,(\tau,\eta,\eta_0,\varphi)$, and consequently we obtained an equation for the function of sources whose solution is related to the required intensity by the simple relationships (8.1.14).Regardless of the cumbersome form, from the mathematical viewpoint equation (8.1.16) is more convenient than the equation for intensity (8.1.9) because it is an integral equation (not integro-differential). Therefore, in the theory of transfer we usually concerned with (8.1.16) [47, 65].

The resultant equation (8.1.16) is the Fredholm integral equation of the second kind. The mathematical theory of these equations has been developed quite sufficiently, in particular, the existence and uniqueness of the solution for these equations have been shown. In the 'operator' form, equation (8.1.16) can be written in the form [46, 53]

$$\tilde{B}=\mathbf{K}\tilde{B}+q,$$

Where $\tilde{B}$ is the sought function of sources $\tilde{B}(\tau,\eta,\eta_0,\varphi)$; **K** is the integral operator of scattering with a kernel

$$\mathbf{K}(\tau,\eta,\varphi,\tau',\eta',\varphi')=\begin{cases}\dfrac{\Lambda(\tau)}{4\pi\eta'}x(\tau,\omega)\exp\left(-\dfrac{\tau-\tau'}{\eta'}\right), \\ \quad 0\le\tau'\le\tau,\ 0\le\eta'\le 1, \\ -\dfrac{\Lambda(\tau)}{4\pi\eta'}x(\tau,\omega)\exp\left(-\dfrac{\tau-\tau'}{\eta'}\right), \\ \quad \tau\le\tau'\le\tau_0, -1\le\eta'\le 0,\end{cases}$$

q is a free term;

$$q(\tau,\eta,\eta_0,\varphi)=\frac{\Lambda(\tau)}{4}Sx(\tau,\omega_0)\exp(-\tau/\eta_0).$$

The formal solution of the Fredholm equation of the second kind is the Neumann series

$$\tilde{B}=q+\mathbf{K}q+\mathbf{K}^2q+\mathbf{K}^3q+\ldots \qquad (8.1.17)$$

The terms of the series have a simple physical meaning: the first (q), as already explained, corresponds to the contribution of single-scattered light; the second ($\mathbf{K}q$) – the application of the scattering operator to single-scattered light, i.e. the contribution of twice-scattered light; similarly, the third one ($\mathbf{K}^2q$) = $\mathbf{K}(\mathbf{K}q)$ – the contribution of light scattered by three times, etc. This means that the Neumann series (8.1.17) is the expansion of the contribution of scattered light in respect of the multiplicity of scattering.

It should be mentioned that the kernel **K** and the free term q are directly proportional to the albedo of single scattering $\Lambda(\tau)$. Let it be that Λ is independent of τ. Then, at the n-th term of the series (8.1.17) we have the coefficient Λ^n which evidently determines the rate of convergence: as Λ approaches unity, i.e. as the absorption decreases in comparison with scattering, the rate of convergence of the series decreases and the multiplicity of scattering which must be taken into account in the calculations, increases. Since q is directly proportional to the parameter S and **K** is independent of S, equation (8.1.17) shows that the function of sources $\tilde{B}$ and according to (8.1.14), the intensity of scattered radiation are directly proportional to S. Consequently, in complete correspondence with the physics of processes, the intensity of scattered light is directly proportional to the flux at the upper boundary of the atmosphere. Therefore, to simplify considerations, in solving the transfer equation it is often assumed that $S = 1$ and the obtained intensity has been multiplied by the specific value of S.

8.2. Analytical methods in radiation transfer theory

Expansion of scattering phase function into a series in terms of Legendre polynomials

In the previous paragraph we obtained the main equations of radiation transfer taking multiple scattering into account. The most important part of the theory of radiation transfer is the mathematical analysis and obtaining solutions in both partial and general cases. The mathematical apparatus of transfer theory has been described in many monographs [47, 64, 65]. We present only the simplest examples of transformation of the transfer equation which are, however, very important for practical problems.

The standard procedure of solving the differential and integral equations is the expansion of their parameters into a series in respect of orthogonal functions. For the transfer equation (8.1.16) we can achieve simplification in expansion of the scattering phase function into a series in terms of the Legendre polynomials. Therefore, the Legendre polynomials will be discussed briefly.*

The Legendre polynomials $P_n(x)$ are determined by the equation

$$P_n(x) = \frac{1}{2^n n!}\frac{d(x^2-1)^n}{dx^n}. \tag{8.2.1}$$

For calculations in practice we can use the recurrent relationship

$$P_n(x) = \frac{2n-1}{n}xP_{n-1}(x) - \frac{n-1}{n}P_{n-2}(x), \tag{8.2.2}$$

(where $P_0(x) = 1$, $P_1(x) = x$), which enables us to calculate in succession P_n, from (8.2.2): $P_2(x) = \frac{1}{2}(3x^2-1)$, $P_3(x) = \frac{1}{2}(5x^3-3x)$. The Legendre polynomials form an orthogonal system of the functions in the range [−1, 1] (main property), and for them

*The Legendre polynomials, spherical functions, Legendre adjoint functions and the addition theorem are described in detail in, for example, a textbook in [63].

$$\int_{-1}^{1} P_n(x)P_m(x)dx = 0, \text{ if } n \neq m; \int_{-1}^{1} P_n^2(x)dx = \frac{2}{2n+1}.$$

Correspondingly, any function $f(x)$ continuous in the range [–1, 1] can be expanded into a series in terms of the Legendre polynomials:

$$f(x) = \sum_{k=0}^{\infty} c_k P_k(x), \tag{8.2.3}$$

where

$$c_k = \frac{2k+1}{2} \int_{-1}^{1} f(x)P_k(x)dx. \tag{8.2.4}$$

Using (8.2.2) and (8.2.3) for the scattering phase function we obtain

$$x(\omega) = \sum_{i=0}^{\infty} x_i P_i(\omega), \tag{8.2.5}$$

where

$$x_i = \frac{2i+1}{2} \int_{-1}^{1} x(\omega)P_i(\omega)d\omega. \tag{8.2.6}$$

It should be mentioned that $x_0 = \frac{1}{2}\int_{-1}^{1} x(\omega)d\omega$, but this is the condition of normalisation of (3.3.13) taking into account the change of the angle to cosine. Consequently, in all cases $x_0 = 1$. The expansion coefficient x_1 is an important characteristic of the phase function:

$$x_1 = \frac{3}{2}\int_{-1}^{1} x(\omega)\omega d\omega.$$

Since ω is the cosine of the scattering angle, $x_1/3$ is the mean cosine of scattering for the given phase function. It characterises its stretching forward: as the mean cosine increases the phase function becomes more stretched. For practical calculations it is interesting to consider finite series, i.e. to cut off (8.2.5) at some number of terms N. Unfortunately, for the strongly elongated aerosol phase functions it is necessary to take into account tens of even hundreds of terms in (8.2.5). Therefore, in calculations in practice using expansion of the indicatrices into a series in terms of the Legendre polynomials, we are faced with the problem of approximating the elongated indicatrices [53]. It should be mentioned that for a Rayleigh phase function $x(\omega)=\frac{3}{4}(1+\omega^2)$ directly $x(\omega)=P_0+\frac{1}{2}P_2(\omega)$.

In the transfer equation, the phase function is a function of the incident angles and scattered radiation. For similar relationships we can use the addition theorem according to which

$$P_i\left(\eta\eta'+\sqrt{(1-\eta^2)(1-(\eta')^2)}\cos(\varphi-\varphi')\right)=$$

$$=P_i(\eta)P_i(\eta')+2\sum_{m=1}^{i}\frac{(i-m)!}{(i+m)!}P_i^m(\eta)P_i^m(\eta')\cos m(\varphi-\varphi'), \qquad (8.2.7)$$

where $P^m_i(x)$ are the Legendre adjoint functions determined by the relationship

$$P_i^m(x)=(1-x)^{\frac{m}{2}}\frac{d^m P_i(x)}{dx^m}. \qquad (8.2.8)$$

For practical calculations of the Legendre adjoint functions we can use the recurrent equations:

$$P_{n-1}^m(x)=\frac{2n+1}{n-m+1}xP_n^m(x)-$$

$$-\frac{n+m}{n-m+1}P_{n-1}^m(x), \qquad \text{at } 0\le m\le n-1, \qquad (8.2.9)$$

$$P_n^{m+2}(x)=2(m+1)\frac{x}{\sqrt{1-x^2}}P_n^{m+1}(x)-$$

$$-(n(n+1)-m(m+1))P_n^m(x), \qquad at\ 0\le m\le n-2,$$

where $P_n^0(x)=P_n(x); P_1^1(x)=\sqrt{1-x^2}$.

Taking into account (8.2.7) the expression for the scattering phase function has the form:

$$x(\omega)=\sum_{i=0}^{N}x_iP_i(\eta)P_i(\eta')+$$

$$+\sum_{i=0}^{N}2x_i\sum_{m=1}^{i}\frac{(i-m)!}{(i+m)!}P_i^m(\eta)P_i^m(\eta')\cos m(\varphi-\varphi').$$

In the double sum we group the terms with the same index m: with $m = 1$ there are terms at all i from 1 to N, with $m = 2$ – at all i from 2 to N, etc. This means that

$$x(\omega)=\sum_{i=0}^{N}x_1P_i(\eta)P_i(\eta')+$$

$$+2\sum_{m=1}^{N}\cos m(\varphi-\varphi')\sum_{i=m}^{N}x_1\frac{(i-m)!}{(i+m)!}P_i^m(\eta)P_i^m(\eta')$$

or in compact form

$$x(\omega)=p^0(\eta,\eta')+2\sum_{m=1}^{N}p^m(\eta,\eta')\cos m(\varphi-\varphi'), \tag{8.2.10}$$

where

$$p^m(\eta,\eta')=\sum_{i=m}^{N}x_i\frac{(i-m)!}{(i+m)!}P_i^m(\eta)P_i^m(\eta'). \tag{8.2.11}$$

For the unknown intensity and the function of sources we write formally expansions identical to (8.2.10)

$$I(\tau,\eta,\eta_0,\varphi)=I^0(\tau,\eta,\eta_0)+$$

$$+2\sum_{m=1}^{N}I^m(\tau,\eta,\eta_0)\cos m\varphi, \tag{8.2.12}$$

$$\tilde{B}(\tau,\eta,\eta_0,\varphi)=\tilde{B}^0(\tau,\eta,\eta_0)+$$

$$+2\sum_{m=1}^{N}\tilde{B}^m(\tau,\eta,\eta_0)\cos m\varphi, \tag{8.2.13}$$

where $I^m(\tau,\eta,\eta_0)$ and $\tilde{B}^m(\tau,\eta,\eta_0)$ are some functions to be determined, $m = 0, \ldots, N$. Substituting (8.2.12), (8.2.13) into the transfer equation (8.1.13) and equating the terms with the same m, we obtain

$$\eta\frac{I^m(\tau,\eta,\eta_0)}{d\tau}=-I^m(\tau,\eta,\eta_0)+\tilde{B}^m(\tau,\eta,\eta_0). \qquad (8.2.14)$$

Let us substitute now (8.2.10), (8.2.12) and (8.2.13) into expression for the function of sources through intensity (8.1.12) and calculate in the obtained equation the integrals in respect of azimuth. The product of the zero terms is independent of the azimuth and the integral is equal to 2π. The other terms in re-multiplication of the series (8.2.10) for the phase function and (8.2.12) for the intensity give the integrals of the type:

$$\int_0^{2\pi}\cos m_1(\varphi-\varphi')\cos m_2\varphi' d\varphi'=\cos m_1\varphi\int_0^{2\pi}\cos m_1\varphi'\cos m_2\varphi' d\varphi'+$$

$$+\sin m_1\varphi\int_0^{2\pi}\sin m_1\varphi'\cos m_2\varphi' d\varphi'.$$

However, $\int_0^{2\pi}\cos m_1\varphi'\cos m_2\varphi' d\varphi'$ is equal to zero if $m_1\neq m_2$, and equal to π if $m_1 = m_2$ and $\int_0^{2\pi}\sin m_1\varphi'\sin m_2\varphi' d\varphi'$ is equal to zero for all m_1 and m_2. Thus, after re-multiplying the series and integration in respect of the azimuth, equation (8.1.12) retains only the terms with equal indices, and at the zero term there will be coefficient 2π, and the remaining terms $4\pi \cos m\varphi$. Equating now the terms with the same m in the right and left parts, we obtain

$$\tilde{B}^m(\tau,\eta,\eta_0)=\frac{\Lambda(\tau)}{2}(\tau,\eta,\eta')I^m(\tau,\eta,\eta')d\eta'+$$

$$+\frac{\Lambda(\tau)}{4}S\,p^m(\tau,\eta,\eta')\exp\left(-\frac{\tau}{\eta_0}\right) \qquad (8.2.15)$$

Finally, for the boundary conditions (8.1.10) we get

$$I^m\ (0,\eta,\eta_0) = 0,\ \text{ if } \eta>0,$$

$$I^m\ (\tau_0,\eta,\eta_0) = 0,\ \text{ if } \eta<0. \qquad (8.2.16)$$

Thus, we have reduced the number of variables in the unknown functions by removing the dependence on azimuth and passing from the transfer equation (8.1.19) to $N+1$ equations (8.2.14)-(8.2.16) which finally have simplified the solution. The obtained equations are mathematically equivalent to the initial ones, in particular, they give the expressions for the intensities through the function of sources similar to (8.1.14):

$$I^m(\tau,\eta,\eta_0)=\frac{1}{\eta}\int_0^{\tau}\tilde{B}(\tau',\eta,\eta_0)\exp\left(-\frac{\tau-\tau'}{\eta}\right)d\tau',\ \text{if}\ \eta>0$$

$$I^m(\tau,\eta,\eta_0)=-\frac{1}{\eta}\int_{\tau}^{\tau_0}\tilde{B}(\tau',\eta,\eta_0)\exp\left(-\frac{\tau-\tau'}{\eta}\right)d\tau',\ \text{if}\ \eta<0,\qquad(8.2.17)$$

whose substitution into (8.2.15) gives the integral equation for the function of sources

$$\tilde{B}^m(\tau,\eta,\eta_0)=\frac{\Lambda(\tau)}{2}\left(\int_{-1}^{0}p^m(\tau,\eta,\eta')\frac{d\eta'}{\eta'}\int_0^{\tau}\tilde{B}(\tau',\eta',\eta_0)\times\right.$$
$$\times\exp\left(-\frac{\tau-\tau'}{\eta'}\right)d\tau'-\int_{-1}^{0}p^m(\tau,\eta,\eta')\frac{d\eta'}{\eta'}\times$$
$$\left.\times\int_{\tau}^{\tau_0}\tilde{B}^m(\tau',\eta',\eta_0)\exp\left(-\frac{\tau-\tau'}{\eta'}\right)d\tau'\right)+\qquad(8.2.18)$$
$$+\frac{\Lambda(\tau)}{4}Sp^m(\tau,\eta,\eta_0)\exp\left(-\frac{\tau}{\eta_0}\right).$$

Expansions (8.2.12) and (8.2.13) are often referred to as expansions by azimuthal harmonics and the functions $I^m(\tau,\eta,\eta_0)$ and $\tilde{B}^m(\tau,\eta,\eta_0)$ – as azimuthal harmonics. Usually in the transfer theory it is preferred to operate with the azimuthal harmonics and the equations for them.

Coefficients of reflection and transmission of the atmosphere

In many problems it is not necessary to calculate the intensity of scattered light in the thickness of the atmosphere, i.e. in relation to τ, and it is sufficient to know the intensity of the radiation outgoing from the atmosphere. For example, problems of this type pear in interpreting the measurements of the intensity of scattered liation from satellites and the brightness of sky from the surface the Earth. In this case, the required intensity can be conveniently presented in the form [47, 65]

$$I(0,-\eta,\eta_0,\varphi) = S\eta_0\rho(\eta,\eta_0),$$
$$I(\tau_0,\eta,\eta_0,\varphi) = S\eta_0\sigma(\eta,\eta_0). \quad (8.2.19)$$

Quantities $\rho(\eta,\eta_0,\varphi)$ and $\sigma(\eta,\eta_0,\varphi)$ are referred to respectively as the coefficients of reflection and transmission of the atmosphere. Since $F = S_0$ is the flux incident on the area parallel to the surface on the upper boundary of the atmosphere, then $\rho(\eta,\eta_0,\varphi) = \pi I(0,-\eta,\eta_0,\varphi)/F$, but this definition in accuracy coincides with the definition (6.3.21) of the coefficient of spectrum brightness.

Expanding $\rho(\eta,\eta_0,\varphi)$ and $\sigma(\eta,\eta_0,\varphi)$ in respect of the azimuthal harmonics

$$\rho(\eta,\eta_0,\varphi) = \rho^0(\eta,\eta_0) + 2\sum_{m=1}^{N}\rho^m(\eta,\eta_0)\cos m\varphi,$$
$$\sigma(\eta,\eta_0,\varphi) = \sigma^0(\eta,\eta_0) + 2\sum_{m=1}^{N}\sigma^m(\eta,\eta_0)\cos m\varphi, \quad (8.2.20)$$

we obtain

$$I^m(0,-\eta,\eta_0) = S\eta_0\rho^m)(\eta,\eta_0),$$
$$I^m(\tau_0,-\eta,\eta_0) = S\eta_0\sigma^m)(\eta,\eta_0). \quad (8.2.21)$$

Reflection of scattered radiation from the orthotropic surface

Previously, we presented equations for the transfer of exclusively scattered radiation. We now take into account the reflection of light from the underlying surface. To simplify considerations, we examine an orthotropic surface with the albedo equal to A. In this case, in solving the transfer equation, the presence of the reflecting surface influences only the zero azimuthal harmonics. To confirm this claim, it is sufficient to use the 'proof by contradiction' principle: actually, if the orthotropic surface affected the non-zero harmonics of intensity then in accordance with (8.2.12) it would also affect its dependence on the azimuth (through cos $m\varphi$), but this contradicts the fact that the surface is orthotropic, i.e. reflecting in the same manner on all azimuths.

Taking this claim into account, we omit the indexes zero at the coefficients of reflection and transmission. In 'addition' reflection from the underlying surface, coefficients of reflection and transmission of the atmosphere change. We introduce the following notations [65]. As previously, the coefficients of reflection and transmission without taking the surface into account are denoted by $\rho(\eta,\eta_0)$ and $\sigma(\eta,\eta_0)$. Identical coefficients, but already with the reflection from the surface taken into account, are denoted as $\bar{\rho}$ (η,η_0) and $\bar{\sigma}(\eta,\eta_0)$. We also need the coefficients of reflection and transmission in illuminating the atmosphere from below in the absence of the surface, i.e. from the point τ_0, and we denote them as $\tilde{\rho}(\eta,\eta_0)$ and $\tilde{\sigma}(\eta_0,\eta)$. The appropriate designations are also introduced for intensity.

It should be mentioned that in a general case, the symmetric relationships hold for the coefficients of reflection and transmission of the atmosphere:

$$\begin{aligned} \rho(\eta,\eta_0,\varphi) &= \rho(\eta_0,\eta,\varphi), \\ \tilde{\rho}(\eta,\eta_0,\varphi) &= \tilde{\rho}(\eta_0,\eta,\varphi), \\ \sigma(\eta,\eta_0,\varphi) &= \tilde{\sigma}(\eta_0,\eta,\varphi). \end{aligned} \tag{8.2.22}$$

We accept that without proof which requires a relatively detailed analysis of the integral equation for the function of sources (8.2.18) [65] which is outside the examined subject of taking into account reflection from the surface. We only note that in Chapter 6 we proved the symmetry of the coefficient of spectral brightness of the surface (6.3.15). This principle is also maintained in complex processes of multiple light scattering.

We determine the flux incident from the atmosphere on the surface

$$\begin{aligned} F^{\downarrow}(\eta_0,\tau_0) = \int_0^{2\pi} d\varphi \int_0^1 I(\tau_0,\eta',\eta_0)\eta' d\eta' + \\ + \pi S\eta_0 \exp\left(-\frac{\tau_0}{\eta_0}\right). \end{aligned} \tag{8.2.23}$$

The first term in (8.2.23) is a hemispherical flux of scattered radiation, according to (3.2.9), the second one is the contribution of direct solar radiation according to the Bouguer law. Direct radiation must be added because we have agreed to examine the equation of transfer and, consequently, determine the intensity and the coefficients of reflection and transmission (according to (8.2.19)) only for scattered radiation. Since according to the definition of the surface albedo (6.3.11) the upward flux $F^{\uparrow}(\eta_0,\tau_0)$ is $F^{\uparrow}(\eta_0,\tau_0) = AF^{\downarrow}(\eta_0,\tau_0)$, then expressing the intensities through the transmission coefficient according to (8.2.21) we obtain

$$F^{\uparrow}(\eta_0,\tau_0) = A\left(2\pi S\eta_0 \int_0^1 \tilde{\sigma}(\eta',\eta_0)\eta' d\eta' + \pi S\eta_0 \exp\left(-\frac{\tau_0}{\eta_0}\right)\right). \quad (8.2.24)$$

The presence of the surface is equivalent to elimination of the atmosphere from below. The calculation of the intensity generated by this illumination at the boundaries of the atmosphere is complicated by the fact that in contrast to illumination from above the light from the bottom comes from different directions. To reduce this case to the already examined case, initially we examine illumination from below only and the one angle with the cosine $\eta' > 0$ (Fig. 8.2), and here $\eta' > 0$ because we deal with the 'inverted' geometry (the pattern should be equivalent to that illuminated from above). We can introduce, at the moment formally, a flux incident on the area normal to the rays and equal to, as in the case of the flux from below, $\pi\tilde{S}(\eta')$. Now, according to (8.2.19), the intensity of scattered radiation at the upper boundary is $\tilde{S}(\eta')\tilde{\sigma}\,\eta'(\eta,\eta')$. However, the atmosphere is illuminated from the bottom not only from one direction η' and, therefore, to determine intensities at the lower and upper boundaries, this expression should be integrated in all directions. In addition to this, for the upper boundary it is important to take into account the direct radiation, coming from the surface, in exactly the same manner as in the case of illumination from above. Consequently, we can write

$$\tilde{I}(0,\eta) = \int_0^{2\pi} d\varphi \int_0^1 \tilde{S}(\eta')\eta'\tilde{\sigma}(\eta,\eta')d\eta' + \pi\tilde{S}(\eta)\exp\left(-\frac{\tau_0}{\eta}\right),$$

$$\tilde{I}(\tau_0,\eta) = \int_0^{2\pi} d\varphi \int_0^1 \tilde{S}(\eta')\eta'\tilde{\rho}(\eta,\eta')d\eta'.$$

It is now easy to determine the intensity at the boundaries of the atmosphere in the presence of the surface. Actually, the intensity at the upper boundary is the sum of intensity resulting only from scattering in the atmosphere $I(0,\eta,\eta_0)$ and the intensity as a result

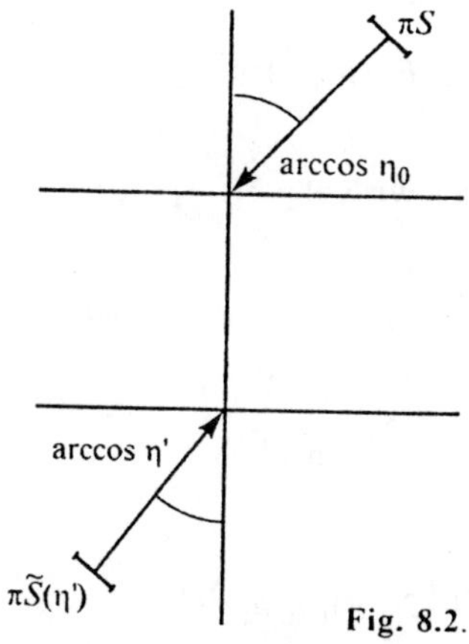

Fig. 8.2. Illumination of the atmosphere from the bottom.

of the illumination of the atmosphere by the light reflected from the surface $\tilde{I}(0,\eta)$. Similarly, the intensity at the lower boundary is the sum of the intensity of the light scattered in the atmosphere $I(\tau_0,\eta,\eta_0)$ and the intensity as a result of illumination by the reflected light $\tilde{I}(\tau_0,\eta)$. Thus,

$$\bar{I}(0,\eta,\eta_0) = S\eta_0\rho(\eta,\eta_0) + 2\pi\int_0^1 \tilde{S}(\eta')\eta'\tilde{\sigma}(\eta,\eta')d\eta' +$$

$$+\pi\tilde{S}(\eta)\exp\left(-\frac{\tau_0}{\eta}\right), \qquad (8.2.25)$$

$$\tilde{I}(\tau_0,\eta,\eta_0) = S\eta_0\sigma(\eta,\eta_0) + 2\pi\int_0^1 \tilde{S}(\eta')\eta'\tilde{\rho}(\eta,\eta')d\eta'.$$

We now determine $\tilde{S}(\eta')$. From the fact that the flux on the area, normal to the rays, is numerically equal to the intensity (Chapter 3), it is concluded that $\pi\tilde{S}(\eta') = I_r(\eta')$, where $I_r(\eta')$ is the intensity of the light reflected from the surface in the direction η'. However, for the orthotropic surface, the reflected intensity does not depend on the direction (Chapter 6), i.e. $I_r(\eta') \equiv I_r$, i.e. I_r is a constant. Consequently

$$\int_0^{2\pi} d\varphi \int_0^1 I_r\eta' d\eta' = F^{\uparrow}(\eta_0,\tau_0),$$

from which

$$I_r = F^{\uparrow}(\eta_0, \tau_0)/\pi \text{ and } \tilde{S}(\eta') = I_r/\pi = F^{\uparrow}(\eta_0, \tau_0)/\pi^2.$$

Substituting the last equation into (8.2.25) we get

$$\bar{I}(0, \eta, \eta_0) = S\eta_0 \rho(\eta, \eta_0) +$$

$$+ \frac{F^{\uparrow}(\eta_0, \tau_0)}{\pi}\left(2\int_0^1 \tilde{\sigma}(\eta, \eta')\eta' d\eta' + \exp\left(-\frac{\tau_0}{\eta}\right)\right),$$

$$\bar{I}(\tau_0, \eta, \eta_0) = S\eta_0 \sigma(\eta, \eta_0) + \frac{F^{\uparrow}(\eta_0, \tau_0)}{\pi} 2\int_0^1 \tilde{\rho}(\eta, \eta')\eta' d\eta'.$$

Dividing both parts of the equalities by $S\eta_0$, according to the definitions (8.2.19), we obtain

$$\bar{\rho}(\eta, \eta_0) = \rho(\eta, \eta_0) +$$

$$+\beta(\eta_0, \tau_0)\left(2\int_0^1 \tilde{\sigma}(\eta, \eta')\eta' d\eta' + \exp\left(-\frac{\tau_0}{\eta}\right)\right), \qquad (8.2.26)$$

$$\bar{\sigma}(\eta, \eta_0) = \sigma(\eta, \eta_0) + 2\beta(\eta_0, \tau_0)\int_0^1 \tilde{\rho}(\eta, \eta')\eta' d\eta',$$

where $\beta(\eta_0, \tau_0) = F^{\uparrow}(\eta_0, \tau_0)/(\pi S\eta_0)$ and according to (8.2.24):

$$\beta(\eta_0, \tau_0) = A\left(2\int_0^1 \bar{\sigma}(\eta', \eta_0)\eta' d\eta' + \exp\left(-\frac{\tau_0}{\eta_0}\right)\right). \qquad (8.2.27)$$

We can now express the coefficients of reflection and transmission of the atmosphere with the accounting surface through identical coefficients for the case with the surface not taken into account. Substituting (8.2.26) into (8.2.27) we get

$$\beta(\eta_0, \tau_0) = A\left(2\int_0^1 \sigma(\eta', \eta_0)\eta' d\eta' +\right.$$

$$\left.+4\beta(\eta_0, \tau_0)\int_0^1 \eta' d\eta' \int_0^1 \tilde{\rho}(\eta', \eta'')\eta'' d\eta'' + \exp\left(-\frac{\tau_0}{\eta_0}\right)\right).$$

Consequently

$$\beta(\eta_0,\tau_0)=\frac{A\left(2\int_0^1\sigma(\eta',\eta_0)\eta' d\eta'+\exp\left(-\frac{\tau_0}{\eta_0}\right)\right)}{1-4A\int_0^1\eta d\eta\int_0^1\tilde{\rho}(\eta,\eta')\eta' d\eta'}.$$

To shorten the equations we introduce the notations [65]

$$\tilde{C}=4\int_0^1\eta d\eta\int_0^1\tilde{\rho}(\eta,\eta')\eta' d\eta',$$

$$\tilde{E}(\eta)=2\int_0^1\tilde{\rho}(\eta,\eta')\eta' d\eta',$$

$$V(\eta_0,\tau_0)=2\int_0^1\sigma(\eta,\eta_0)\eta d\eta+\exp\left(-\frac{\tau_0}{\eta_0}\right), \tag{8.2.28}$$

$$\tilde{V}(\eta,\tau_0)=2\int_0^1\tilde{\sigma}(\eta,\eta')\eta' d\eta+\exp\left(-\frac{\tau_0}{\eta_0}\right)=$$

$$=2\int_0^1\sigma(\eta',\eta)\eta' d\eta'+\exp\left(-\frac{\tau_0}{\eta_0}\right)=V(\eta,\tau_0),$$

where in deriving the equality $\tilde{V}(\eta,\tau_0) = V(\eta,\tau_0)$ we took into account the symmetry ratio (8.2.22). The equations (8.2.26) and (8.2.27) now give

$$\bar{\rho}(\eta,\eta_0,\tau_0)=\rho(\eta,\eta_0,\tau_0)+\frac{AV(\eta,\tau_0)V(\eta_0,\tau_0)}{1-A\tilde{C}},$$

(8.2.29)

$$\bar{\sigma}(\eta,\eta_0,\tau_0)=\sigma(\eta,\eta_0,\tau_0)+\frac{A\tilde{E}(\eta)V(\eta_0,\tau_0)}{1-A\tilde{C}}.$$

It should be mentioned that for the coefficient of reflection of the atmosphere the symmetry property is also maintained when reflection from the surface is taken into account

$$\bar{\rho}(\eta,\eta_0,\tau_0) = \bar{\rho}(\eta_0,\eta,\tau_0).$$

Thus, solving the equation of transfer and finding the coefficients of reflection and transmission without considering the presence of the surface, from (8.2.28) and (8.2.29) we can easily determine them already taking into account the orthotropic surface. Of course, it is now necessary to solve the equation for the 'inverted' atmosphere in order to find $\tilde{\rho}(\eta,\eta_0)$, but only for zero harmonic and the equation of transfer is the simplest. In a general case of calculating the intensity inside the atmosphere and also calculating the intensity in reflection from a non-orthotropic surface, using the same method of summation of the intensities in illumation from top and bottom, it is also possible to express the parameters of the 'atmosphere plus surface' system through the parameters of the atmosphere 'without the surface' [65]. Thus, in the modern theory of transfer, the problem of reflection from the surface is solved analytically and we can examine the transfer of exclusively scattered radiation without taking direct and reflected radiation into account.

8.3. Numerical methods in the theory of radiation transfer

Specific features of numerical methods

A practical problem of the theory of transfer radiation is the development of numerical algorithms of computer calculations of the intensity of radiation with multiple scattering taken into account. The importance of the problem results from the need for interpreting the results of measurements of the intensity of scattered light and also calculating the hemispherical fluxes of solar radiation which determine the energy regime of the atmosphere.

It should be mentioned that the complexity and specific features of the transfer equation create problems in numerically solving this equation by standard computing methods (substitution of derivatives by finite differences, substitution of integrals by finite sums). Therefore, special algorithms have been developed for this equation and the concepts used as a basis for these algorithms greatly differ. At present, there are more than ten methods of numerical solution of the transfer equation which greatly differ in their ideology [53]. We select only four of them illustrating this difference of concepts using these methods as an example. It should be mentioned that because of the previously mentioned special features of the transfer equation, the development of numerical methods of solving this equation is closely connected with the analytical methods and it is often difficult to separate these two approaches, as will be shown on examples of the methods of spherical harmonics and composition of the layers.

The method of spherical harmonics

This method is used for determining the intensities from the integro-differential equation (8.1.9), but we shall clarify its concept using a simpler integral equation for the function of sources (8.2.18) [65]. We find the unknown functions $B^m(\tau,\eta,\eta_0)$ in the form of an expansion into a series in terms of the Legendre adjoint functions identical with (8.2.11), separating thus the variables η and a τ:

$$\tilde{B}^m(\tau,\eta,\eta_0)=\sum_{i=m}^{N} c_i^m(\tau)P_i^m(\eta)\tilde{B}_i^m(\tau,\eta_0), \tag{8.3.1}$$

where $c_i^m(\tau)=x_i(\tau)\dfrac{(i-m)!}{(i+m)!}$. The functions to be determined are $\tilde{B}_i^m(\tau,\eta_0)$. Substituting (8.2.11) and (8.3.1) into the integral equation (8.2.18) and equating the coefficients at $\tilde{B}_i^m(\tau,\eta_0)$ with equal m and i, we obtain

$$\tilde{B}_i^m(\tau,\eta_0)=\frac{\Lambda(\tau)}{2}\sum_{i=m}^{N} c_i^m(\tau)\left(\int_0^1 P_i^m(\eta')P_i^m(\eta')\frac{d\eta'}{\eta'}\times\right.$$

$$\times\int_0^1 \tilde{B}_i^m(\tau,\eta_0)\exp\left(-\frac{\tau-\tau'}{\eta'}\right)d\tau'-$$

$$\left.-\int_{-1}^0 P_i^m(\eta')P_i^m(\eta')\frac{d\eta'}{\eta'}\int_1^{\tau_0}\tilde{B}_i^m(\tau,\eta_0)\exp\left(-\frac{\tau-\tau'}{\eta'}\right)d\tau'\right)+ \tag{8.3.2}$$

$$+\frac{\Lambda(\tau)}{4}SP_i^m(\eta_0)\exp\left(-\frac{\tau}{\eta_0}\right),$$

where $m = 0, \ldots, N$; $i = m, \ldots, N$.

The relationships (8.3.2) are a system of integral equations for determining the unknown functions $\tilde{B}_i^m(\tau,\eta_0)$. It would appear that we have not gained anything by obtaining, instead of N independent equations (8.2.18), a system of $\frac{1}{2}(N+1)(N+2)$ equations. However, firstly, the system (8.3.2) as may easily be seen is 'triangular': the equation with $m = N$ includes only one unknown function $\left(\tilde{B}_N^N(\tau,\eta_0)\right)$, the equation with $m = N-1$, $i = N$ only two $\left(\tilde{B}_N^{N-1}(\tau,\eta_0) \text{ and } \tilde{B}_N^N(\tau,\eta_0)\right)$ and so on. Thus, the system should be solved by the 'method of inverse course': initially, to determine $\tilde{B}_N^N(\tau,\eta_0)$, then $\tilde{B}_N^{N-1}(\tau,\eta_0)$ and so on. Secondly, exchanging integrations in respect of η' and τ' and taking into account the explicit expressions for the Legendre adjoint functions according to (8.2.9), integration in respect of η' may be carried out analytically. Its result is expressed in the form of a series of the already known special functions – integral exponents $E_n(x)$ (see paragraph 7.5) of the argument $(\tau,-\tau')$. In a general case, the derivation of formulas for the coefficients of the series and the results are very combersome and we shall therefore not discuss this and only confirm the possibility of retaining integration only in respect of τ in equations (8.3.2). Therefore, thirdly, the obtained integrals are one-dimensional and it is now easy to solve them by the standard numerical methods of solving integral Fredholm equations of the second kind: reduction to the system of algebraic linear equations by representing the function at discrete values τ or the iteration methods, based on expansion into a Neumann series (8.1.17).

The method of discrete ordinates

This method is close to the 'standard' schemes of numerical solution of differential equations. It is based on replacement in the integro-differential equation (8.2.18) of the integral in respect of the angles by the Gauss quadrature equation, i.e. transition to the discrete grid for the scattering angles (thus the name of the method). Consequently, equation (8.2.18) transfer to a system of ordinary differential equations of the first order. The method of solving these systems are well known and based on the determination of the eigennumbers and vectors of their matrices. In the case of the transfer equation, this matrix may be converted to a special form greatly simplifying the solution procedure [37].

The method of addition of layers

We examine a partial problem of determination of the intensity of radiation at the boundaries of the atmosphere. The method of addition of layers is based on the method of addition of intensities if they are known for every layer separately, in combining two atmospheric layers. Of course, this is the simplest and at the same time sufficiently effective method of calculating intensity [53]. If we add to the layer with multiple scattering a very thin layer whose parameters may be regarded as constant in respect of τ and the scattering in this layer is only single, the expressions for the intensities at the boundaries of the combined layer are obtained in the explicit form. Dividing the entire atmosphere into such thin layers, we can added them successfully. To describe the method, we use the equations of intensities for the layer with single scattering illuminated from different directions. All considerations will be made for the expansion of intensities in respect of the azimuthal harmonics.

In the case of single scattering of light and constant parameters of the layer, the function of sources (8.2.15) has the form

$$\tilde{B}^m(\tau,\eta,\eta_0)=\frac{\Lambda}{4}Sp^m(\eta,\eta_0)\exp(-\tau/\eta_0).$$

Substituting this equation into the expression for intensity through the function of sources (8.2.17) at the boundaries of the layer ($\tau = 0$ and $\tau = \Delta\tau$) and to stress the approximation of single scattering, denoting the intensities by letter i, we obtain

$$i^m(0,-\eta,\eta_0)=\frac{1}{\eta}\frac{\Lambda}{4}Sp^m(\eta,\eta_0)\int_0^{\Delta\tau}\exp(-\tau'/\eta_0)\exp(-\tau'/\eta)d\tau'=$$

$$=\frac{1}{\eta}\frac{\Lambda}{4}Sp^m(\eta,\eta_0)\frac{1-\exp(-\Delta\tau(1/\eta_0+1/\eta))}{1/\eta_0+1/\eta},$$

$$i^m(\Delta\tau,\eta,\eta_0)=\frac{1}{\eta}\frac{\Lambda}{4}Sp^m(\eta,\eta_0)\int_0^{\Delta\tau}\exp(-\tau'/\eta_0)\times$$

$$\times\exp\left(-\frac{\Delta\tau-\tau'}{\eta}\right)d\tau'=$$

$$=\frac{1}{\eta}\frac{\Lambda}{4}Sp^m(\eta,\eta_0)\exp(-\Delta\tau/\eta)\frac{1-\exp(-\Delta\tau(1/\eta_0-1/\eta))}{1/\eta_0-1/\eta}.$$

Finally,

$$i^m(0,-\eta,\eta_0) = \frac{\Lambda}{4} S\eta_0 p^m(\eta,\eta_0) \frac{1-\exp(-\Delta\tau(1/\eta_0 + 1/\eta)}{\eta+\eta_0},$$

$$i^m(\Delta\tau,\eta,\eta_0) = \tag{8.3.3}$$

$$= \frac{\Lambda}{4} S\eta_0 p^m(\eta,\eta_0) \frac{\exp(-\Delta\tau/\eta) - \exp(-\Delta\tau/\eta_0)}{\eta-\eta_0}.$$

Evidently, because of the constancy of the parameters of the layers if the layer is illuminated from below $\tilde{i}^m(0,-\eta,\eta_0) = i^m(0,-\eta,\eta_0)$ and $\tilde{i}^m(\Delta\tau,\eta,\eta_0) = i^m(\Delta\tau,\eta,\eta_0)$.

We now examine a layer not illuminated with direct rays but illuminated with radiation with the intensity $I_0^m(\eta)$, and assuming that the coefficient of reflection and transmission of the layer $\rho^m(0,\eta,\eta_0)$ and $\sigma^m(\Delta\tau,\eta,\eta_0)$ are given, we determine the intensities on the upper and lower boundaries of the layer $I^m(0,-\eta)$, $I^m(\Delta\tau,\eta)$. It should be noted that, generally speaking, the scattering in the layer is multiple. This problem has already been solved in the previous paragraph but the results were not given in the explicit form. We examine only one direction of the radiation η' incident on the layer and introduce formally for it the intensity of the incident flux $\pi S^m(\eta')$. Then, according to (8.2.21) we have

$$I^m(0,-\eta,\eta') = S^m(\eta')\eta'\rho^m(\eta,\eta'),$$

$$I^m(\Delta\tau,\eta,\eta') = S^m(\eta')\eta'\sigma^m(\eta,\eta').$$

To determine the total intensity we integrate with respec to all directions of incidence and for transmission we take into account direct radiation

$$I^m(0,-\eta) = \int_0^{2\pi} d\varphi \int_0^1 S^m(\eta')\eta'\rho^m(\eta,\eta')d\eta',$$

$$I^m(\Delta\tau,-\eta) = \int_0^{2\pi} d\varphi \int_0^1 S^m(\eta')\eta'\sigma^m(\eta,\eta')d\eta' + \tag{8.3.4}$$

$$+\pi S(\eta)\exp(-\Delta\iota/\eta).$$

Further, taking into account that $\pi S(\eta',\varphi) = I_0(\eta',\varphi)$, and expanding both parts in respect of the azimuthal harmonics, we obtain

$S'''(\eta') = \frac{1}{\pi} I_0'''(\eta')$. Substitution of this expression into (8.3.4) gives the required result:

$$I'''(0,-\eta) = 2\int_0^1 I_0'''(\eta')\eta'\rho'''(\eta,\eta')d\eta',$$

$$I'''(\Delta\tau,\eta) = 2\int_0^1 I_0'''(\eta')\eta'\sigma'''(\eta,\eta')d\eta' + \qquad (8.3.5)$$

$$+ I_0'''(\eta)\exp(-\Delta\tau/\eta).$$

For the method of composition of layers it is however convenient to write (8.3.5) not through the coefficients of reflection and transmission but through the intensity of radiation emerged from the layer illuminated by direct rays with the cosine of the angle η'. From (8.2.21) we obtain

$$\rho'''(\eta,\eta') = \frac{1}{S\eta'} I'''(0,-\eta,\eta'),$$

$$\sigma'''(\eta,\eta') = \frac{1}{S\eta'} I'''(0,-\eta,\eta'),$$

and therefore

$$I'''(0,-\eta) = \frac{2}{S} I_0'''(\eta')I'''(0,-\eta,\eta')d\eta',$$

$$I'''(\Delta\tau,\eta) = I_0'''(\eta)\exp(-\Delta\tau/\eta) + \qquad (8.3.6)$$

$$+ \frac{2}{S}\int_0^1 I_0'''(\eta')I'''(\Delta\tau,\eta,\eta')d\eta'.$$

It should be mentioned that the left part in (8.3.6) contains the intensitie for the case of lightning by radiation $I_0'''(\eta')$ and under the integrals – in illumination by direct rays from the direction η'. For the latter we haven't introduced any special notation, and they will be distinguished on the basis of the number of arguments.

In the case of lighting the layer from below all considerations remain the same. However, in order to remain in the system of co-ordinates similar to the case of direct radiation, it is necessary to 'invert' the co-ordinate axes (to exchange $\Delta\tau$ and 0) and all intensities should be denoted by $\tilde{I}$.

$$\tilde{I}^m(\Delta\tau,\eta) = \frac{2}{S}\int_0^1 \tilde{I}_0^m(\eta')\tilde{I}^m(\Delta\tau,\eta,\eta')d\eta',$$

$$\tilde{I}^m(0,-\eta) = \tilde{I}_0^m(\eta)\exp(-\Delta\tau/\eta) + \tag{8.3.7}$$

$$+\frac{2}{S}\int_0^1 \tilde{I}_0^m(\eta')\tilde{I}^m(0,\eta,\eta')d\eta'.$$

We now pass directly to the method of composition of layers (Fig. 8.3). Let us assume that the first layer has the optical thickness τ and is illuminated by direct solar radiation with the cosine of zenith angle η_0. Assume that we have obtained (taking into account multiple scattering) the intensities outgoing from the layer for the case of illuminating it from above and below $I_1^m(0,-\eta,\eta_0)$, $I_1^m(\tau,\eta,\eta_0)$, $\tilde{I}_1^m(0,-\eta,\eta_0)$ and $\tilde{I}_1^m(\tau,\eta,\eta_0)$. To this we add from the bottom a thin layer with a thickness $\Delta\tau$ in which scattering is assumed to be single, the layer parameters are assumed to be independent of τ and, consequently, the intensities leaving this layer are determined by equations (8.3.3). It is required to find the intensities of radiation outgoing from the combined layer: $I^m(0,-\eta,\eta_0)$, $I^m(\tau+\Delta\tau,\eta,\eta_0)$, $\tilde{I}^m(0,-\eta,\eta_0)$ and $\tilde{I}^m(\tau+,\Delta\tau,\eta,\eta_0)$.

The key to solving the problem is the determination of the intensities at the contact boundaries of the layers. We determine $I^m(\tau,-\eta,\eta_0)$ – the intensity of radiation from the first into second layer. The second layer is now illuminated from above by direct radiation which, however, is weakened in accordance with Bouguer's law during passage through the first layer, i.e. instead of S for the second layer we should now take $S\exp(-\tau/\eta_0)$. In addition to this, the second layer is illuminated from above by the radiation from the first layer $I_0^m(\eta) = I_1^m(\tau,\eta,\eta_0)$. Generally speaking, this intensity should change because of the scattering in the first layer of radiation incoming from the second layer. However, if this change is taken into account, the scattering in the second layer will no longer be single scattering (we take into account the scattering of radiation whose change has already been caused by the scattering in the second layer). Thus, remaining in the framework of approximation of single scattering in the second layer, for the intensity $I^m(\tau,-\eta,\eta_0)$ we obtain not an equation but

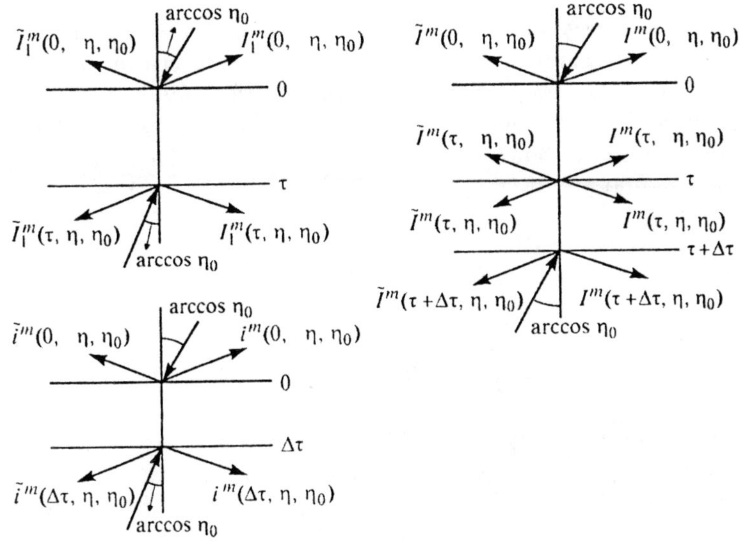

Fig. 8.3. Method of composition of layers.

an explicit relationship

$$I^m(\tau,-\eta,\eta_0) = i^m(0,-\eta,\eta_0)\exp(-\tau/\eta_0) + \\ +\frac{2}{S}\int_0^1 I_1^m(\tau,\eta',\eta_0)i^m(0,-\eta,\eta')d\eta'. \tag{8.3.8}$$

The first term is the intensity of direct radiation reflected by the second layer, taking into account the extinction of the incident direct radiation by the first layer, the second term is the reflected radiation according to (8.3.6).

We find $\tilde{I}^m(\tau,-\eta,\eta_0)$ – the intensity of radiation passed through the second layer in the case of illuminating from below. The second layer is now illuminated from above below by direct radiation incident directly on its boundary (without extinction), and from above by the radiation of the first layer with the intensity $\tilde{I}_0^m(\eta) = \tilde{I}_1^m(\tau,\eta,\eta_0)$. The required intensity is the sum of the radiations transmitted in the case of lighting from below and reflected in illumination from above:

$$\tilde{I}^m(\tau,-\eta,\eta_0) = i^m(\Delta\tau,-\eta,\eta_0) + \\ +\frac{2}{S}\int_0^1 \tilde{I}_1^m(\tau,\eta',\eta_0)i^m(0,-\eta,\eta')d\eta'. \tag{8.3.9}$$

At the boundary of the layers we now determine the intensities of radiation outgoing from the first layer. To determine $I^m(\tau,\eta,\eta_0)$ we take into account the illumination of the layer from above, giving $I_1^m(\tau,\eta,\eta_0)$, it is necessary to add the illumination from below with the intensity determined previously for which the reflection should be taken into account in accordance with (8.3.7). Consequently

$$I^m(\tau,\eta,\eta_0) = I_1^m(\tau,\eta,\eta_0) + \\ + \frac{2}{S}\int_0^1 I^m(\tau,-\eta,\eta_0)\tilde{I}_1^m(\tau,\eta,\eta')d\eta'. \tag{8.3.10}$$

To determine $\tilde{I}^m(\tau,\eta,\eta_0)$ it is necessary to take into account the direct radiation incident on the first layer from the bottom (however, this radiation was attenuated during the passage through the second layer by $\exp(-\Delta\tau/\eta_0)$ times, and the illumination of the first layer from the same side with intensity $\tilde{I}^m(\tau,-\eta,\eta_0)$. In both cases, we have a reflection. Consequently

$$\tilde{I}^m(\tau,\eta,\eta_0) = \tilde{I}_1^m(\tau,\eta,\eta_0)\exp(-\Delta\tau/\eta_0) + \\ + \frac{2}{S}\int_0^1 \tilde{I}^m(\tau,-\eta,\eta_0)\tilde{I}_1^m(\tau,\eta,\eta')d\eta'. \tag{8.3.11}$$

Using the previously determined intensities on the level of the contact of the layers we find the required intensities at the boundaries of the already combined layer. Here there are again four cases.

For $I^m(\tau+\Delta\tau,\eta,\eta_0)$ we must take into account the transmission, by the second layer, of the incident direct radiation attenuated by the first layer plus the transmission of the radiation with $I_0^m(\eta) = I^m(\tau,\eta,\eta_0)$. We obtain

$$I^m(\tau+\Delta\tau,\eta,\eta_0) = i^m(\Delta\tau,\eta,\eta_0)\exp(-\tau/\eta_0) + \\ + I^m(\tau,\eta,\eta_0)\exp(-\Delta\tau/\eta) + \\ + \frac{2}{S}\int_0^1 I^m(\tau,\eta',\eta_0)i^m(\Delta\tau,\eta,\eta')d\eta'. \tag{8.3.12}$$

For $\tilde{I}^m(\tau+\Delta\tau,\eta,\eta_0)$ we have the direct radiation incident directly on the second layer. For this radiation we taken into account the reflection to which it is necessary to add the transmission from radiation with $I^m_0(\eta) = I^m(\tau,\eta,\eta_0)$:

$$\tilde{I}^m(\tau+\Delta\tau,\eta,\eta_0) = i^m(0,-\eta,\eta_0) + \\ +\tilde{I}^m(\tau,\eta,\eta_0)\exp(-\Delta\tau/\eta) + \\ +\frac{2}{S}\int_0^1 \tilde{I}^m(\tau,\eta',\eta_0)i^m(\Delta\tau,\eta,\eta')d\eta'. \tag{8.3.13}$$

For $I^m(0, -\eta,\eta_0)$ when the layer is illuminated from above and transmission in the case of illuminating from below with $\tilde{I}^m_0(\eta) = \tilde{I}^m(\tau,-\eta,\eta_0)$:

$$I^m(0,-\eta,\eta_0) = I^m_1(0,-\eta,\eta_0) + \\ +I^m(\tau,-\eta,\eta_0)\exp(-\tau/\eta) + \\ +\frac{2}{S}\int_0^1 I^m(\tau,-\eta',\eta_0)\tilde{I}^m_1(0,\eta,\eta')d\eta'. \tag{8.3.14}$$

Finally, for $\tilde{I}^m(0,-\eta,\eta_0)$ we have illumination from below, attenuated by the second layer, and the illumination also from below with $\tilde{I}^m_0 = I^m(\tau,-\eta,\eta_0)$:

$$\tilde{I}^m(0,-\eta,\eta_0) = \tilde{I}^m_1(0,-\eta,\eta_0)\exp(\Delta\tau/\eta_0) + \\ +\tilde{I}^m(\tau,-\eta,\eta_0)\exp(-\tau/\eta) + \\ +\frac{2}{S}\int_0^1 \tilde{I}^m(\tau,-\eta',\eta_0)\tilde{I}^m_1(0,\eta,\eta')d\eta'. \tag{8.3.15}$$

Equations (8.3.3), (8.3.8)–(8.3.15) include the algorithm of the method of composition of the layers. In fact, using the thin layer of the atmosphere as the first layer and finding the intensities of this layer at its boundaries in the approximation of single scattering (8.3.3) we can gradually add thin layers to this layer until the entire atmosphere is exhausted. It should be taken into account that the method gives accurate results only for the boundaries; the determined intensities inside the atmosphere are of auxiliary nature and cannot be used for determining the vertical dependence of intensity. In fact, for this purpose it would be necessary to recalculate all intermediate intensities when adding every new layer and this is not done (and cannot be done in the frames of this method). It should also be mentioned that thin layers with single scattering are usually represented by layers with $\Delta\tau$ of the order of 0.01–0.05.

Monte Carlo method*

This is one of the most powerful computing methods of the transfer theory which makes it possible to solve numerically the problems which are 'not solvable' by other methods [46]. However, we examine the simplest realisation of the Monte Carlo method with an example of calculating hemispheric fluxes. Finally, it is possible to calculate flows after calculating intensities, integrating them in respect of angles, but in transfer theory there are also simpler methods which make it possible to calculate hemispherical flows directly. One of these is the Monte Carlo method.

The concept of the Monte Carlo method is the representation of radiation transfer in the atmosphere in the form of a random process and modelling of this process in a computer. For statistical computer-based modelling it is necessary to have a device playing the role of 'a blind case'. Similar algorithms, referred to as 'random number generators' are well known at the present time and we shall not discuss them. The application of the Monte Carlo method requires a generator of uniformly distributed random numbers in a range [0, 1]. We agree to denote these random numbers by α and when it appears in the text it will denote a new random number.

In the Monte Carlo method radiation transfer is regarded as movement through a medium of photons. To clarify the concept of the method we discuss a simpler case. Let it be that an atmosphere whose optical thickness τ_0 is illuminated by the Sun at the angle with cosine η_0 by the flux πS on an area normal to the rays. From Bouguer's law we know that the extinction of direct radiation in passage through such an atmosphere is equal to $P = \exp(-\tau_0/\eta_0)$. However, quantity P can also be interpreted as the probability p of the photon passing through the atmosphere without any interaction with it. We examine gradually the movement of the photons through the atmosphere. For each photon we use a random number of a =

* The other, more accurate term is the method of statistical modelling but the 'Monte Carlo method' is preferred. The name comes from the name of the Mediterranean resort.

α (every time a new number) and if $a < p$ the photon has passed through the atmosphere without interaction (if $a > p$ then not). We count the number of the transmitted photons: let it be that their number is $N(\tau_0)$ and the total number of photons is N. Because of the uniformity of the generator of random numbers it may be asserted that at high N the fraction of transmitted photons is equal to p but this means that $N(\tau_0) = N \exp(-\tau_0/\eta_0)$, i.e. $N(\tau_0)$ is the flux of direct (not interacting with the atmosphere) radiation at the lower boundary, but only in the units of the number of photons N. To reduce to energy units it is necessary to multiply it by the expression of the flux at the upper boundary for a single photon, i.e. by $\pi S\eta_0/N$. Thus, for direct radiation we have

$$F^{\downarrow}(\tau_0) = \frac{N(\tau_0)}{N}\pi S\eta_0. \tag{8.3.16}$$

Finally, the result (8.3.16) was known previously. We used it to explain that, simulating the transfer of radiation through the atmosphere as a random process and counting the photons, we may obtain the required values of fluxes if the number of photons is large. To extend this result to the case of multiple scattering it is necessary to model all three process: free path of the photon, its interaction with the atmosphere (absorption and scattering) and its interaction with the surface (absorption and reflection).

The free path of the photon is its displacement without interaction with the atmosphere. We have already mentioned this. Let it be that the photon is on the level τ_1 and moves at the angle with cosine η. The probability of the passage of the path $\Delta\tau$ is $\exp(-\Delta\tau/|\eta|)$. However, this relationship can also be interpreted differently: we use the random number $a = \alpha$, which will be regarded as the given probability and from this number we determine $\Delta\tau$: $\exp(-\Delta\tau/|\eta|) = a$; $\Delta\tau = -\eta \ln a$. It is precisely a random model of the free path. Now, the new position of the photon in the atmosphere τ_2 is determined from the equation

$$\tau_2 = \tau_1 - \tau \ln a \tag{8.3.17}$$

and taking into account that $\ln a < 0$, equation (8.3.17) also holds in motion downwards ($\eta > 0$) and upwards ($\eta < 0$). If in modelling using (8.3.17) $\tau_2 < 0$, the photon has left the atmosphere and it is necessary to 'launch' the next one, if $\tau_2 > \tau_0$, the photon has reached the surface and it is necessary to model interaction of the photon with the surface; otherwise ($0 < \tau_2 < \tau_0$) – photon has remained in the atmosphere and it is necessary to model its absorption or scattering in the atmosphere.

We now examine the interaction of the photon with an orthotropic surface. According to the definition of the albedo, it is the fraction of reflected radiation. Transferring this claim to the language of probabilities, identical with that made previously for the fraction of direct radiation, we immediately obtain that the albedo A is the probability of reflection of the photon from the surface. Then, if $\alpha < A$, the photon has been reflected, otherwise it is absorbed by the surface and it is necessary to model a new photon. In the reflection process a photon obtains a new direction arccos η_2 and φ_2. Since the surface is orthotropic, they are all of equal probability and are independent of the direction of incidence on the surface η_1 and φ_1. However, in this case they can be modelled quite simply using the uniform distribution

$$\eta_2 = -\cos\left(\frac{\pi}{2}\alpha\right), \varphi_2 = 2\pi\alpha,$$

where the letters α denote the different random numbers.

The case of interaction with the atmosphere is slightly more complicated. Let us assume that this interaction takes place on level τ_1 and the direction of the photon is (η_1, φ_1). The meaning of the albedo of single scattering $\Lambda(\tau_1)$ as already shown in explaining this term in paragraph 8.1 is completely identical with the meaning of the albedo of the surface: $\Lambda(\tau_1)$ is the probability of scattering of the photon. It appears that if $\alpha < \Lambda(\tau_1)$ scattering takes place, otherwise the photon is absorbed in the atmosphere and it is necessary to start modelling the trajectory of the next photon.

In scattering it is necessary to determine the angle and azimuth of scattering. We should mention the probability interpretation of the scattering phase function (see Chapter 3): the phase function $x(\gamma)$ is the probability density of light scattering at the angle γ. According to the definition of the probability density, the probability of scattering in the angle range from 0 to γ is

$$p = \frac{1}{2}\int_0^{\gamma} x(\gamma)\sin\gamma \, d\gamma,$$

where 1/2 is taken from the normalisation condition (3.3.13). We 'invert' this relationship using as the probability the random number $a = \alpha$ and pass from the angles to cosines (see paragraph 8.1). We obtain an equation for determining the cosine of the angle of scattering η in random modelling

$$\int_{-1}^{\eta} x(\tau_1,\omega)d\omega = 2\alpha. \tag{8.3.18}$$

Usually, in the Monte Carlo method the phase function is given in the form of a table in respect of argument ω: $x(\omega_i)$, $i = 1, \ldots, M$, $\omega_1 = -1$, $\omega_M = 1$. Consequently, writing integral (8.3.18) on this grid through the quadrature trapezidal formula we easily obtain an explicit expression for η. Actually, we examine a table of quantities

$$S_i = \int_{-1}^{\omega_i} x(\tau_1,\omega')d\omega' = \sum_{j=1}^{i-1}\frac{1}{2}(x(\tau_1,\omega_{j+1}) + x(\tau_1,\omega_j))(\omega_{j+1} - \omega_j).$$

We determine number k from the condition $S_k \le 2a \le S_{k+1}$. Evidently, the required point η is between ω_k and ω_{k+1} and

$$2a = S_k + \int_{\omega_k}^{\eta} x(\tau_1,\omega')d\omega' = S_k + \frac{1}{2}(x(\tau_1,\eta) + x(\tau_1,\omega_k))(\eta - \omega_k).$$

In the trapezoid rule, the subintegrand is approximated by a linear function and, consequently, using linear interpolation, we have

$$x(\tau_1,\eta) = x(\tau_1,\omega_k) + \frac{x(\tau_1,\omega_{k-1}) - x(\tau_1,\omega_k)}{\omega_{k-1} - \omega_k}(\eta - \omega_k)$$

and to determine η we obtain the quadratic equation

$$d_k(\eta - \omega_k)^2 + 2x_k(\eta - \omega_k) + (2S_k - 4a) = 0,$$

with the notations

$$d_k = \frac{x(\tau_1,\omega_{k-1}) - x(\tau_1,\omega_k)}{\omega_{k-1} - \omega_k},\quad x_k = x(\tau_1,\omega_k).$$

The solution of the equation is:

$$\eta = \omega_k + \frac{-x_k + \sqrt{x_k^2 + d_k(4a - 2S_k)}}{d_k}, \tag{8.3.19}$$

with the plus sign in front of the root selected from the condition $\omega_k < \eta \le \omega_{k+1}$. As regards the azimuth of scattering φ, since the phase function is independent of the azimuth, as in the case of reflection from the surface, then

$$\varphi_2 = 2\pi\alpha. \qquad (8.3.20)$$

After modelling the angle and azimuth of scattering by (8.3.19), (8.3.20), it is necessary to find the new direction of the photon (η_2, φ_2). It is determined from the well known equations of spherical trigonometry [67]:

$$\eta_2 = \eta_1\eta - \sqrt{(1-\eta_1^2)(1-\eta^2)}\cos\varphi, \qquad (8.3.21)$$

$$\varphi_2 = \varphi_1 + \arccos\left(\frac{\eta - \eta_1\eta_2}{\sqrt{(1-\eta_1^2)(1-\eta_2^2)}}\right). \qquad (8.3.22)$$

Thus, we have learnt how to model all processes of interaction of the photon in the atmosphere. Now, we gradually simulate N trajectories of photons (usually for fluxes we use N of the order of thousands). At the beginning of the trajectory, every photon has the co-ordinate $\tau = 0$, $\eta = \eta_0$, $\varphi = 0$. We simulate its free path and an interaction with the atmosphere or the surface, and then a new free path, and so on, until the photon emerges through the upper boundary of the atmosphere or is absorbed by the atmosphere on the surface. To calculate the fluxes on the level τ, we count the number of photons passing through the given level, i.e. the number of cases when $\tau_2 < \tau < \tau_1$ at $\eta > 0$ for $F^{\downarrow}(\tau)$ and when $\tau_2 > \tau > \tau_1$ at $\eta < 0$ for $F^{\uparrow}(\tau)$. After modelling, the required fluxes are finally determined from the relationships identical to (8.3.16). It should be mentioned that in calculating the fluxes above the orthotropic surface and with the phase function independent of the azimuth, the azmith co-ordinate may be ingored. Physically, this follows from the fact that the model of transfer and calculation of the fluxes (integrals in respect of the azimuth) does not contain processes with azimuthal anisotropy. Mathematically, we have clearly proved this statement, showing that the zenith angles of scattering in the atmosphere and reflection on the surface do not depend on the azimuth φ_1 of the photon prior to interaction, and only the zenith angle takes part in the modelling of the path of the photon and influences the counting of photons.

8.4. Algorithms and programmes for calculating radiation characteristics of the atmosphere (radiation codes)

Here we examine various methods of calculating the transmittance function, intensities and radiation fluxes (thermal and solar) in the atmosphere. These methods were used as a basis for the developing in recent years a large number of algorithms and calculation programmes (radiation codes). In many cases these codes are available for mass application and are used widely for solving different applied and scientific problems – calculating the transmittance functions for different paths of propagation of radiation, calculating the intensity and fluxes for different measurement geometry, etc. We describe the most widely used codes briefly. It should be stressed that we do not aim to provide a complete review of the radiation codes used at present because the number of these codes is very large, and we shall only illustrate them on typical examples [93, 97].

The direct method of calculation was used as a basis for developing codes in many scientific institutions (for example, St. Petersburg State University, A.I. Vavilov State Optical Institute, Institute of Atmospheric Optics of the Russian Academy of Sciences, Institute of Physics of the Atmosphere of the Russian Academy of Sciences, Kurchatov Institute of Atomic Energy, The Geophysical Laboratory of US Air Force, etc.). We describe typical examples of the currently available radiation codes.

The code FASCOD-3 (The Fast Atmospheric Signature Code – version 3) uses an effective algorithm for direct calculating transmittance functions (for homogeneous and inhomogeneous conditions) and calculating the intensity of radiation in the spectral range from ultraviolet to microwave range (0–50 000 cm^{-1}) in the altitude range from 0 to 120 km. In calculating radiation we can take into account the processes of multiple scattering of radiation, the effects of line mixing, deviation from LTE, absorption of oxygen and ozone in ultraviolet and visible ranges of the spectrum. In addition to this, calculations can be carried out for the so called weight functions used when solving inverse problems of atmospheric optics (Chapter 10). Calculations are carried out assuming a Voigt contour of spectral lines. Deviations from Lorentz lines are taken into account using continual adsorption of H_2O and a correction function for CO_2. The parameters of the spectral lines are selected on the basis of the HITRAN database. The radiation code includes different models of aerosol and clouds. The models of the atmosphere includes six climatological regions of the Earth (including changing profiles of the content of H_2O, O_3, CH_4, CO and NO_2) and mean global profiles of 20 other gases of the atmosphere. Calculations may take into account the spectral characteristics of reflection from the underlying surface. The quality of calculations using the FASCOD-3 radiation code was verified by comparison with calculations using other codes and experimental data.

RADTRAN code is designed for calculating atmospheric attenuation and the brightness temperature of thermal radiation in the spectral range 1–300 GHz. Calculations are carried out taking into account the molecular adsorption of oxygen, water vapour (lines and continuum), and extinction in clouds and precipitation. Atmospheric models were taken from LOWTRAN (see later) or are introduced into the code by the user. The geometry of measurements and the emitting properties of the surface are also specified by the user. In addition to this, the code contains the emissivities of nine types of surface (taking into account their polarisation characteristics). Multiple scattering is taken into account when calculating the radiation characteristics in precipitation.

GENLN-2 code is used to calculate the transmittance function and the intensity of thermal radiation for different measurement geometries. The refraction in the atmosphere and the specific nature of the atmosphere itself are taken into account. The shape of the spectral lines is of the Voigt type. The deviation of the shape from the Voigt shape for the wings of the lines is taken into account. The parameters of the lines are taken from the HITRAN database. The phenomena of line mixing in the adsorption bands of CO_2 are taken into account. The latest versions of GENLN-2 code can also be used to carry out calculations for a non-equilibrium atmosphere.

SHARC code was specially developed for calculating non-equilibrium radiation in the infra-red range of the spectrum (2–40 μm) for the atmospheric altitudes from 60–300 km. The code takes into account the radiation of the five most importance atmospheric gases for the examined altitudes – CO_2, NO, O_2, H_2O and CO. The code is used for calculating the population of the excited states of these molecules. In addition to this, the code contains a module for calculating the characteristics of the gas composition of the atmosphere. The selectivity of adsorption is taken into account using a model of the isolated line which restricts the limiting spectral resolution of the calculations to the value of 0.1 cm^{-1}. The shape of the lines of absorption is of the Voigt type. The initial parameters of the lines are taken from the HITRAN database.

SPbGU code is designed for calculating the transmittance functions and intensities of radiation in the infrared range of the spectrum taking into account deviations from the LTE conditions. The measurement geometry is on the horizon of the planet. The parameters of the spectral lines are taken from the HITRAN base, the line shape is of the Voigt type. CO_2 line mixing is taken into account. The code contains a block for calculating partial derivatives of the intensity of radiation using different parameters of the physical state of the atmosphere (kinetic temperature, vibrational temperatures, the content of adsorbing gases). Calculations for non-equilibrium conditions are performed using the profiles of vibrational transitions of the appropriate levels of the molecules. The SPbGU code is used mainly for solving inverse problems in the conditions of LTE breakdown.

The GOMETRAN code is designed for calculating the intensity of outgoing reflective and scattered solar radiation in the spectrum range 240–270mm. The last versions of the code, for example SCIATRAN, can be used to calculate the outgoing radiation in a longer wavelength range. In particular, GOMETRAN is designed for interpreting the measurements of GOME satellite multi-channel spectrometer (European satellite ERS-2) and, therefore, it can be used to calculate the partial derivatives of outgoing radiation in respect of different parameters of the atmosphere – the content of absorbing gases and the optical characteristics of the aerosol. To solve the equation of radiation transfer in the plane-parallel model of the atmosphere taking multiple scattering into account it is necessary to use the method of discrete ordinates. The approximate calculations of the sphericity of the atmosphere is carried out only taking into account the sphericity when calculating direct solar radiation. The initial version of the code is designed for use in a cloudless atmosphere. This was followed by the development of versions for a cloudy atmosphere. The k method is used to take into account selective molecular adsorption. Additionally, the authors of code GOMETRAN (Bremen University, Germany) developed codes for calculating the outgoing solar radiation for a spherical model (examination of the horizon of a planet).

The two following calculation codes do not use the direct method of calculating the radiation characteristics of the atmosphere.

LOWTRAN code. There are various versions of this code which have been gradually improved. We shall describe version LOWTRAN-7. The code is used to calculate the transmittance functions and the intensity of radiation for a relatively low spectral resolution (starting at 20 cm^{-1}) in a wide spectral range from 0 to 50 000 cm^{-1}. The intensities of radiation (thermal and solar) can be calculated taking into account multiple scattering. The calculations are carried out using the transmittance functions for finite spectral intervals in the form of exponential functions. Selective and continual molecular absorption, molecular scattering, aerosol scattering and adsorption, absorption and scattering on clouds and precipitation are taken into account. Calculation can also be carried out for different measurement geometries in a spherical atmosphere taking refraction effects into account. Calculations can be carried out using six mean climatic models of the atmosphere or arbitrary models introduced by the user. Molecular adsorption is taken into account using the transmittance function in the form

$$P_{\Delta\nu}(\nu) = \exp\left[-C_{\nu} u \left(\frac{p}{p^{0}}\right)\left(\frac{T_{0}}{T}\right)^{m}\right]^{a}, \tag{8.4.1}$$

where the parameters C, n, m and a are determined for every spectral range 20 cm^{-1} wide on the basis of approximation of the calculation of the transmittance function for different pressures, temperatures and the content of absorbing gases by the direct method (code FASCOD) for the same spectral ranges.

Aerosol models represent a four-layer atmosphere – boundary layer (0–2 km), troposphere (2–10 km), stratosphere (10–30 km) and mesosphere (30–120 km). For the boundary layer there are various types of aerosol models – village, city, sea, tropospheric, desert and 'naval military'. The models of the clouds includes five types of water clouds and also a number of crystalline clouds. Because of the relatively simple approximation of molecular absorption and approximate consideration of the effect of multiple scattering the code LOWTRAN-7 is characterised by medium accuracy and is usually used for applied calculations.

The MODTRAN code was developed as an improved version of the LOWTRAN code and it differs from the latter mainly by the possibility of carrying out calculations with a higher spectral resolution (starting at 2 cm^{-1}). In addition to this, the code uses a more advanced approach when taking into account molecular absorption (use of the Voigt line shape, the temperature dependence of the transmission functions, etc.). As regards its quality and possibilities, this code occupies an intermediate position between LOWTRAN and FASCOD codes.

CHAPTER 9

RADIATION ENERGETICS OF THE ATMOSPHERE–UNDERLYING SURFACE SYSTEM

The influx of heat in the form of radiant energy (energy of electromagnetic radiation) is the most important compound part of the total influx of heat whose effect changes the thermal regime of the atmosphere and of the underlying surface [20, 32, 33, 37, 43, 51, 92, 102, 103]. The radiation balance or the balance of radiant energy of the system is the difference between the radiation absorbed by the system and its natural radiation. In meteorology and physics the atmosphere we examine the radiation balance of difference systems: the surface, atmosphere, the atmosphere-underlying surface system. We start with the analysis of an important component of radiation balance (RB) – the radiation from the Sun on the upper boundary of the atmosphere.

9.1. Solar insolation at the upper boundary of the atmosphere

Insolation Q is the flux of solar radiation incident on the unit horizontal area during the given period of time $(t_2 - t_1)$ [32, 37, 79]:

$$Q = \int_{t_1}^{t_2} F^{\downarrow}(t)dt. \qquad (9.1.1)$$

These fluxes may be examined in the vicinity of the underlying surface, on different levels in the atmosphere, and so on. It should be stressed that these fluxes relate to the entire spectrum of solar radiation.

It is natural to start the study with insolation on the upper boundary of the atmosphere because this value determines the energy received from the Sun at different latitudes and different periods of the year. The solar radiation flux at the upper boundary of the atmosphere is determined by the equation

$$F^{\downarrow}(t) = F_0^{\downarrow} \cos\theta(t), \qquad (9.1.2)$$

where $F_0^{\downarrow}$ is the flux on the unit area, normal to the propagation of solar radiation, at the upper boundary of the atmosphere; θ is the zenith angle of the Sun at the given point and time. If it is taken into account that the distance between the Earth and the Sun changes during movement of the Earth around the orbit, we may write that

$$F_0^{\downarrow} = \frac{r_0^2}{r^2} S_0, \tag{9.1.3}$$

where r_0 and r are the mean and instantaneous distances of the Earth from the Sun; S_0 is the flux of solar radiation, corresponding to the mean distance (the solar constant for the Earth). The relative changes of the solar flux at the upper boundary of the atmosphere $d = \frac{F_0^{\downarrow} - S_0}{S_0}$ are presented in Table 9.1 for different months of the year. It should be noted that in winter the amount of solar energy received by the Earth in the Northern hemisphere is almost 7% larger than in summer.

The total solar energy incoming every day on the unit area may be determined by integration in equation (9.1.3) in respect of 'light' time of the day, i.e. from sunrise to sunset:

$$Q = S_0 \frac{r_0^2}{r^2} \int_{sunrise}^{sunset} \cos\theta(t)dt. \tag{9.1.4}$$

In equation (9.1.4) we ignore the variation of the ratio d during the day. The zenith angle of the Sun may be expressed through other angles – the inclination of the Sun, the time angle h and latitude φ:

$$\cos\theta = \sin\varphi \cdot \sin\delta + \cos\varphi \cdot \cos\delta \cdot \cos h. \tag{9.1.5}$$

Substituting equation (9.1.5) into equation (9.1.4) and denoting the angular velocity of rotation of the Earth by $\omega = 2\pi/24 = \pi/12$ rad/hour), we obtain

Table 9.1. Relative changes (d, %) of the flux $F_0^{\downarrow}$ at the upper boundary of the atmosphere in relation to the month N

N	1	2	3	4	5	6	7	8	9	10	11	12
d	3.4	2.8	1.8	0.2	−1.5	−2.8	−3.5	−3.1	−1.7	−0.3	1.6	1.8

$$Q = S_0\left(\frac{r_0}{r}\right)^2 \int_{-H}^{H} (\sin\varphi\cdot\sin\delta + \cos\varphi\cdot\cos\delta\cdot\cos(t\omega))\cdot dt, \qquad (9.1.6)$$

where H is the half of the light period of the day, i.e. the time from sunrise or sunset to the solar midday. Integrating equation (9.1.6), we obtain

$$Q = \frac{S}{\pi\omega}\left(\frac{r_0}{r}\right)^2\left(\sin\varphi\cdot\sin\delta\cdot H + \frac{1}{\omega}\cos\varphi\cdot\cos\delta\cdot\sin(H\,\omega)\right). \qquad (9.1.7)$$

In equation (9.1.7) the value of H in the second term on the right is expressed in radians (180 = π rad). The results of calculations by equation (9.1.7) of the daily sums of the solar energy, incoming on the unit area at the upper boundary of the atmosphere, in relation to latitude and the day of the year are presented in Fig.9.1.

Since the Sun is closest to the Earth in January (the winter in the northern hemisphere), the distribution of daily sums of the solar energy is far from uniform. The southern hemisphere receives more radiation than the northern one. The maximum insolation takes place in the summer at poles because of the duration of the light time of the day (24 h). The minimum amount, equal to zero during polar nights, is at both poles. After integrating equation (9.1.7) in respect of the yearly time period it can be shown that the total annual insolation is the same for the appropriate latitudes of the northern and southern hemispheres.

9.2. Radiation balance of the surface

According to definition, the radiation balance of the surface R_s is the difference between the radiation absorbed by the surface F_a and its natural radiation F_e:

$$R_s = F_a - F_e. \qquad (9.2.1)$$

Absorbed radiation consists of two components – absorbed solar radiation $Q(1-A)$ and absorbed outgoing long-wave radiation of the atmosphere $F_a^{\downarrow}$ (Fig.9.2):

$$F_a = Q(1 - A) + F_a^{\downarrow}, \qquad (9.2.2)$$

where Q is the total solar radiation consisting of direct Q_d and scattered Q_s solar radiation (Q=Q_d+Q_s); A is the surface albedo. If the surface is 'absolutely black' it absorbs all descending thermal

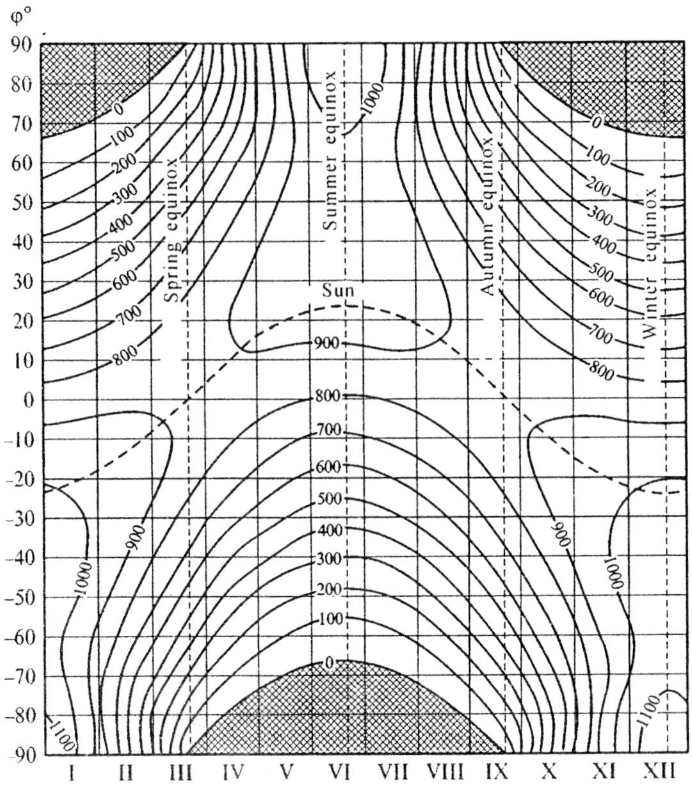

Fig.9.1. Daily sums of the solar energy incident on the unit area on the upper boundary of the atmosphere in relation to latitude and the day of the year [37].

radiation of the atmosphere $F_a^{\downarrow}$ (it is often referred to as the downward radiation of the atmosphere) i.e. $F_a^{\downarrow} = F^{\downarrow}$. If the emissivity of the surface is ε, the surface reflects part of the downward radiation of the atmosphere $(1-\varepsilon)F^{\downarrow}$ and consequently $F_a^{\downarrow}=\varepsilon F^{\downarrow}$. Thus, for the 'non-black' surface

$$F = Q(1-A) + \varepsilon F_a^{\downarrow}. \qquad (9.2.3)$$

Since in equations (9.2.2) and (9.2.3) we consider the quantities that are integral in respect of the spectrum, the surface albedo and emissivity also relate to the entire spectrum of solar and thermal radiation, respectively.

The flux of natural radiation of the surface $F_s^{\downarrow}$ is determined by its temperature and emissivity and can be represented as $\varepsilon\sigma_B T_s^4$,

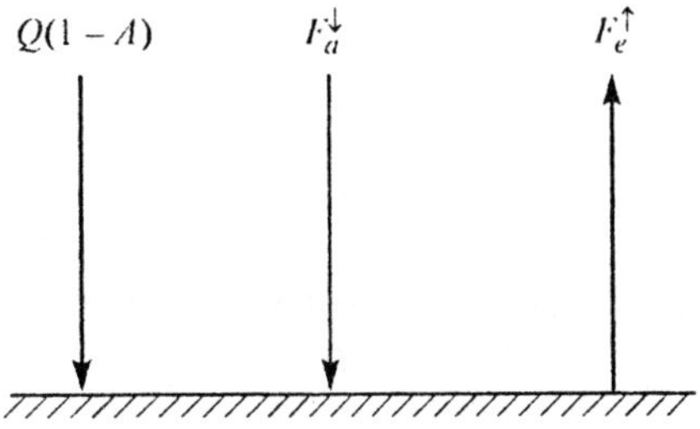

Fig.9.2. Radiation balance of the underlying surface.

where T_s is the temperature of the underlying surface. The descending thermal radiation of the atmosphere may be written in the following approximate form

$$F^{\downarrow} = \sigma_B \bar{T}_a^4 (1 - P_a), \tag{9.2.4}$$

where $\bar{T}_a$ is the mean temperature of the atmosphere; P_a is the integral transmittance function for the entire thickness of the atmosphere (for the radiation flux).

In some cases, the radiation balance is subdivided into its short-wave R_{s-w} and long-wave R_{l-w} parts. If we introduce the concept of the effective thermal radiation on the surface

$$F_{eff} = F_s^{\uparrow} - F^{\downarrow}, \tag{9.2.5}$$

Then the radiation balance for the non-black surface may be represented as follows

$$R_s = R_{s-w} + R_{l-w} = Q(1 - A) - \varepsilon F_{eff} = \\ = Q(1 - A) - \varepsilon\sigma_B T_s^4 + \varepsilon\sigma_B \bar{T}_a^4 (1 - P_a). \tag{9.2.6}$$

Calculations and measurements of surface radiation balance

The radiation balance of the underlying surface strongly affects its temperature, the distribution of temperature in soil, the ground temperature of the atmosphere, the processes of evaporation and snow melting, formation of mist and light frost, and the processes of transformation of the properties of air masses. The radiation balance of the underlying surface greatly varies and depends on the latitude, the time of year and day, climate conditions, and the properties of the underlying surface. The radiation balance is calculated for different time periods – hours, days, months, seasons and years. The methods of calculating the quantities included in the determination of the radiation balance of the underlying surface were described in chapters 7 and 8.

As an example, the diurnal variation of the radiation balance, and its short-wave and long-wave components according to the results of observations in a steppe, are presented in Fig.9.3. In daytime, the radiation balance is positive, at night it is negative, i.e. during the day the surface is heated by solar radiation, and during the night it cools down because of thermal long-wave radiation – radiation cooling of the surface. According to the results of observations, the radiation balance passes through zero at a height of the Sun of 10–15°. The long-wave component of the radiation balance is always negative, i.e. the radiation of the underlying surface is always greater than the radiation of the atmosphere absorbed by the surface (downward radiation of the atmosphere). This is caused by, in particular, by the presence of the 'transparency window' in the central infrared part of the spectrum in the atmosphere at a height of 8–12 μm with low values of the absorption coefficient and, consequently, the low downward radiation of the atmosphere. The downward radiation of the atmosphere depends on the content of the absorbing components of the atmosphere (mainly on the content of water vapour), the profile of temperature and most markedly on the cloud conditions of the atmosphere. The maximum downward radiation of the atmosphere (and consequently, the minimum value of the longwave component of the radiation balance) is found in the

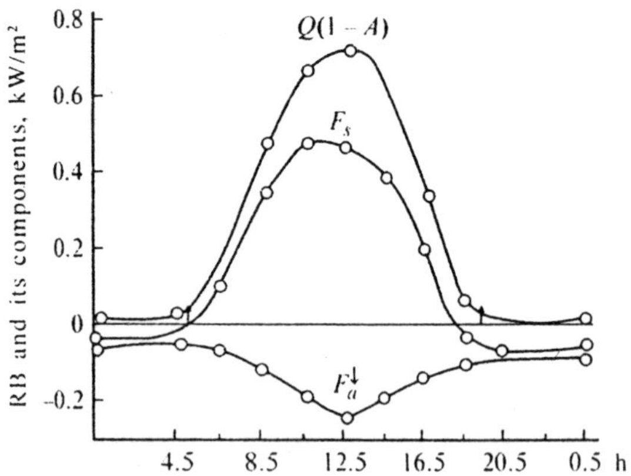

Fig.9.3. Daily course of radiation balance (RB), its short- and longwave components, according to the results of observations in a steppe [43].

presence of lower stratum clouds.

The measurements and calculations of the radiation balance of the underlying atmosphere have been continuing for a long time. As an example, Table 9.2. shows the calculations of the mean latitude distribution of the annual sums of the radiation balance for various surfaces (dry land, ocean) and for the entire surface of the globes [43].

Table 9.2 shows that the annual average values of the radiation balance of the underlying surface are positive, i.e. radiation heating of the surface takes place. The maximum annual values of the radiation balance of the underlying surface are found in the tropical regions of the surface of the ocean.

The results of a large number of measurements and calculated data were used to construct maps of the geographic distribution of the radiation balance of the surface. An example of these calculations is shown in Fig. 9.4. Figure 9.4. shows that the annual radiation balance is positive on the entire territory of the globe and changes from values close to zero in the central Arctic (10 kcal/cm^2·year) in the vicinity of the boundary of permanent ice to 80–120 kcal/cm^2·year in tropical latitudes. Nevertheless, the annual sums of the radiation balance may also be negative in regions with constant or very long-term icy or snow covers, i.e. in Arctic and

Table 9.2. Mean latitude distribution of the radiation balance of the underlying surfaces (kcal/cm^2·year)

Latitude, deg	Ocean	Dry land	Mean
70 – 60 N	23	20	21
60 – 50	29	30	30
50 – 40	51	45	48
40 – 30	83	60	73
30 – 20	113	69	96
20 – 10	119	71	106
10 – 0	115	72	105
0 – 10 S	115	72	105
10 – 20	113	73	104
20 – 30	101	70	94
30 – 40	82	62	80
40 – 50	57	41	56
50 – 60	28	31	28
Entire globe	82	49	72

Antarctic regions. Analysis of the map shown in Fig. 9.4 shows that the variation of the radiation balance in going from dry land to ocean takes place abruptly and this is expressed in a break in the isolines in the vicinity of the shorelines. This is caused mainly by the rapid change of the albedo of the underlying surfaces. The values of the albedo of the oceans are considerably lower than the albedo of the dry land. This increases the value of the absorbed total solar radiation and the radiation balance of the oceans. The second reason for the large changes in the radiation balance in going from ocean to dry land is differences in the temperature of these surfaces leading to differences in the long-wave radiation of the surfaces.

9.3. Radiation balance of the atmosphere

The incoming part of the radiation balance of the atmosphere R_a forms as a result of the absorption by the atmosphere of the longwave radiation of the underlying surface $F^{\uparrow}_{s,a}$ and the absorbed direct and scattered solar radiation Q_a (Fig. 9.5). Strictly speaking, the incoming part of the radiation balance should include the solar radiation reflected from the surface and absorbed by the atmosphere. Therefore, Q_a denotes the éntire solar radiation absorbed by the atmosphere.

The outgoing part of the radiation balance is determined by the long-wave atmospheric radiation the direction of the surface (downward radiation of the atmosphere) $F_a^{\uparrow}$ and into cosmic space $F^{\uparrow}_{a,\infty}$. Thus, the equation for the radiation balance of the atmosphere may be written in the form

$$R_a = Q_a + F^{\uparrow}_{s,a} - F^{\downarrow}_a - F^{\uparrow}_{a,\infty}. \tag{9.3.1}$$

If the integral transmittance function of the entire thickness of the atmosphere is denoted by P_a, then

$$F^{\uparrow}_{s,\infty} = (1 - P_a)F^{\uparrow}_s = (1 - P_a)\varepsilon\sigma_B T_s^4. \tag{9.3.2}$$

The sum $F^{\uparrow}_{\infty} = F^{\uparrow}_s P_a + F^{\uparrow}_{a,\infty}$ is the outgoing radiation of the atmosphere–surface system. Taking this into account, we can write the following equation for the radiation balance of the atmosphere:

$$R_a = Q_a + \underbrace{F^{\uparrow}_s - F^{\downarrow}}_{F_{eff}} - \underbrace{F^{\uparrow}_s P_a - F^{\uparrow}_{a,\infty}}_{F^{\uparrow}_{\infty}} = Q_a + F_{eff} - F^{\uparrow}_{\infty}, \tag{9.3.3}$$

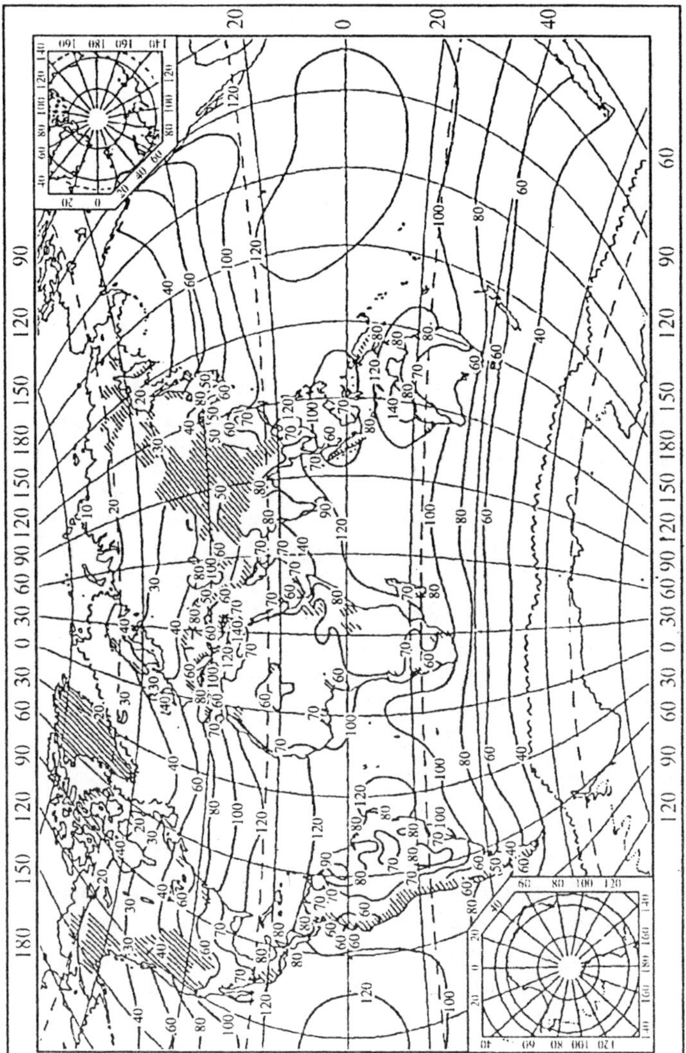

Fig. 9.4. Example of calculations of the map of the geographic distribution of the radiation balance of the surface [103].

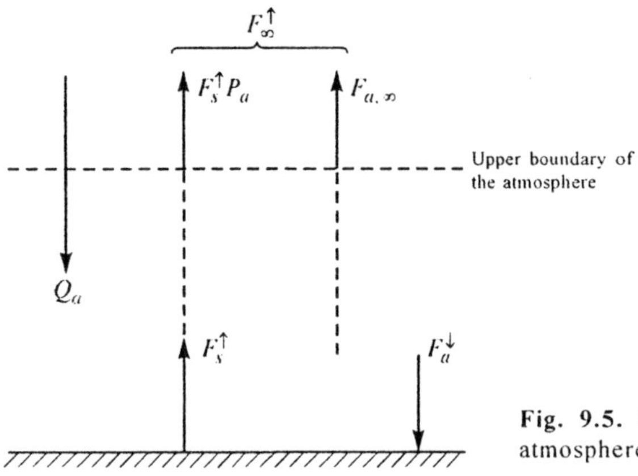

Fig. 9.5. Radiation balance of the atmosphere.

Calculations show that the mean yearly radiation balance of the atmosphere in all latitudes on average per year is negative, i.e. the atmosphere on average 'cools down' as a result of radiation transfer of radiation. This is associated with the fact that the absorption of solar radiation by the atmosphere is relatively small and the absorption of the thermal radiation of the surface by the atmosphere does not compensate the atmospheric downward radiation and the radiation outgoing into space. The variation of the radiation balance of the atmosphere with the latitude for the northern hemisphere is shown by the data in Table 9.3.

As indicated by Table 9.3, the maximum cooling is observed in tropics, the minimum – in the mean latitudes. This radiation cooling of the atmosphere is compensated by the turbulent influx of heat from the Earth's surface, and the main, – by the influx of heat as a result of condensation of the water vapour.

Radiation influx

In examining the energetics of the atmosphere it is important to study the energy absorbed and radiated not only by the atmosphere as a whole but also by its individual layers. The radiation balance of the individual layers (or levels) of the atmosphere is expressed

Table 9.3. The radiation balance of the atmosphere (annual average) for different latitudes zones (W/m²)

Latitude	0–10	10–20	20–30	30–40	40–50	50–60	60–70
R_a	–101	–110	–109	–92	–80	–80	–93

in the rate of radiation heating because of the absorption of solar radiation or in the rate of radiation cooling as a result of atmospheric radiation.

Figure 9.6 shows the results of calculating the profiles of the rates of radiation heating in the cloudless atmosphere as a result of the absorption of solar radiation. Calculations were carried out taking into account the absorption of O_3, H_2O, O_2 and CO_2, multiple molecular scattering and the reflection from the underlying surface (albedo A = 0.15). It may be seen that in the tropical model of the atmosphere the rate of heating of the troposphere is higher than in the model of the mean latitudes. This is explained by the higher content of water vapour in the tropics and by the correspondingly higher absorption of solar radiation. The maximum rate of radiation heating is found at an altitude of approximately 3 km and equals 4 deg/day. For the model of the mean latitude it exceeds 3 deg/day at this altitude. Above this altitude, the rate of radiation heating rapidly decreases as a result of a rapid decrease of the concentration of water vapour with the increase of altitude, and reaches the minimum value at an altitude of 10–20 km. In higher layers of the atmosphere the rates of radiation heating increase as a result of the absorption of solar radiation in the ozone absorption bands. The ozone concentration reaches the maximum

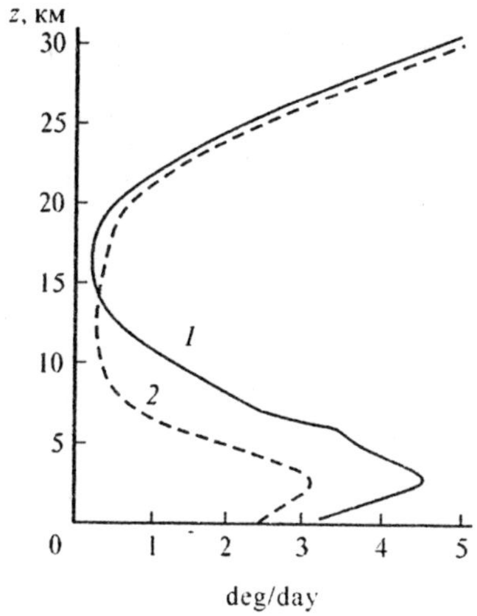

Fig.9.6. The results of calculations of the profiles of the rate of radiation heating in a cloudless atmosphere as a result of absorption of solar radiation [37]. 1) tropics; 2) mean latitudes, winter.

value approximately at an altitude of 25–30 km.

The rate of radiation heating as a result of the absorption of solar radiation varies greatly and depends on many parameters, mainly on the condition of the atmosphere (the content of absorbing gases, the presence of clouds of different types, the content and nature of aerosol particles), the zenith angle of the Sun (latitude, season, daytime), and surface albedo. To characterise the spatial variability of the rates of radiation heating as a result of absorption of solar radiation by ozone and molecular oxygen, we give Fig. 9.7. The graph shows that meridional distribution of the rates of heating of the upper atmosphere from 20 to 100 km. The graph shows, for example, that these rates in the summer months have two regions of maximum values – close to the altitude of 50 km, where they reach 18 deg/day, and close to 100 km where they reach 40 deg/day. In the former case, heating is caused by the absorption of ozone, in the second case – by the absorption of oxygen. In winter, the rate of radiation heating is considerably lower than in summer.

If the solar radiation heats the entire Earth atmosphere, the role of infrared radiation is more complicated. Thermal radiation mainly cools down the atmosphere because it carries the energy to the

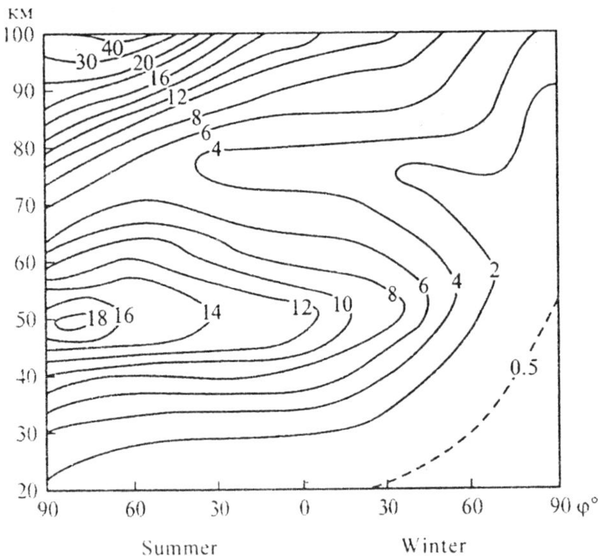

Fig.9.7. Meridional distribution of the rate of heating of the upper atmosphere from 20 to 100 km [8].

cosmic space by outgoing radiation. However, under certain conditions, and at certain levels in the atmosphere and in certain regions infrared radiation may heat the atmosphere.

Radiation cooling of the atmosphere is caused mainly by the radiation of water vapour, carbon dioxide and ozone. Figure 7.5 showed the examples of calculations of the rates of radiation changes of the temperature of the atmosphere for the absorption bands of ozone and carbon dioxide. The zone-averaged meridional distributions of the rates of radiation changes of temperature for January and July are shown in Fig. 9.8. Maximum cooling is found in the summer stratosphere and is determined exclusively by the transfer of radiation in the bands of ozone and carbon dioxide. This is caused by the fact that the contribution of water vapour at these altitudes is very small because of the low concentration of water vapour at these altitudes. In addition to this, the ozone is responsible for slight heating of the stratosphere by infrared radiation. This effect is clearly visible in tropical and sub-tropical latitudes above the tropopause in both summer and winter. This heating is caused by the increase of the ozone concentration with the altitude in the stratosphere and by low temperatures in the region of the tropopause.

The rates of radiation changes of temperature as a result of the transfer of infrared radiation depend mainly on the thermal structure of the atmosphere, the amount of absorbing gases, and the presence of clouds of different type. Figure 9.9 shows the meridional distribution of the zone-averaged total rates of radiation changes of temperature (as a result of solar and infrared radiation) for January and July. Calculations were carried out taking into account climate models of the Earth atmosphere with clouds also taken into account. Infrared cooling exceeds solar heating in almost the entire atmosphere. In the upper stratosphere, above approximately 25 km, there is strong cooling (4–5 deg/day) as a result of radiation of CO_2 and O_3 whose magnitude is higher than heating as a result of the absorption of solar radiation by ozone. Cooling in the troposphere is associated mainly with the radiation of water vapour and is maximum in the tropics. The presence of clouds (see later) results in slow cooling at the lower levels of the atmosphere. The effect of the clouds changes with the latitude and season because of their variability. The region of slow heating extends from the summer polar regions to tropical latitudes in the winter hemisphere on the levels of approximately 5 km and is associated with the absorption of solar radiation by water vapour.

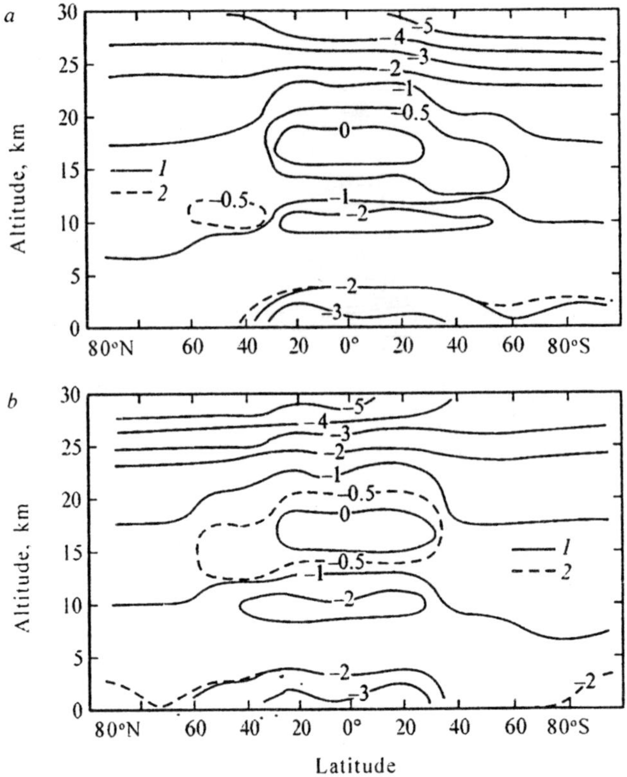

Fig.9.8 Zone-averaged meridional distribution of rates of radiation variations of temperature in the absence (1) and presence (2) of aerosols [37] for January (a) and July (b).

This special feature is also caused by heating as a result of the presence of clouds. The maximum values of heating are found in the effect caused by the continuous solar illuminated of the atmosphere is stronger than the effect of large solar zenith angles. Infrared cooling in the same regions is relatively small because of the low temperature of the atmosphere and the effect of clouds. In both hemispheres, the maximum cooling is found in the boundary layer in the region of the tropics in the winter. This is associated with a high content of water vapour in the vicinity of the surface and partially with the relatively small amount of clouds in the tropics in winter in comparison with the amount of clouds in summer.

Effect of cloudiness on radiation balance

The clouds in the Earth's atmosphere cover on average on 50% of

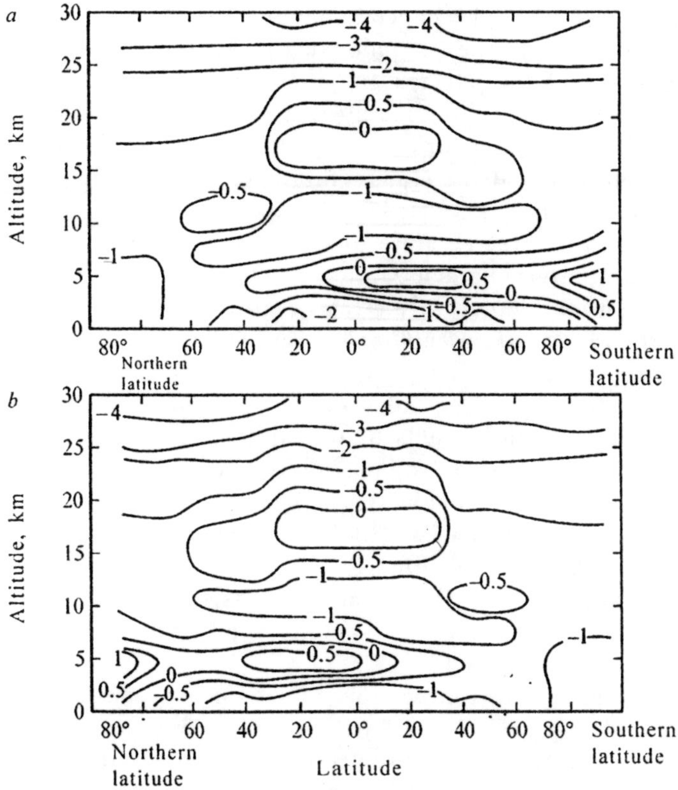

Fig.9.9. Meridional distribution of the zone-averaged total rates of radiation changes of temperature (as a result of solar and infrared radiation) for January (a) and July (b) [37].

the surface of the globe. They are a powerful 'regulator' of the radiation balance of the surface and of the atmosphere itself. The presence of clouds greatly increases the reflection of solar radiation into space. Usually, the albedo of the clouds is greater than the albedo of the oceans and land. An exception is surfaces covered with snow and ice. This effect, referred to in the English literature as the effect of solar albedo, decreases the amount of solar energy available for transformations in the atmosphere–surface system and results in cooling of this system.

On the other hand, the clouds reduce the amount of thermal radiation outgoing into space as a result of the absorption of thermal radiation by the Earth's surface and the layers of the atmosphere below the clouds. This decrease is associated with the fact that in this case the outgoing radiation forms at lower temperatures detected at the upper boundaries of the clouds. This effect is referred to as the greenhouse effect and increases the radiation balance of the surface and the atmosphere and causes heating of the atmosphere–surface system. The resultant effect of clouds on radiation balance strongly depends on the horizontal and vertical characteristics of the clouds, their phase state, the content of liquid or solid phase of water, and the size distribution function of cloud particles and temperature of the clouds.

For the quantitative characterization of the effect of clouds on the rate of radiation change of the temperature of the atmosphere we present the results of calculations (Figs. 9.10 and 9.11) Figure 9.10 shows the vertical profiles of the rates of radiation cooling as a result of transfer of infrared radiation for a cloudless atmosphere and atmosphere with the clouds of the upper, middle and lower strata for the standard model of the atmosphere. The clouds are distributed between the levels: for the upper stratum – 200–450 mbar for the middle one – 450–735 mbar, and for the lower one – 745–950 mbar. The position of the clouds are shown in the figure. The absorption of H_2O, CO_2 and O_3 was taken into account. The rates of radiation cooling for the cloud conditions are given for the cloud amounts of up to 100%.

In the cloudless atmosphere, the rate of infrared cooling is maximum (~2 deg/day) near the Earth surface. The rate decreases with increase of altitude to ~12 km and then increases because of absorption and emission in the bands of carbon dioxide and ozone. In the presence of the clouds of a higher stratum the maximum rate of cooling (~4 deg/day) is found at the upper boundary of the clouds. There is also small heating in the vicinity of the base of the clouds.

In the case of the clouds of the middle belt there is strong cooling (~11 deg/day) at the upper boundary of the clouds and intensive heating (~4 deg/day) at their base. Slow heating also takes place in the adjacent layer of the atmosphere in the vicinity of the lower boundary of the clouds of the middle stratum.

A similar pattern is also detected for the clouds of the lower stratum. The graphs show clearly that the presence of the clouds greatly changes the vertical profiles of infrared radiation changes of temperature. In this case, the main special features are the strong cooling of the upper part of the clouds and intensive heating of the base of the clouds.

The rates of heating as a result of the absorption of solar radiation is shown in Figure 9.11. Calculations were carried out for

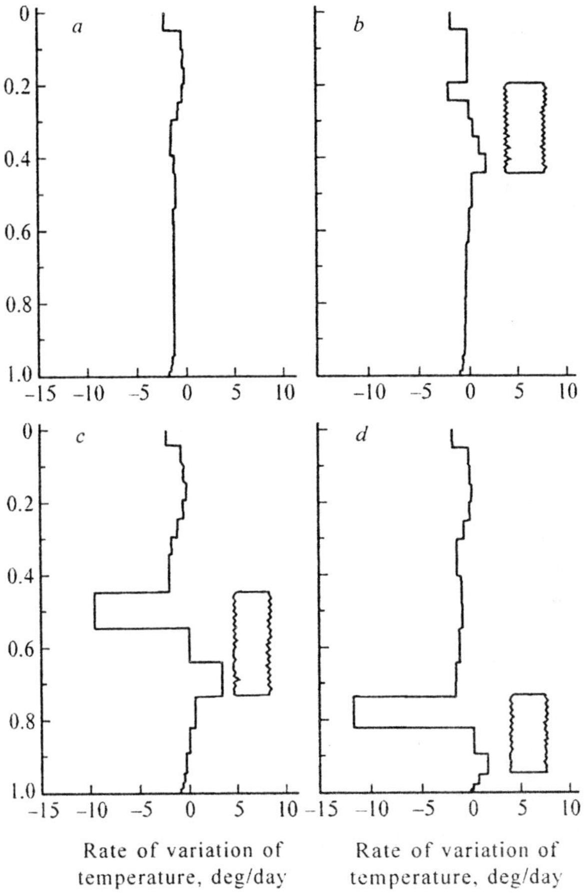

Fig.9.10. Vertical profiles of the rates of radiation cooling as a result of transfer of infrared radiation for a cloudless atmosphere (a) and an atmosphere with clouds of the upper (b), middle (c) and lower (d) stratum for the standard model of the atmosphere [103].

a solar constant of 1365 W/m^2, the surface albedo of 0.2 and the cosine of the solar zenith angle of 0.8. The heating rates in the cloudless atmosphere are on average smaller than ~1.5 deg/day. In the presence of the clouds of the upper belt the rate of heating is ~2 deg/day at the upper boundary of the clouds. For the clouds of the middle and lower stratums, these rates are ~8 deg/day at the upper boundary of the clouds.

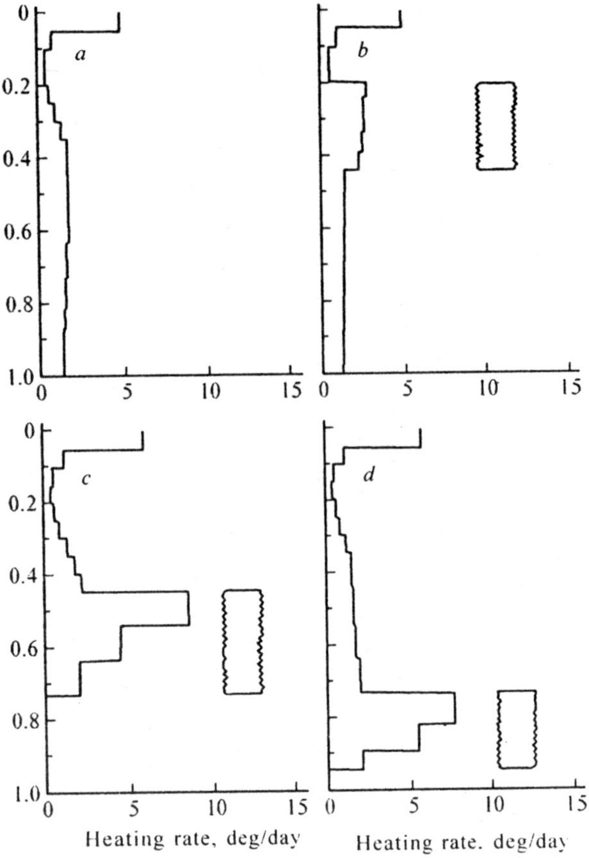

Fig.9.11 Vertical profiles of the rates of radiation heating as a result of the absorption of solar radiation [103]. (a) Cloudless atmosphere; (b) clouds of the upper stratum; (c) clouds of the middle stratum; (d) clouds of the lower stratum.

9.4. Radiation balance of the planet

Study of the radiation balance of the entire atmosphere–underlying surface system is of considerable interest. This balance characterises the balance of the radiation energy in the vertical column, including the active layer of the surface and the entire atmosphere. In other words, the radiation balance of the Earth as the planet is discussed here. 'The input part' of this balance consists of the direct and scattered solar radiation absorbed by the Earth's surface and the atmosphere, and 'output' part is the outgoing longwave radiation:

$$R = Q(1 - A) + Q_a - F_\infty^\uparrow. \tag{9.4.1}$$

Equation (9.4.1) is obtained as the sum of radiation balances of the surface (9.2.6) and the atmosphere (9.3.3).

The equation for R can also be written in the form

$$R = Q_\infty(1 - A_n) - F_\infty^\uparrow, \tag{9.4.2}$$

where Q_∞ is the mean (per annum) flux of the direct solar radiation at the upper boundary of the atmosphere; A_n is the albedo of the Earth as the planet. The value of the mean flux of the solar radiation Q_∞ is easy to calculate. Infact, the amount of solar energy coming to the Earth in unit time is equal to the product of the solar constant S_0 by the area of the cross section of the Earth πR^2 (R is the mean radius of the Earth). This value is distributed over the entire surface of the globe and equal to $4\pi R^2$. Thus, the mean value of the flux of solar radiation per unit of the horizontal surface of the Earth at the upper boundary of the atmosphere is 1/4 S_0. The first term in the right-hand part of equation (9.4.2) is the solar radiation absorbed by the entire atmosphere–underlying surface system. It should be stressed that here A is not the albedo of the surface and it is the albedo of the entire atmosphere–underlying surface system, i.e. the albedo of the planet.

Investigations of the radiation balance (RB) of the planet and of its components have been carried out using calculation method and special apparatus on board of satellites. These satellite devices can be used to measure the solar constant S_0, the component of solar radiation reflected and scattered by the atmosphere–underlying surface system, characterised by the albedo of the planet A, and outgoing long-wave radiation. Investigations of the components of the radiation balance of the Earth using artificial satellites have been conducted for a long period of time starting in 1959 (TIROS satellite). The polar satellites make it possible to study, during a relatively short period of time, the entire globe and the geostationary satellites can be used for the almost continuous studying of the radiation balance in specific geographical regions (tropical and middle latitudes).

A relatively high stability of the climate of the Earth indicates that the heat losses averaged over the globe as a result of the outgoing long-wave radiation balance approximately the absorbed solar radiation. The mean global albedo, according to current estimates, is approximately 30%, the mean flux of the outgoing long-wave radiation is 229 W/m^2.

Figure 9.12 shows seasonal variations of the global albedo of the planet, the outgoing thermal radiation and radiation balance of the Earth. These data were obtained on the basis of 48 average monthly maps constructed using satellite measurements. The mean monthly albedo of the planet has a maximum value (approximately 32%) during the December solstice. The seasonal variations are partially connected with differences in the surfaces of land and the ocean, in the cloud cover and in distribution of the snow and icy covers for northern and southern hemispheres. Seasonal variations $F_\infty^\uparrow$ show a maximum in the close to June–July. This is associated with the large areas of the dry land ion the northern hemisphere and with high temperatures of the dry land in the summer months. Thus, the global mean temperature of the surface of the globe is maximum in July (~16.7 °C) and minimum in January (~13.1 °C). The seasonal variations of the surface temperature control the seasonal dependence of the outgoing thermal radiation. In regions with great humidity, another factor determining the seasonal dependence of $F_\infty^\uparrow$

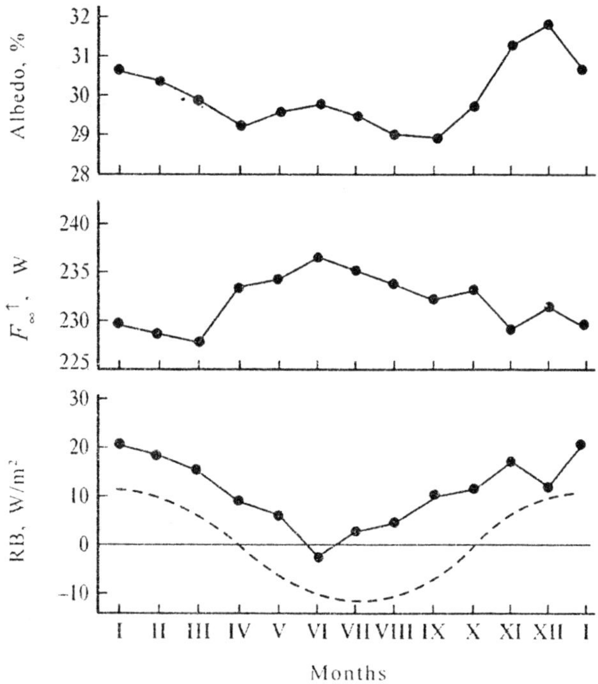

Fig.9.12. Seasonal variations of global albedo of the planet, outgoing thermal radiation and Earth's radiation balance [103]. Dotted curve – average monthly deviations from average yearly insolation for S_0 = 1.376 W/m².

is the seasonal variation of the cloud amount.

The lower part of Figure 9.12 shows the seasonal behaviour (in the form of deviations from the mean annual value) of solar insolation at the upper boundary of the atmosphere caused by changes in the distance between the Earth and the Sun. The seasonal variation of the radiation balance of the Earth repeats the seasonal dependence of solar insolation. It should be mentioned that the maximum difference in the solar insolation (January–July) reaches 22 W/m^2. The seasonal variations of solar radiation, absorbed by the atmosphere–surface system are approximately half of the variations of solar insolation. This is associated with the increase of the albedo of the planet with increase of solar insolation. The seasonal variations of the outgoing thermal radiation of the same order as the variations of the absorbed solar radiation, but they are shifted by 180° in phase.

The latitude variations of the mean annual and seasonal (winter and summer) radiation balances of the atmosphere–underlying surface system are shown in Fig. 9.13. The mean annual radiation balance is approximately symmetric in the northern and southern hemispheres with the maximum value at the equator. This maximum is associated with the minimum values of the outgoing long-wave radiation in the tropics caused by the presence of heavy cumulus clouds, and with the strong absorption of solar radiation in the equatorial regions. This special feature is most clearly visible in Fig. 9.14 which shows the latitude distributions of the zone-averaged values of absorbed solar radiation and outgoing long-wave radiation. The negative radiation balance in the polar regions is associated with the fact that the values of the outgoing thermal radiation are considerably greater than the absorbed solar radiation. The low values of absorption are caused by the high values of the albedo of snow and ice. The latitude variations of the radiation balance are also caused by changes of the mean zenith angles of the Sun with latitude. In the winter period the radiation balance has a maximum in the vicinity of 30°S, whereas in the summer in the vicinity of 15–20°N.

The global maps of the mean values of the outgoing long-wave radiation, the albedo of the planet and the radiation balance of the atmosphere–surface system, obtained on the basis of satellite measurements over a period of five years (1979–1983) are shown in Fig.9.15 [103]. The values of the albedo show large variations at the dry land–ocean boundaries in the latitude band 30°N–30°S as a result of the presence of high convective clouds with a high

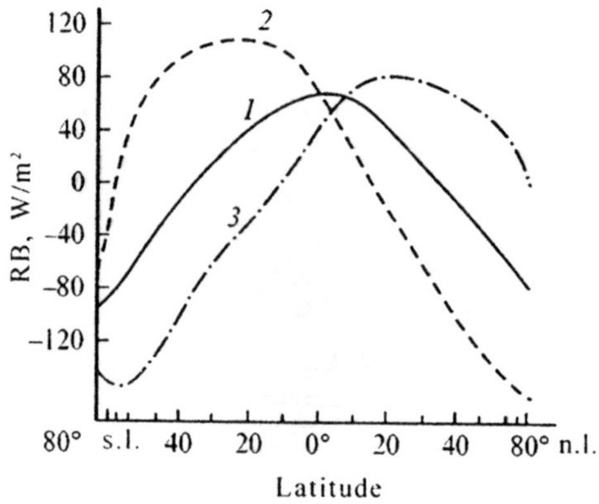

Fig.9.13. Latitudinal variations of the mean annual (1) and seasonal (winter and summer) (2,3) radiation balances of the atmosphere–underlying surface system [103].

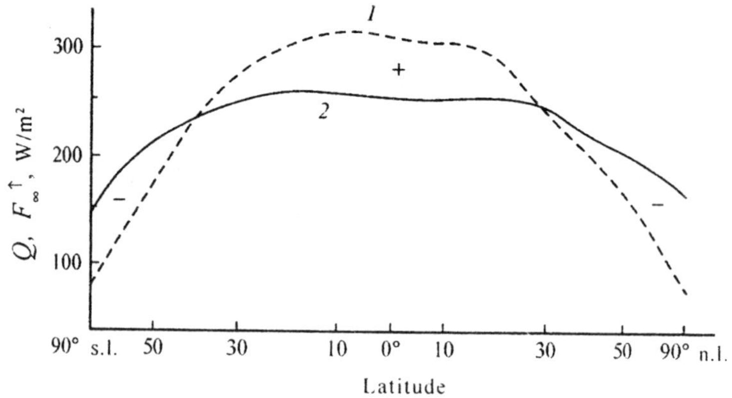

Fig.9.14. The latitudinal distributions of the zone-averaged values of absorbed solar radiation (1) and outgoing long-wave radiation (2) [37].

albedo and low values of thermal radiation. In the ranges extending from 30°N and 30°S to geographic poles, the radiation balance is relatively zone-homogeneous, especially in the southern hemisphere. In low latitudes there are individual regions of the input and output of radiation energy. Large variations in the radiation balance are detected in tropical and subtropical regions where the region of desserts in Africa and Arab lands is characterised by negative or slight positive anomalies. In the regions close to Asia with high intensity convective processes there are high positive values of the radiation balance. On the whole, the correlation coefficient between the albedo of the system and the outgoing long-wave radiation is

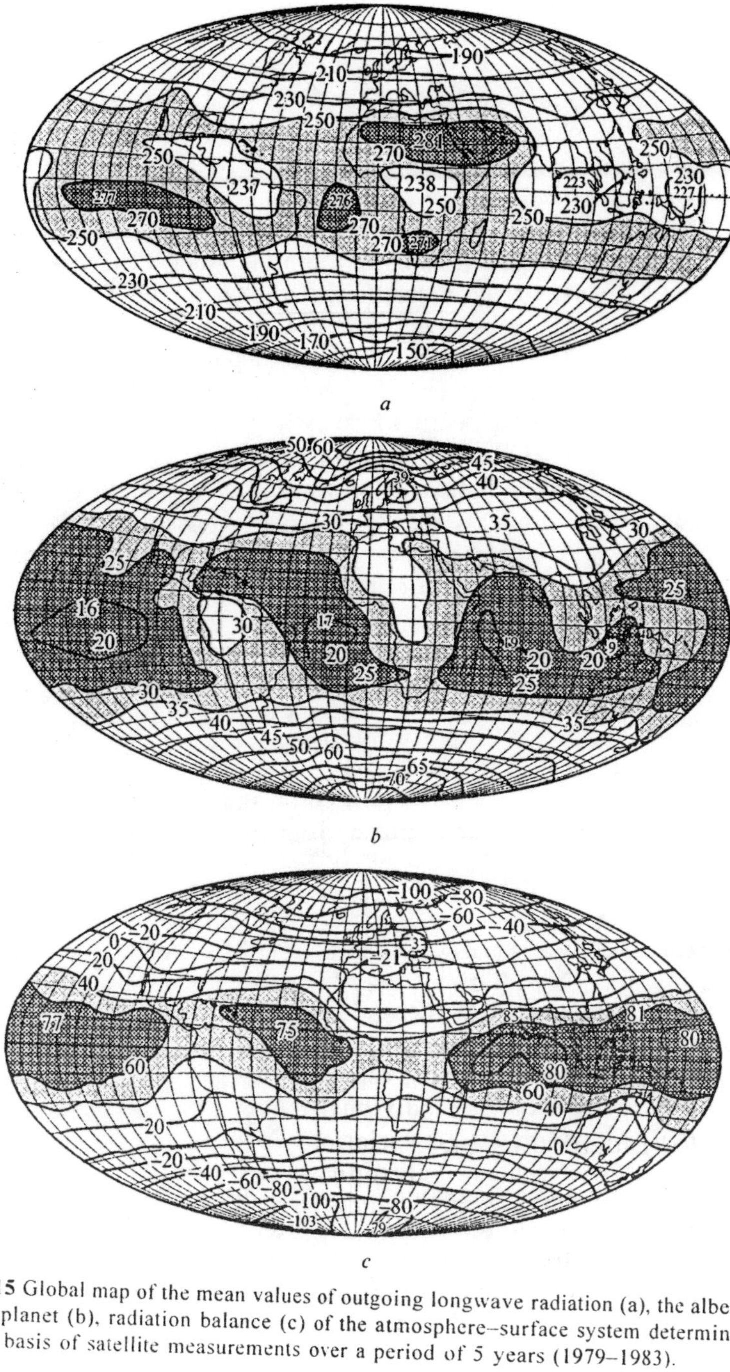

Fig.9.15 Global map of the mean values of outgoing longwave radiation (a), the albedo of the planet (b), radiation balance (c) of the atmosphere–surface system determined on the basis of satellite measurements over a period of 5 years (1979–1983).

negative mainly as a result of the effect of clouds on these values. An exception is the regions of the desert where the number of clouds is minimum and the surface is strongly reflective and is warm. For the latitude greater than 40° the radiation balance of the planet is basically negative. This sink of radiation energy increases at approach to the poles. The radiation balance of the planet is determined mainly by the fields of temperature and cloud cover.

The components of the radiation balance depend on the time of day. The adsorbed solar radiation and the albedo change during the day as a result of the dependence of the processes of scattering and absorption on the zenith angle of the Sun and also on the atmospheric state – mainly of the changes with time on the type and amount of clouds. The outgoing long-wave radiation changes during the day as a result of daily variations of the amount and types of clouds, humidity and the content of absorbing gases and the aerosol and the temperature stratisification of the atmosphere. The quantitative characterisation of the effect of clouds is carried out using the special parameter of the cloud radiation effect [103]. If the outgoing long-wave radiation of the atmosphere–underlying surface system in the conditions of partial cloud cover is represented as the sum of the radiation for the cloudless F_1 and cloudy F_2 atmosphere:

$$F = (1 - n)F_1 + nF_2. \tag{9.4.3}$$

where n is the amount of the clouds, the parameter of the cloud radiation effect C is determined from the equation

$$C = F - F_1 = n(F_2 - F_1).$$

In some cases, this parameter is determined separately for solar C_s and thermal C_e radiations, respectively. The global mean annual values of C_s and C_e are equal to 48 and 31 W/m^2, respectively, which correspond to the parameter of the total radiation effect of the clouds of $C = -17$ W/m^2.

Global radiation balance

We study the global radiation balance of the atmosphere– underlying surface system determined by calculations. The main input parameters in the calculations of the global radiation balance of the atmosphere–underlying surface system are the vertical profiles of the different characteristics of the atmosphere (temperature, gas

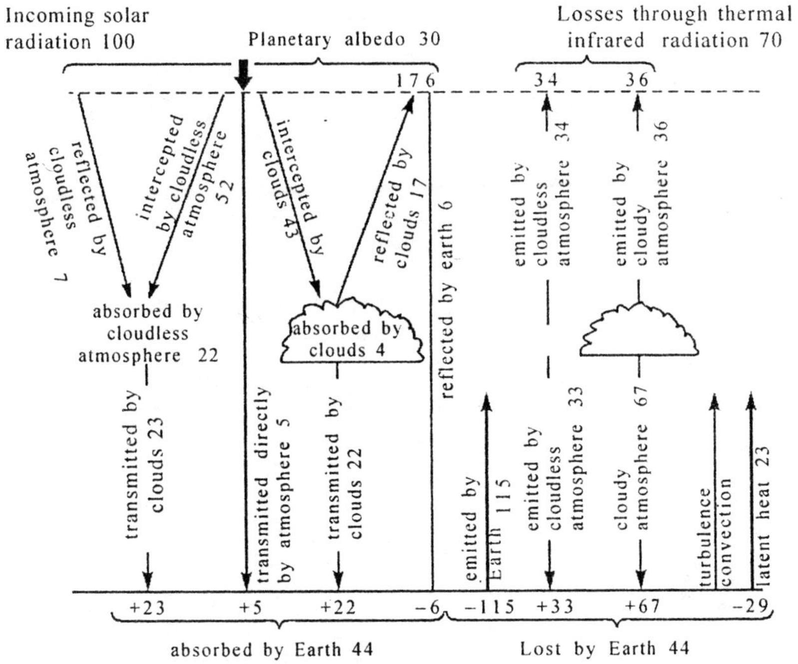

Fig.9.16 Analysis of the global radiation balance of the atmosphere–underlying surface system determined by calculations [37].

and aerosol composition), the geometrical and physical properties of the cloud cover, the global amount of every cloud type (lower, middle and upper strate), the albedo of the Earth's surface, the duration of solar radiance, and the zenith angle of the Sun. The results of analysis for the global mean state of the atmosphere are presented in Fig.9.16.

Figure 9.16 shows firstly the distribution of solar radiation in the atmosphere, secondly the distribution of thermal infrared radiation, thirdly the contribution of non-radiation processes to energy transfer (turbulence, convection and condensation of water vapours). The radiation, averaged per annum, coming at the upper boundary of the atmosphere from the Sun, is regarded as 100 %. This solar radiation is sub-divided into three components: the component propagating in the cloudless atmosphere (52%), the component propagating in the cloudy atmosphere (43%) and the component directly arriving on the underlying surface (5%). From the incoming solar radiation, 26% is absorbed by the atmosphere, and 22% – in the conditions of the cloudless atmosphere (absorption by atmospheric gases and aerosol) and 4% – by clouds. The Earth's surface absorbs 44%. The total absorption of solar radiation by the atmosphere–underlying surface system is 70% and 30% of solar radiation is reflected back into the cosmic space, including 7% reflected by the cloudless atmosphere, 17% reflected by clouds, 6% – by the surface of the Earth.

At the same time, the atmosphere and the underlying surface generate their own thermal radiation. The outgoing thermal radiation is 34% for the cloudless atmosphere 36% for the cloudy atmosphere which gives a total of 70%. Thus, the losses of energy by the system as a result of outgoing thermal radiation are equal to the absorbed solar radiation. The upward flux of thermal radiation on the level of the Earth's surface in the units used here is 115%. The downward radiation of the atmosphere =100% (downward radiation of the cloudless atmosphere 33%, downward radiation of the cloudy atmosphere 67%). Thus, the effective radiation flux at the surface is 15%. Adding up together (taking the sign into account) the thermal radiation fluxes coming in and leaving the atmosphere we obtain that as a result of thermal radiation the atmosphere losses 55% of radiant energy incoming from the Sun. If it is taken into account that the atmosphere absorbs only 26% of the arriving solar radiation, the radiation losses of the atmosphere are 29%. These loses are compensated by the generation of heat as a result of the condensation of water vapour evaporated from the surface (the latent heat flux) and by the heat flux from the underlying surface as a result of turbulence and cellular convection. These upward heat fluxes equal 23 and 6%, respectively.

9.5. Radiation factors of climate changes

Climate and its changes

The most convincing confirmation of changes in the climate of the Earth is Ice Ages – gradual oncoming and disappearance of ice cover in moderate latitudes observed in the last million years. It should be mentioned that the last Ice Age ended approximately 10 thousands years ago. The past climates of the Earth are studied in a special section of climatology – paleoclimatology.

Climate change factors

Changes in the climate detected in the last decades and manifested in different manners – the increase of the mean global temperature of the Earth, the decrease of the polar 'ice caps' and the rising of the level of the world's oceans – have a strong effect on different aspects of the life of humans and functioning of various branches of economy. This has been the main reason for special interest paid to the problems of the climate of the Earth in the last couple of decades. The task of forecasting climatic changes is one of the priority tasks in modern science. To solve this problem, new models of the climate have been developed and are being developed on the basis of the theory of the climate which is very complicated because of a large number of factors which may influence the climate of the planet and the complexity of the quantitative description of different physical, chemical, and biological processes, controlling the state of the planet. In the group of the factors of climate changes one can mention astronomical reasons (Chapter 1), the drift of the continents, changes in the position of the geomagnetic pole, and many others. For the relative short-term (several hundreds of years) forecasting of climate changes, the most important factor, according to the current views, is the radiation factors – special features of the distribution of radiation balance over the Earth surface. As shown previously, they are related with special features of the composition of the atmosphere: the content of the gases and aerosols absorbing radiation, characteristics and amount of clouds, the properties of the underlying surface.

In particular, it should be stressed that the climate of the Earth is determined to a large degree by the natural greenhouse effect (GE) [95]. This effect is caused by the relatively high transparence of the atmosphere for short-wave solar radiation and considerable absorption of the long-wave radiation by the surface of the Earth by various atmospheric gases such as water vapour, carbon dioxide, ozone and clouds. Because of this absorption, a large part of the radiation of the surface does not escape to the cosmic space and heats the atmosphere. If there was no absorption of long-wave radiation then, according to simple estimates, the mean temperature of the Earth would be approximately 255 K. The mean temperature of the Earth's surface is 20–30 K higher than this temperature.

When examining the Earth's climate there are two important problems – the time variation of the natural GE and the effect of antopogenic activity on the natural GE. In analysis of these problems it is conventional to sub-divide the processes determining the GE of the atmosphere–surface system, into two classes:

– the forcing processes regarded as the external effects on the climatic system;

– feedback processes in the climatic system.

The processes of the effect include changes in the content of optically active gases (CO_2, methane, etc), natural and antopogenic aerosols, the products of volcanic eruptions and the solar constant. Direct consequences of these effects are changes of the temperature in the system and its circulation. The temperature changes may lead to other changes of the radiation properties of the atmosphere and the surface. For example, an increase of temperature can increase the evaporation of the water vapour from the surface of oceans. An increase of the content of water vapour in the atmosphere (the most important greenhouse gas) can result in a further increase of the temperature of the system. This example of the forcing illustrates a positive feedback. Other feedbacks may be associated with changes of the amount, height and type of clouds, the planetary albedo with a result of changes in the snow and ice cover, and with changes of the vegetation cover and the albedo of land. Finally, the relatively slow but very large potential changes in the heat fluxes and the accumulation of energy by the oceans should be considered as a feedback. It should be mentioned that the separation of the processes into the forcing processes and the feedback processes is not always unambiguous. For example, changes in the albedo of land may be interpreted as the feedback processes and the processes of the effect on the climatic system, taking into account, for example, changes in the albedo of land as a result of forest cutting.

Suitable examples of different effects and also feedbacks in the climatic system are numerous and this in particular explains the difficulties in the simple evaluation of current changes in the Earth's climate. Another reason for difficulties and indeterminacy of these estimates is the fact that many processes, including radiation processes, have been studied insufficiently. For example, insufficient attention has been given to examining the processes of absorption of solar radiation by different types of clouds. Estimates of this component of the radiation balance of the atmosphere vary from 3 to 10%.

Effect of greenhouse gases

Changes in the composition of the atmosphere were discussed in Chapter 2 where it was noted that they may be caused by both natural and antropogenous reasons. For example, the agricultural activity of mankind increases the content of carbon dioxide in the atmosphere. Thus, the problem of climate forecasting is complicated by the need to separate natural reasons for changes in the climate from antropogeneous reasons which may be controlled.

In the atmosphere of the Earth there are also changes of other atmospheric optically (greenhouse) gases in the infrared range of the spectrum – methane, N_2O, different freons. An increase in the content of these gases will lead to the greenhouse effect. Table 9.4 gives the characteristics of contributions of different atmospheric gases to the greenhouse effect.

In addition to the greenhouse effect (per one molecule in relation to one molecule of CO_2) , Table 9.4. gives information of the mean mixing ratios of gases (in fractions of 10^{-9}) in 1992, and also the possible effect on the absorption of radiation and the gas composition of the atmosphere. The data show that many trace gases are more effective in the influence on the greenhouse effect in the atmosphere (per molecule!) than the 'conventional' greenhouse gas CO_2.

To characterise the effect of changes in the composition of the greenhouse gases in the atmosphere, we present estimates of changes in the surface temperature caused by corresponding changes of the content of these gases (Table 9.5). These estimates were obtained using a relatively simple one-dimensional model of the atmosphere disregarding feedbacks in the climatic system.

For comparison, Table 9.5 also gives the changes in the surface temperature caused by a decrease of the ozone content, a 2% increase of the solar constant and the addition of the stratospheric layer of the aerosol with the optical thickness of 0.15. The data in Table 9.5 show that the main greenhouse gas (with the exception of water vapour) is the carbon dioxide and the change of the content of carbon dioxide may lead to considerable climatic changes.

Water vapour, clouds and precipitation

As already mentioned, the water vapour is the main greenhouse gases in the Earth's atmosphere and an increase of the temperature of the atmosphere–surface system can increase the extent of evaporation of water from the surface of the oceans, the humidity of the atmosphere and the greenhouse effect of water vapour. However, in addition to this, an increase of atmospheric humidity may result in an increase of the amount of clouds because of the condensation of water vapour in the atmosphere. The increase of the amount of clouds influences the radiation balance in two ways: on the one hand, the reflection of the solar incoming radiation increases and, on the other hand, the outgoing thermal radiation of

Table 9.4. Efficiency of greenhouse effect induced by various gases [99]

Gas	Mixing ratio in 1992, ppb	Efficiency of GE*	Comments
O_3	10–200		Absorbs UV and IR radiation
CO_2	356000	1	Absorbs IR radiation, affects stratospheric ozone
CH_4	1714	21	Absorbs IR radiation, affects tropospheric O_3 and OH, stratospheric O_3 and H_2O
N_2O	311	206	Absorbs IR radiation, affects stratospheric ozone
$CFCl_3$ (CFC–11)	0.268	12400	As above
CF_2Cl_2 CFC–12)	0.503	15800	As above
CH_2HCl (HCFC–22)	0.105	10600	As above
CH_3CCl_3	0.160	2730	As above
CF_3Br (H–1301)	0.002	16000	As above

* Greenhouse effect is given in relation to the GE of carbon dioxide

Table 9.5. Variations of surface temperature [99]

Mechanism of the effect	*T*, deg
Increase of content of CO_2 (300 – 600 ppm)	1.31
Increase of content of CH_4 (0.25 – 0.56 ppm)	0.16
Increase of content of N_2O (0.16 – 0.32 ppm)	0.27
Increase of content of CFC-11 (0 – 1 ppb)	0.07
Increase of content of CFC-12 (0 – 1 ppb)	0.08
50% decrease of ozone content at all heights	–0.38
Increase of solar constant by 2%	1.35
Addition of the stratospheric layer of aerosol with the optical thickness of 0.15	–0.99

the atmosphere–surface system decreases. We discussed these effects in paragraph 9.4 and it was shown that the overall effect in the mean may result in cooling of the climatic system. However, in individual situations, for example, in regions with high values of the albedo of the underlying surface (snow or ice) the overall effect is equal to zero or may even result in heating the climatic system.

The processes of condensation of water vapour and of precipitation on the surface of the Earth are important factors of climate regulation. An increase in the evaporation from the surface of the ocean, and an increase in the intensity of the processes of condensation of water vapour and the formation of clouds change the amount of heat coming into the atmosphere. Precipitation, especially in the form of snow, can greatly change the albedo of the surface and, consequently, its radiation balance.

Atmospheric aerosol

Study of the data obtained in the ground-based measurements of direct and scattered solar radiation shows that the amount of short-wave radiation coming on the surface in the conditions of a cloudless atmosphere on the surface, greatly changes from year to year. The main reason for these changes is the large changes in the content of aerosol particles in the atmosphere. Aerosol attenuates the solar radiation incident on the surface, increases the intensity of the processes of scattering in the atmosphere, including backscattering, i.e. can increase the component of the solar radiation reflected into space. If the aerosol is absorbing, this increases the absorption of solar radiation in the atmosphere. To a lesser degree but still noticeably the aerosol can influence the thermal radiation fluxes. For example, the scattering and, especially, the absorption of infrared radiation by the aerosols reduce the intensity of outgoing thermal radiation. The overall effect of the aerosol is greatly changeable and this is associated with large changes in the concentration of aerosol particles and their microphysical and, consequently, optical properties. On the average, according to current views, an increase of the content of aerosol in the atmosphere results in cooling the climatic system.

The stratospheric aerosol plays an important role in impacting the climate (Table 9.5). Its concentration in the stratosphere can increase by several orders of magnitude as a result of strong volcanic eruptions. It is therefore assumed that the eras of increased volcanic activity of the Earth correspond to the eras of cooling down.

Changes in the characteristics of the underlying surface

Changes in the radiation properties of the underlying surfaces can also greatly affect the radiation balance of both the surface and the entire planet. Consequently, natural and antropogenous changes of the surfaces can be responsible for changes in the climate of the Earth. These changes can be manifested both as the forcing processes (for example, antropogenous – cutting down forests, agricultural activities, growth of large cities) and feedbacks. In both cases, the albedo of the underlying surfaces changes. The processes of feedback in the climatic system include processes such as the changes of the area covered by snow and ice as a result of the warming of the climate, increasing the size of deserts, rising the level of the world oceans and flooding the areas of land. An increase of the albedo of the underlying surface or of the planetary albedo results in the increase of the reflection of incidence solar radiation into space and, consequently, may result in cooling the climate of the Earth.

CHAPTER 10

RADIATION AS A SOURCE OF INFORMATION ON THE OPTICAL AND PHYSICAL PARAMETERS OF ATMOSPHERES OF PLANETS

10.1. Direct and inverse problems of the theory of transfer of radiation and atmospheric optics

In the theory of transfer of radiation and atmospheric optics we are concerned with two types of problems which are referred to direct and inverse problems [47, 71]. The direct problems include determination of characteristics of radiation. In inverse problems the available characteristics of radiation are used to determine the optical or physical parameters of the atmosphere or underlying surfaces. It should be mentioned that the concepts of direct and inverse problems are typical of many sections of mathematical physics. The concepts of direct and inverse problems of mathematical physics are based on the directivity of the investigated causal connections.

Direct problems are oriented in the direction of the course of the causal connection, i.e. they represent problems of determining the effect of known causes. They include the problems of determination of space–time fields for given sources, calculation of the device response to a known signal at the input, and so on.

The inverse problems are associated with the conversion of the causal connection, i.e. finding unknown causes from known effects – determination of the characteristics of the sources of the field in some points or regions of the space on the basis of the results of measurement of the parameters of the fields, retrieval of the input signal on the basis of a response at the output of a device, and so on [75]. The simplest example of an inverse problem will be discussed. The Bouguer law describes the processes of extinction of radiation during its propagation in the atmosphere. The simplest characteristic of such extinction is the optical thickness of the investigated layer. Knowing the incoming I_0 and outgoing I radiation for this layer, it is easy to determine an important optical characteristic – the optical thickness of the layer:

$$\tau = \ln\left(\frac{I_0}{I}\right). \tag{10.1.1}$$

Equation (10.1.1) is the simplest example of solving inverse problems of the theory of radiation transfer.

If we know the physical reasons for the observed weakening of radiation, for example, it is caused by natural absorption, we can write an explicit expression for the optical thickness of the layer: $\tau = ku$, where k is the absorption coefficient, u is the amount of absorbing substance. If we know the values of the absorption coefficient, for example, if these values were measured previously in laboratory conditions, the next simple step is the determination of the amount of absorbing substance u. This is an example of solving an inverse problem of atmospheric optics.

Remote methods have been used successfully for a long time in the study of the atmosphere of the Earth and other planets. In his monograph Twilight [61], G.V. Rosenberg described an interesting example, possibly one of the first remote measurements of the atmospheric characteristics. The Middle Ages philosopher and scientist El Hasan determined the vertical dimensions of the Earth's atmosphere (52 thousand steps) analysing twilight phenomena.

At present, in investigations of the Earth the remote methods are used both in observation from space and from the surface of the Earth or using different flying systems. The advanced remote methods of measurement use the measurement of radiation in a very wide spectral range – from ultraviolet to radiowave range. All remote methods of measurements are inverse problems of atmospheric optics. However, we show that the inverse problems of atmospheric optics is a more general concept [71]. In atmospheric optics in solving direct and inverse problems we are concerned with the following quantities and characteristics (Table 10.1):

a) characteristics of the radiation field (it will be denoted by symbol J);

b) parameters of the physical state of the medium (X);

c) parameters of the interaction of radiation with the medium (optical parameters) (A);

d) boundary conditions (G);

e) geometry of the examined medium (S).

Table 10.1. Main characteristics used on atmospheric optics [71]

Parameter or charactristic	Examples	Notation
Characteristics of the field of electromagnetic radiation	Stokes vector – parameter, radiation intensity, etc	J
Parameters of the physical state of the medium	Temperature of atmosphere and underlying surface, concentration of absorbing, scattering and emitting molecules and aerosol particles, speed of wind, moisture content of soil, etc	X
Parameters of the interaction of radiation with the medium (optical parameters)	Einstein coefficients, coefficients of absorption, scattering and radiation, scattering phase function, etc	A
Boundary conditions	Solar radiation at the upper boundary of the atmosphere, radiation of the underlying surface, etc	G
Geometry of the medium	Plain-parallel atmosphere, spherical atmosphere, etc	S

When solving the direct problems of atmospheric optics it is assumed that the parameters of the physical state of the medium and the parameters of interaction of radiation with the medium are known together with the boundary conditions and geometry of the system, and it is required to determine some characteristics of the radiation field. This means that schematically the direct problem may be represented as follows:

$$X+A+G+S\rightarrow J.$$

The inverse problems of atmospheric optics may be formulated in different ways. The classification of different inverse problems of the atmospheric optics is given in Table 10.2. It is assumed that the geometry of the medium is given for all these problems, as it is usually the case in practice.

The first type of inverse problems is different remote methods of measurement. They can be represented by the scheme:

$$J+A+G\rightarrow X$$

The measured characteristics of the radiation field and the given parameters of the interaction of radiation with the medium and the boundary conditions are used to determine different parameters of the physical state of the atmosphere and the underlying surface. The simplest example of a problem of this type (determination of

Table 10.2 Classification of inverse problems of atmospheric optics [71]

Type	Given	To be determined	Comments
1	J, A, G	X	Remote methods of measuring physical parameters of medium
2	J, X, G	A	Determination of optical parameters of medium
3	J, X, A	G	Determination of boundary conditions

the content of the absorbing gas) was presented previously.

The second type of inverse problems of atmospheric optics is directed to determining optical characteristics of the atmosphere. This type of inverse problems can be represented by the scheme:

$$J+X+G \rightarrow A.$$

It should be noted that this approach is traditional in the laboratory investigations of the optical characteristics of atmospheric gases. His approach is also used when determining the optical properties of a real atmosphere.

The third type of inverse problems of atmospheric optics is formulated in relation to the boundary conditions as follows:

$$J+X+A \rightarrow G.$$

The best known example of this type is the problem of determination of the integral or spectral solar constant on the basis of the ground-based or balloon measurements of direct solar radiation.

10.2. Remote measurement methods

Two types of methods are used for measuring the parameters of the atmosphere and the underlying surface: contact and remote measurements. Contact measurements are measurements of a parameter at a specific 'point' (limited volume) of the atmosphere or surface (limited area). In contact measurements a sensitive sensor is in direct contact with the investigated object. An example – measurement of the temperature of the atmosphere using a thermometer. Contact measurements are used for a large number of measurements of different parameters of the atmosphere and the surface. However, it is almost impossible to obtain by these measurements the information on the state of the atmosphere in the regional or global scale. To a great degree, this relates to other planets. Therefore, remote measurements are very important in atmospheric sciences.

In a general case, remote measurements are based on the recording of the characteristics of different fields – gravitational, electrical, magnetic, electromagnetic, acoustic. In these methods, the characteristics of the medium are measured at a distance from the investigated volume of the atmosphere or surface area. These distances can very large, for example, for satellite measurements or in examination of planets from the surface of the Earth. In this monograph, we examine remote methods based on recording the characteristics of the electromagnetic field (radiation). The processes of generation of radiation or its transformation depend on the optical and physical parameters of the medium. These dependencies are also the physical basis of the examined remote measurement methods.

Figure 10.1 shows the diagram of remote measurements. The most important element of these measurements is the measuring device. The input of the device receives electromagnetic radiation. The devices measure different characteristics of the radiation field – angular, spectral, polarization – depending on time and often on the point in space. At the output of the device the signals are proportional to some functionals of the radiation field. For example, in spectral measurements we obtain information on the spectral dependence of radiation with a specific spectral resolution:

$$I_{\Delta\nu}(\nu) = \int_{\Delta\nu} I(\nu)\varphi(\nu - \nu')d\nu, \qquad (10.2.1)$$

where $\Delta\nu$ is the resolved spectral range; $\varphi(\nu-\nu')$ is the spectral slit function of the device which describes quantitatively the reaction (response) of the device to radiation with different frequency.

The device also carries out angular and time averaging of intensity of radiation described by the appropriate slit functions. It is important to stress that for subsequent interpretation of measurements, i.e. obtaining important optical or physical parameters of the medium, it is necessary to know different characteristics of the device. These characteristics are usually examined in advance in a laboratory.

To solve the inverse problems of atmospheric physics, including problems in remote sounding of the medium, as indicated by Table 10.2 and Fig.10.1, it is necessary not only to measure the specific

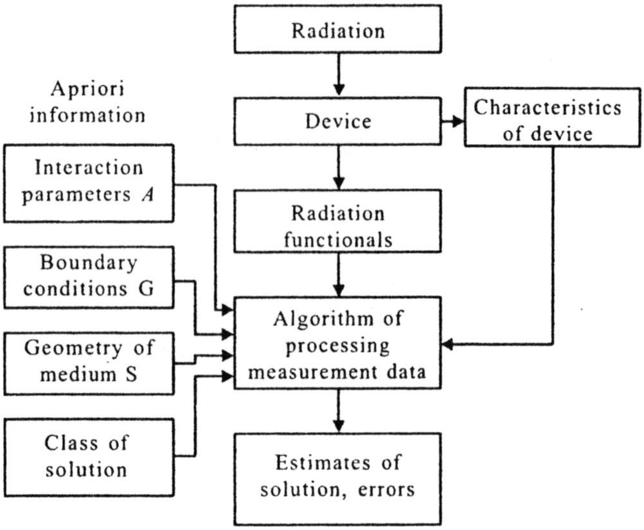

Fig.10.1 General block diagram of remote measurements [71].

characteristic of radiation but also provide a certain amount of information which is referred to as *a priori*. The *a priori* information differs depending on the type of inverse problems. Thus, the remote methods of measurement of the parameters of the atmosphere and the underlying surface include not only the measuring device but also a specific amount of the *a priori* information and also the algorithm of retrieving the required parameter. Therefore, the accuracy of remote sounding by a medium depends not only on the accuracy of measurements of some characteristics of the radiation field but also on the accuracy of *a priori* information.

Direct measurements of the device are only the initial stage of realization of remote measurements, they do not solve the given task of determination of the parameters of the medium. This is carried out in the algorithm of processing the measurement data (algorithm of interpretation) – a special system of calculation codes for a computer. This algorithm includes the *a priori* information on the model of radiation transfer, the parameters of interaction of radiation with the medium, the boundary conditions and the geometry of the medium. This information, included in the equation of radiation transfer, forms the physical–mathematical model of remote methods of measuring the parameters of the atmosphere and the underlying surface. As mentioned previously, this model should also contain different characteristics of the measurement device – spectral and angular resolution, appropriate slit functions, different measurement errors, and so on.

It may be shown that the majority of the inverse problems of atmospheric physics are reduced, from the mathematic viewpoint, to solving integral equations of a special type – Fredholm integral equation of the first kind [75]:

$$y(x) = \int_a^b K(x,y) f(y) dy, \tag{10.2.2}$$

where $y(x)$ is the known (measured) function; $K(x,y)$ is the kernel of the integral equation; $f(y)$ is the function to be determined. In this case $y(x)$ is the characteristics of radiation; $f(y)$ is the required parameters of the atmosphere or surface. The relationship (10.2.1) also belongs to this type of equation. Infact, if we want to obtain, from the measurements of the radiation intensity in finite spectral intervals $I_{\Delta\nu}(\nu)$, the monochromatic intensity at the input of the device $I(\nu)$ at the known kernel of the equation $K(x, y) = \varphi(\nu-\nu')$, it is necessary to solve the equation of type (10.2.1). A simple example relating to atmospheric optics is the example presented previously for the optical thickness of the atmosphere. In a general case, for an inhomogeneous atmosphere the optical thickness is expressed as follows

$$\tau(\nu) = \int k(\nu,z)\rho(z)dz \tag{10.2.3}$$

If we know the optical thickness, then to determine the vertical distribution of the density of the absorbing gas (with the given absorption coefficient $K(x,y)$) it is necessary to solve the Fredholm integral equation (10.2.2)

A general special feature of these equations is their incorrectness in the classic sense (according to Hadamard) which leads in particular to the need to specify another type of *a priori* information – the class of the solution to be found. An exception is represented by the inverse problems of remote refractometry reduced to solving the integral Volterra equations of the first kind which are correct according to Hadamard. These special features of the solution of the inverse problems require the application of special algorithms of interpretation of the measured data.

Like any other measurement methods, the remote measurements

are characterised by the errors (random and systematic) of the measurement of the sought parameter of the atmosphere or underlying surface. In addition to this, measurements in atmospheric physics, meteorology, climatology, etc (including remote measurements) are characterised by the spatial (horizontal and vertical) resolution, periodicity of measurements, and the rate of transfer of the results of measurements to users. Since remote measurements represent a set of the measuring device (with its measurement errors), the specific amount of *a priori* information and the interpretation algorithm, all these components determine the accuracy characteristics of remote measurements.

10.3. Classifications of remote measurement methods

Because of the existence of a large number of remote measurement methods in atmospheric optics it is convenient to consider their general classification. The remote measurement methods of the parameters of the environment are classified according to different criteria:

– the type of radiation used (nature of radiation, source of radiation);

– the main processes of interaction of radiation with the investigated medium;

– illumination conditions (the time of day);

– the spectral range;

– the parameter to be determined;

– the carrier used.

Firstly, the remote measurement methods are subdivided into passive and active (depending on the nature of radiation used). The passive methods, using the measurements of the characteristics of natural radiation fields, include:

1) Methods of atmospheric radiation (equilibrium and non-equilibrium);

2) Methods of scattered radiation (solar and reflected from the moon);

3) Methods of attenuation and absorption (transmittance) mainly of solar radiation, but also the radiation of the moon and stars;

4) Refraction methods;

5) Methods of reflected radiation.

6) In many cases, the second and fifth passive methods are combined into a single method – the method of scattered and reflected radiation.

The active methods of sounding, using artificial sources of electromagnetic radiation, are:

- lidar sounding;
- radar sounding;
- the refraction method;
- the extinction and absorption method.

The classification of the remote measurement methods on the basis of the main processes of interaction of radiation with the investigated medium is close to that described previously. This classification specifies:

- scattering methods (various types – molecular (Rayleigh), aerosol, Raman, etc);
- extinction (absorption) methods;
- atmospheric radiation methods;
- refraction methods, etc.

In this classification, the methods of absorptions, scattering and refraction are used in both passive and active measurement methods.

According to the illumination conditions (the time of day) the remote methods can subdivided into: 1) day methods (above the illuminated side of the planet); 2) night methods; 3) methods used in the terminator region (the region of transition from the day to night side of the planet). The latter methods, especially in gound-based measurements, are often referred to as the methods of twilight sounding. The first and third passive methods are associated with the application of solar radiation as a source of information on the medium state and are used for the day side of the planet. At night, the remote methods can also be based on the measurement of radiation of stars and of the solar radiation reflected from the Moon, and also on the measurement of various glows of the atmosphere.

In principle, the active sounding methods can also be used at any time of day. However, the presence during the day of a high level of reflected and scattered solar radiation complicates the application of, for example, lidar methods in the visible and near infra-red ranges of the spectrum above the day side of the planet.

From the viewpoint of the carrier used, the remote methods are subdivided into ground-based, aircraft, balloon, rocket and space.

According to the geometry of measurements, the cosmic methods can be sub-divided into the methods of nadir and limb (on the horizon of the planet) sounding. Figure 10.2 shows different types

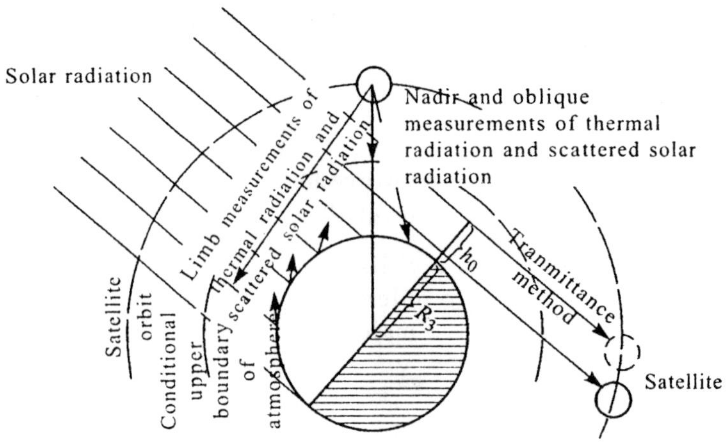

Fig.10.2 Different types of geometry and methods of space measurements.

of geometry and different methods of space measurements. In nadir geometry of measurements (or close to those – oblique), the outgoing radiation is recorded in directions in the vicinity of the local vertical. The majority of currently available satellite devices carries out angular scanning in the vicinity of the nadir (in the majority of cases normal to the plane of the orbit) so that the horizontal fields of the investigated characteristics can be determined. The identical result may be obtained using special receivers – linear or matrix type. The range of scanning angles, the angular aperture of the devices, the type of scanning and the height of the space carrier determine the spatial range of measurements and the horizontal resolution of remote measurements.

The remote methods of measuring the parameters of the atmosphere and underlying surface can be classified by the parameter to be determined. In this classification there are remote methods for determination of:

– temperature, density and pressure of the atmosphere;

– characteristics of the clouds – amount, the height of upper and lower boundary (vertical structure), temperature of the upper boundary (UB), water content, phase composition, and microphysics of clouds;

– intensity of precipitation;

– the content of absorbing gases – water vapour, ozone and other trace gase (TG);

– wind fields;

– aerosol characteristics (optical and microphysical);

– properties of the underlying surface – temperature, moisture content and the optical characteristics of the underlying surfaces (reflectivity and emissivity).

10.4. Remote methods of measurement based on measurements of attenuation (absorption) of radiation

These methods were already discussed when we considered the concepts of the inverse problems of the theory of transfer and atmospheric physics. In the majority of cases, these methods use the radiation of the sun and they are applicable during the day time, although a large number of examples in recent years have been associated with the utilization of radiation of stars and of the solar radiation reflected from the moon. The two latter cases enable this remote method to be used at night.

The considered methods are most efficient in studying the characteristics of the gas and aerosol atmospheric state, and also in a number of cases of clouds. In addition to this, using the temperature dependence of molecular absorption, it is possible to determine the temperature of the atmosphere. Selecting different spectral ranges and spectral intervals from the ultraviolet range to the radio range in measurements, for example of solar radiation, we can obtain information on the content of tens of different gases in the atmosphere. These experiments have been conducted from the surface of the Earth for a very long time. It is interesting to note that these remote methods have been used to detect the presence of ozone and many other gases in the Earth's atmosphere. The discussed remote methods are often referred to as the transmittance methods because the information on the physical parameters of the physical state is extracted in fact from the transmittance functions of the atmosphere (transmissivity of the atmosphere).

Taking into account different factors attenuating the radiation, the optical thickness may be represented in the form:

$$\tau(\nu,s_1,s_2)=\int_{s_1}^{s_2}\sum_{i=1}^{N}k_i(\nu,s)\rho_i(s)ds+\tau_R(\nu,s_1,s_2)+\tau_a(\nu,s_1,s_2), \quad (10.4.1)$$

where $k_i(\nu,s)$ is the coefficient of the absorption of the i-th gas component; $\rho_i(s)$ is its density; $\tau_R(\nu,s_1,s_2)$ and $\tau_a(\nu,s_1,s_2)$ are the optical thicknesses of Rayleigh and aerosol extinction. It should be mentioned that equation (10.4.1) was written for the case of the monochromatic optical thickness. In actual experiments it is necessary to take into account the finite spectral resolution of the devices. Thus, the examination carried out here is idealized because the measurements of strictly monochromatic radiation are not possible. However, there are devices with super high spectral resolution (for example, heterodyne spectrometers) for which the study of the monochromatic case is of practical interest. Equation (10.4.1) shows that the radiation extinction in a general case is determined by different physical reasons – true absorption by atmospheric gases, extinction as a result of molecular (Rayleigh) scattering, the absorption and scattering by atmospheric aerosols. Using different spectral dependences of different extinction mechanisms, we can define different components of this extinction. Depending on the spectral range of measurements, the characteristics of the device and the aims of interpretation, equation (10.4.1) represents the basis of different remote methods:

- determination of the characteristics of the gas composition of the atmosphere;
- determination of the temperature of the atmosphere;
- measurement of the velocity of wind;
- determination of the density of atmosphere;
- measurements of different characteristics of atmospheric aerosols.

We consider different remote methods of measurement assuming for simplicity it is possible by some means to separate different components of the radiation extinction in equation (10.4.1).

Determination of the characteristics of the gas composition of the atmosphere by the transmittance method

Firstly, it should be mentioned that in the ground-based experiments there is a problem of determination of the intensity of the extra-atmospheric solar radiation $I_0(\nu)$. This intensity is determined using the long or short Bouguer methods. For the model of a plane-parallel horizontally homogeneous atmosphere we can write the equation for the optical thickness:

$$\tau_\rho(\nu) = \frac{1}{\cos\theta}\int_0^\infty k(\nu,z)\rho(z)dz. \qquad (10.4.2)$$

The success of solving the inverse problem – determination of the vertical profile of the density of the absorbing gas $\rho(z)$ from equation (10.4.2) – depends on the behaviour of the kernel of the equation (the absorption coefficient $k(\nu,z)$ in this case) in relation to altitude at different frequency. Let us assume that, for example, the absorption coefficient does not depend on altitude. We can therefore carry out evident transformations:

$$\tau_\rho(\nu) = \frac{k(\nu)}{\cos\theta}\int_0^\infty \rho(z)dz = \frac{k(\nu)}{\cos\theta}u, \quad u = \int_0^\infty \rho(z)dz. \qquad (10.4.3)$$

Equations (10.4.3) show that in the examined case we can not obtain information on the vertical profile of the density of the absorbing gas but can obtain information only on its integral content in the entire thickness of the atmosphere. It should be mentioned that the determination of the value of the total content u of different gases is often of considerable practical interest. For example, ground-based measurements of the absorption of ultraviolet solar radiation enable us to determine the integral (total) content of ozone with a high accuracy (1–3%). This is associated with the fact that the absorption coefficient of ozone in the ultraviolet range of the spectrum does not depend on the pressure of the atmosphere and depends only slightly on the temperature of the atmosphere. Ground-based stations for measuring the total ozone content operate in many countries of the world. For example, in CIS countries measurements of this type are now carried out at ~40 ozone measuring stations. In addition to this there is a network of stations carrying out surface spectroscopic measurements of the total content of CH_4, CO, H_2O, CO_2, N_2O in the infrared range of the spectrum. In the infrared range of the spectrum, the dependencies of the molecular absorption of the atmospheric gases greatly differ in the majority of cases. As shown previously (see Chapter 4), the molecular coefficient of absorption in the spectral line $k(\nu,z)=k(\nu,p(z), T(z))$. The relationships for the Lorentz line (derived from analysis of the equations (4.5.20) and (4.5.24)), presented in Chapter 4, show that in the centre and the wing of the line the dependences of the absorption coefficient on pressure and, consequently, altitude in the atmosphere, greatly differ. The absorption coefficient in the wing of the line is directly proportional to pressure, whereas in the centre of the line it is inversely proportional. Thus, the contributions to the optical thickness of the molecular absorption of different layers of the atmosphere in different sections of the Lorentz line greatly differ and it is intuitively clear that solution of the integral equation (10.4.2) may provide information in particular on the vertical profile of the density of the absorbing gas. This information can be obtained from the measurements with high spectral resolution enabling us to scan different spectral lines. This is realized in the microwave range of the spectrum. At middle spectral resolution the majority of the interpretation methods are based on the application of the model (mean) vertical profile of the content of the absorbing gas and the problem of determination of the total gas content is solved.

We consider another example of the application of the measurements of atmospheric transmittance. Figure 10.3 shows the geometry of space measurements of solar radiation. In movement of a satellite on the orbit, the sun rises or sets beyond the horizon of the planet in relation to the satellite. At these moments, the spectral devices installed on the satellite, can be used for measurements of solar radiation both during passage of the radiation through the atmosphere and outside it. In this case we can determine the transmittance function of the atmosphere on slant paths, i.e. the ratio $I(\nu)/I_0(\nu)$. These methods of sounding the atmosphere are often referred to as twilight methods because they are carried out during the 'darkening' of radiation by the Earth (the Sun in the present case).

The Bouguer law for the considered case can be written in the form

$$I(\nu,h_0)=I_0(\nu)\exp\left(-2\int_{h_0}^{\infty}w(z,h_0)k(\nu,z)\rho(z)dz\right), \qquad (10.4.4)$$

where h_0 is the tangent height, i.e. the minimum distance of the trajectory of propagation of a light ray from the surface of the Earth (Fig.10.3); $w(z,h_0)$ is the Jacobian of transition from coordinate f along the path of propagation of radiation to altitude z. The multiplier 2 forms as a result of division of the path into two identical integration sections which holds for the model of the spherically homogeneous (layer-stratified) atmosphere.

It should be mentioned that the measurements of solar radiation may be carried out as a function of the frequency and trajectory of propagation of radiation characterised by, for example, tangent height h_0. It is also important to mention that in space experiments (in opposite to the ground-based measurement method when there is a problem of determination of the extra-atmospheric value of solar radiation $I_0(\nu)$, solved by special methods) we can measure directly the extra-atmospheric radiation at high values of tangent

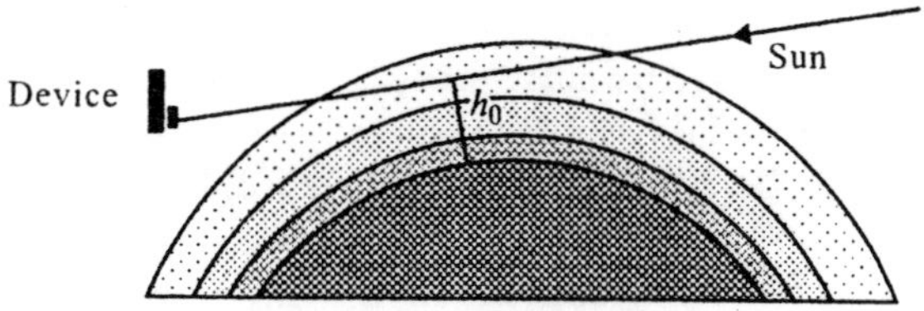

Fig.10.3. Geometry of cosmic measurements of solar radiation.

height h_0.

Passing in the monochromatic case to the optical density of molecular absorption we obtain

$$\tau(\nu, h_0) = 2\int_{h_0}^{\infty} w(z, h_0)k(\nu, z)\rho(z)dz. \tag{10.4.5}$$

Here the kernel of the integral equation is the product $w(z,h_0)k(\nu,z)$, which is assumed to be known, and the function to be determined is the density of the absorbing gas $\rho(z)$. Equation (10.4.5) is the Volterra equation of the first type since the integration limit is variable. Formally, we can solve this equation analytically differentiating the optical density in respect of tangent height h_0

$$\rho(h_0) = -\frac{d\tau(\nu, h_0)}{dh_0}\frac{1}{2w(h_0, h_0)k'(\nu, h_0)}. \tag{10.4.6}$$

The problems of a stable numerical solution of equation (10.4.5) remain because the numerical differentiation of the measured function also belongs to ill-posed (in the classic sense) problems.

It should be mentioned that our examination was carried out for the intensity of radiation. Just as devices cannot measure monochromatic radiation, and measure the radiation energy in finite spectral range (this is taken into account by integration over the spectrum taking into account the slit function of the device), they measure the radiation energy infinite solid angles. This is taken into account by integrating the intensity of radiation in finite solid angles taking the angular sensitivity of the devices into account.

In the discussed method, measurements at different tangent heights give a possibility to obtain information on the vertical structure, and the spectral extinction dependences make it possible to separate contributions to the extinction of different components, in particular, the absorption of different gases.

The possibility of extracting actual information on the vertical profile of the density of the absorbing gas is associated with the behaviour of the kernel of the equation (10.4.5) as a function of altitude at different tangent heights. Figure 10.4 gives the kernels of the integral equation (10.4.5) for the mixing ratio of ozone $q(z)$ in measurements of solar radiation in the infra-red region of the spectrum at a wave length of 285 nm in the region of the Hartley–Huggins ozone absorption band at different tangent heights in the atmosphere. It is seenthat at different tangent heights, the kernels are placed at different atmospheric altitudes. This behaviour of the kernels is also a physical basis of the possibility of determining by space measurements atmospheric transmittance the information on the vertical profile of the mixing ratio or the density of the absorbing gas. The character of the dependence of the kernels of equation (10.4.5) on altitude in the atmosphere is determined by the specific features of the geometry of space measurements, in particular by the fact that at the given tangent height h_0 solar radiation is not absorbed by the layers situated below this tangent height.

The spaxe scheme of measurements of the characteristics of the gas composition in the atmosphere was used in the last couple of decades in many experiments of different ranges of the spectrum.

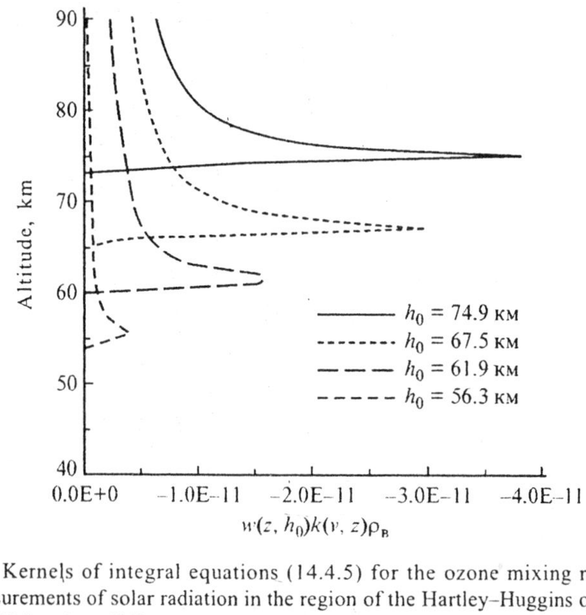

Fig.10.4. Kernels of integral equations (14.4.5) for the ozone mixing ratio $q(z) = \rho_{O3}/\rho_a$ in measurements of solar radiation in the region of the Hartley–Huggins ozone absorption band (λ=285 nm) at different tangent heights in atmosphere.

The devices with high spectral resolution and the selectivity of the spectra of molecular absorption enables us to study the content of many atmospheric gases in the upper troposphere, stratosphere and mesosphere. For example, in experiments with the infrared interferometer 'ATMOS' with high spectral resolution, the vertical profiles of more than 30 atmospheric gases were measured simultaneously.

The results of measurements of the Sun radiation in the ultraviolet, visible and near-infrared ranges of the spectrum by spectrometers SAGE-I, SAGE-II, POEM, OZON-MIR and others were used to determine the content of ozone, NO_2, H_2O and aerosol extinction. Figure 10.5 shows an example of the retriveal of the vertical profile of the ozone content from the results of measurements with OZON–MIR spectrometer (long-term orbital station MIR). The same graph gives the ozone profile determined using a different remote method (limb method) based measuring the thermal radiation of the horizon of the planet (EMS equipment). The majority of space experiments of this type use the measurements of the radiation of the sun.

However, in this case the number of these measurements per day is not large. Recently, similar measurements have been taken using the radiation of different stars. In this case it is possible to increase greatly (by an order of magnitude or more) a number of measurements per day. It is also possible to increase greatly the spatial coverage of different regions of the Earth by measurements

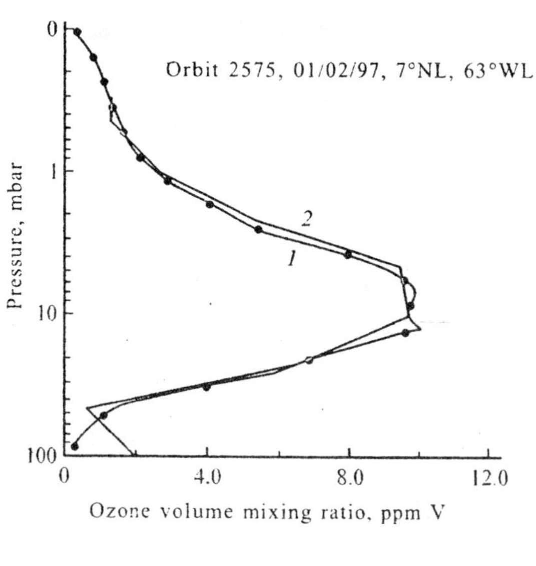

Fig.10.5. Example of restoration of the vertical ozone profiles from measurements with OZON–MIR spectrometer (1) and limb thermal radiation measurements (2) of the horizon of the planet [56] (equipment MLS).

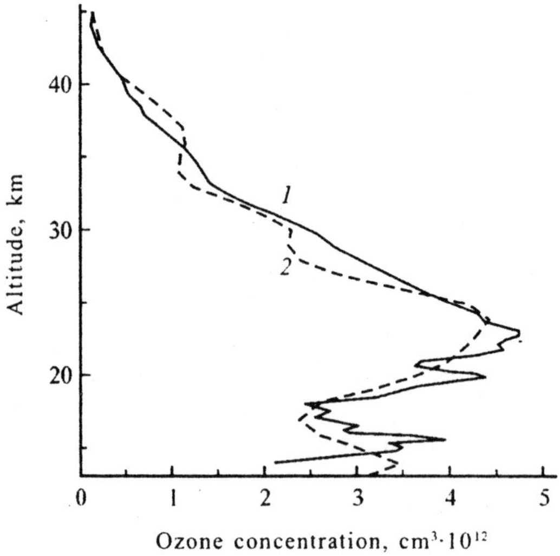

Fig.10.6. Example of retrieval of the vertical ozone profile from measurements of star radiation (experiments with UVISI equipment on board USA MSX satellite). 1) lidar sounding; 2) occultation experiment."

in comparison with the use of solar radiation. Figure 10.6 gives an example of the retrieval of the vertical profile of the ozone content from the measurements of radiation of stars (experiments with equipment UVISI on board of the American satellite MSX). To verify the results of remote satellite measurements, Fig.10.6 also gives the profile of the ozone content determined using ground-based lidar sounding.

Determination of the vertical profile of temperature by the transmittance method

We now return to the expression for the optical density of molecular absorption for the space geometry of measurements – equation (10.4.5). Previously it was assumed that the density of the absorbing gases is the required function. However, we assume that this function is known. This assumption in the Earth's atmosphere is fulfilled with high accuracy and in a wide range of latitudes for various atmospheric gases such as oxygen and carbon dioxide. In this case we can formulate the inverse problem in relation to the vertical profile of the temperature of the atmosphere $T(p)$. Actually, as mentioned previously, the intensity of the spectral lines (and also half-widths but in most cases to a lesser extent) depends on temperature (Chapter 4). For example, for the intensity of vibrational–rotational lines of linear molecules we can write to a first approximation:

$$S(T) = S_0\left(\frac{T}{T_0}\right)\exp\left(k_B E''\left(\frac{1}{T_0}-\frac{1}{T}\right)\right), \qquad (10.4.7)$$

where S_0 is the intensity of the line at reference temperature T_0; E'' is the energy of the low vibration level of the quantum transition of the molecules; k_B is the Boltzmann constant.

As shown by the relationship (10.4.7), depending on the values of energy E'' the dependence of the intensity of the line on temperature of the atmosphere differs. For example, for a CO_2 absorption line situated close to the frequency of 668.60867 cm^{-1}, the value of E'' is equal to 464.1717 cm^{-1}. Simple calculation shows that for this line the derivative $dS(T)/dT$ at T_0 = 296 K is approximately equal to $0.01S_0$ [K^{-1}]. Thus, when the temperature of the atmosphere varies by 1 K, the intensity of the line, and consequently, optical thickness $\tau(\nu)$ change by 1% which may be recorded by space devices. This means that in the examined case, equation (10.4.5) is a non-linear integral Volterra equation of the first kind in relation to the vertical profile of the temperature of the atmosphere.

Determination of the vertical profile of the density of the atmosphere by the transmittance method

The density of the atmosphere can be determined on the basis of different physical principles. We shall therefore discuss two remote methods:

- the method of molecular absorption;
- the method of extinction as a result of Rayleigh scattering.

The first method is based on relationship (10.4.5). It is used to determine, for example, for the cosmic measurement scheme, the vertical profile of the density of the i-th gas component of the atmosphere. The determination of the density of gas components such as O_2 or CO_2 whose mixing ratios in the Earth's atmosphere are known and are constant in time and in space, enables us to determine the total density of the atmosphere on the basis of the relationship

$$\rho_a = \rho_i / q_i, \qquad (10.4.8)$$

where ρ_a is the total density of the atmosphere. Substituting the relationship $\rho_i(z) = \rho_a(z)q_i(z)$ into equation (10.4.5) gives the explicit form of the integral equation for determining the total density of the atmosphere

$$\tau(\nu,h_0) = 2\int_0^{\infty} w(z,h_0)k(\nu,z)q_i(z)\rho_a(z)dz, \qquad (10.4.9)$$

where $K(z,h_0,\nu)=w(z,h_0)\ k(\nu,z)q_i(z)$ is the kernel of the integral equation.

To illustrate the second approach we assume for simplicity that the only component of extinction of radiation is the optical thickness of Rayleigh (molecular) scattering. The data presented previously (Chapter 5) show that the Rayleigh scattering coefficient depends on concentration of air molecules. Consequently, we can formulate the appropriate integral equation for the density of air.

Determination of the characteristics of atmospheric aerosols by the transmittance method

Different inverse problems in relation to the characteristics of the atmospheric aerosols have also been formulated for the transmittance methods. To simplify considerations, it is assumed that the extinction of radiation takes place only as a result of scattering and absorption on atmospheric aerosols. Therefore, the optical density in relationship (10.4.1) is determined by the extinction of radiation on aerosols. For the case of extinction on spherical homogeneous particles for polydispersed aerosols particles we can write the following expression for the optical thickness of aerosol extinction $\tau_a(\lambda)$ (5.3.2):

$$\tau_a(\lambda) = \sec\theta \int_0^{\infty} \alpha_a(z)dz =$$

$$= \sec\theta \int_0^{\infty}\int_0^{\infty} N_a(z)\pi r^2 Q_e(r,m,\lambda,z)n_a(r,z)dzdr, \qquad (10.4.10)$$

where $Q_e(r,m,\lambda,z)$ is the extinction factor of particles with radius r with the complex refractive index (CRI) m; $n_a(r,z)$ is the size distribution function of the particles. It should be mentioned that in equation (10.4.10) we took into account the dependence of Q_e and n_a on altitude in the atmosphere. For the case of an aerosol–homogeneous atmosphere, i.e. for Q_e and n_a independent of height, we can formulate the following integral relationship in relation to the normalized distribution function $f_a(r)$:

$$\tau_a(\lambda) = \sec\theta N_0 \int_0^\infty \pi r^2 Q_e(r,m,\lambda) f_a(r) dr, \tag{10.4.11}$$

where N_0 is the total number of the aerosol particles. Measuring the spectral dependence of the optical density of aerosol extinction, we can determine $f_a(r)$ [29,84]. The possibilities of the examined methods are associated with the behaviour of the kernel of this equation as a function of the particle radii at different wavelengths. Figure 10.7 shows the kernel of this equation for the volume distribution function of the particles for the case of particles with the complex refractive index $m = 1.60-0.02i$.

Equation (10.4.11) shows that to calculate the kernels of the equation it is necessary to know the physical–chemical characteristics of atmospheric aerosol – its complex refractive index m. This situation is realised, for example, for water aerosol, i.e. for clouds and precipitation. In a general case, the problem of remote determination of the aerosol characteristics should be formulated as a problem of simultaneous determination of $f(r)$ and m.

10.5. Remote methods using measurements of atmospheric radiation

A large group of inverse problems of atmospheric optics for atmospheric radiation has been formed. These problems should be sub-divided into the methods of thermal radiation, the infrared non-equilibrium radiation and methods using the measurements of glow of the atmosphere. These methods are based on the integral form of the equation of radiation transfer (Chapter 3).

Thermal radiation methods

For the outgoing thermal radiation for a plane–parallel horizontal homogeneous atmosphere with nadir geometry of measurements we

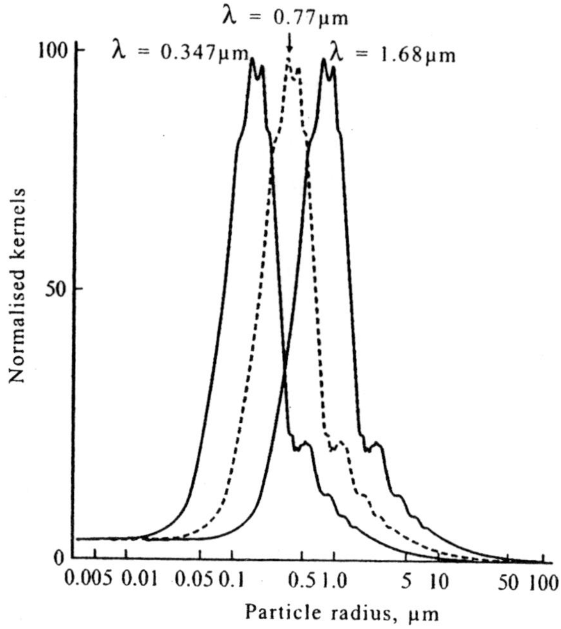

Fig.10.7. The kernels of equation (10.4.11) for the volume distribution function of the particles for the case of particles with the complex refractive index m = 1.60–0.02i [117] for different wave lengths.

can write:

$$I(\nu,\theta) = B_\nu(T_s)\exp\left(-\int_0^\infty k_\nu(z)\rho(z)dz\right) + \\ + \int_0^\infty B_\nu(T(z))k_\nu(z)\rho(z)\exp\left(-\int_z^\infty k_\nu(z')\rho(z')dz'\right)dz. \qquad (10.5.1)$$

When writing equation (10.5.1) we have used the assumptions on the establishment of local thermodynamic equilibrium, on the negligible effects of scattering of thermal radiation and the absolutely black underlying surface. Equation (10.5.1) is a physical–mathematical basis for formulating different inverse problems of the thermal region of the spectrum for space and ground-based measurement schemes. In the later case, in infrared spectral range, the integrated term describing the contribution of the underlying surface is absent.

For the microwave range, this integrated term takes into account the relic cosmic radiation falling on the upper boundary of the atmosphere. We discuss examples of these inverse problems.

The integral form of the equation of transfer of thermal radiation (equation 10.5.1) shows that thermal radiation depends on the temperature and characteristics of the content of the absorbing and emitting components (for example, different gases). In a general case we can write

$$I(\nu,0) = I[T_0, T(z), q_i(z), a, b, c,...], \qquad (10.5.2)$$

where a, b, c, etc, characterise the optical properties of the atmosphere and the underlying surface. As already mentioned, in remote measurement methods it is assumed that these parameters are known. However, in this case equation (10.5.2) shows that thermal radiation is determined by the entire set of the functions describing the physical state of the atmosphere – vertical profiles of temperature, and the mixing ratios of different gases, in a general case by the characteristics of clouds and aerosols, etc. Even if we can measure the spectral and angular dependences of thermal radiation, it is difficult to hope to extract from these measurements all required the information on the parameters of the atmosphere and the underlying surface. Here 'help' comes again from the spectral selectivity of the optical properties of the atmosphere. In the entire thermal range of the spectrum from 3–4 μm to radio waves there are 'spectrally localized' absorption bands of different gases and 'transmittance windows' – the range in which atmospheric extinction is relatively small. Although the absorption bands of different gases overlap and there are no absolutely 'clean' transmittance windows, this property does make it possible to a certain degree to 'separate the variables' in the relationship (10.5.2).

Determination of the temperature of the underlying surface

If it is assumed that the atmosphere does not absorb radiation, then equation (10.5.1) retains only the integrated term describing the radiation of the underlying surface for the space measurement scheme. In real conditions, the absorption (and radiation) of the atmosphere is also detected in the transmittance windows. However, it is important that, for example, in the 8–12 μm transmittance window, the outgoing thermal radiation depends mainly on the temperature of the surface or the characteristics of clouds (if they are located in the field of view of the satellite device).

When solving the given problem, in addition to the previously mentioned range 8–12 μm (or narrower spectral ranges in this region) we utilize the transmittance window in the near infrared range of the range (3.7.μm) and also the optically transparent intervals in the microwave range. The accuracy of determination of the temperature of the underlying surface depends on the accuracy of taking into account the effect of the atmosphere, the accurate assignment (or independent determination) of the emitting properties of the surfaces, and the effect of clouds (in the infra-red region of the spectrum). At present, this remote method makes it possible to measure the temperature of the surfaces of oceans with the accuracy of 0.5–1.0 K.

Determination of the vertical profile of temperature (nadir measurement geometry)

It is now assumed that we have determined the temperature of the underlying surface. It is also assumed that the measurements of the spectral angular dependence of the outgoing thermal radiation are performed in the absorption bands of atmospheric gases whose concentrations are known and remain constant with time, and in space. Such gases in the atmosphere of the Earth are CO_2 and O_2. Consequently, equation (10.5.2) retains one unknown function – the vertical temperature profile. Methods have been developed and applied for determining the vertical profile of temperature from measurements of the spectral dependence of outgoing radiation in the infrared absorption bands of carbon dioxide at 15 and 4.3 μm and in the absorption band of oxygen in the microwave range at 0.5 cm. A separate line of absorption of oxygen at 0.25 cm can also be used for this.

The possibility of remote determination of the vertical temperature profiles is connected with the fact that the outgoing radiation in the spectral ranges with different optical densities is generated by different altitude layers of the atmosphere. This mechanism is shown in Fig. 10.8 which shows the so-called weight functions of the investigated inverse problem characterising the regions of formation of the outgoing thermal radiation in different spectral intervals of 4.3 and 15 μm CO_2 absorption bands.

The weight functions are the derivatives of the transmittance functions of the atmosphere in the integral form of the transfer equation which 'weigh' the altitude distribution of the Planck function or the temperature in the microwave range. For the microwave range they are very close to the kernels of the integral

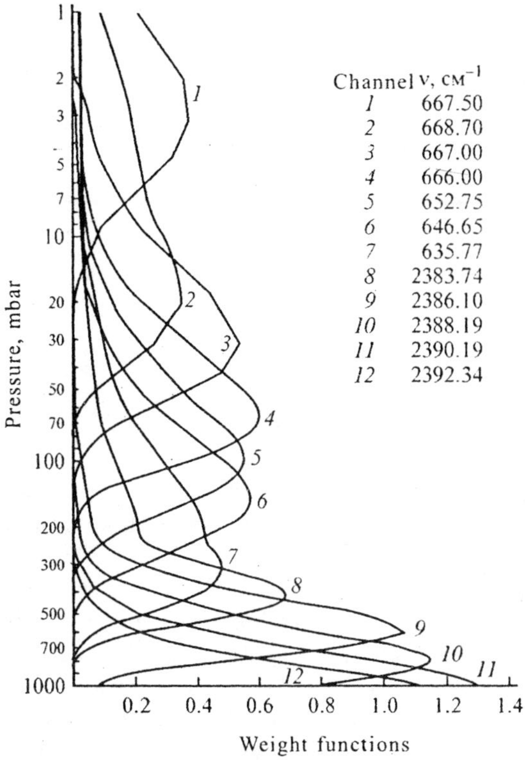

Fig.10.8. Weight functions characterising the regions of formation of outgoing thermal radiation in different spectral intervals of the absorption bands of CO_2 at wavelengths of 4.3 and 15 μm [105].

equation of variational derivatives, if the temperature dependence of the transmittance function is not taken into account. For the infrared range the kernels differ from the weight functions by the multiplier – the derivative of the Planck function in respect of temperature. The weight functions together with the kernels of the integral equations and variational derivatives are often used for analysis and illustration of the possibilities of remote measurements.

Determination of vertical profile of the temperature (limb geometry of measurements)

The outgoing radiation can be measured directing the device, mounted on board a satellite, to the horizon of the planet. In this case, we deal with the limb geometry of measurements or the measurement of the radiation of the planet horizon. In this measurement scheme the device should have a sufficiently high angular resolution or, in other words, it should be capable of recording the outgoing radiation in a narrow range of tangent heights (Fig.10.2).

The transfer equation for the considered geometry of the measurements can be written in the form:

$$I(\nu,h_0)=\int_0^{\infty}B_\nu(T(z))w(z,h_0)\times\frac{\partial}{\partial z}\left(\exp\left(-\int_z^{\infty}k_\nu(z')\rho(z')w(z',h_0)dz'\right)+\right. \tag{10.5.3}$$

$$\left.+\exp\left(-\int_{h_0}^{z}k_\nu(z')\rho(z')w(z',h_0)dz'-\int_{h_0}^{\infty}k_\nu(z')\rho(z')w(z',h_0)dz'\right)\right)dz,$$

where $w(z,h_0)$ is the Jacobian of transition from integration along the path of formation of radiation to the integration with respect to the vertical coordinate. Equation (10.5.3) holds for the model of a spherical layered-homogeneous atmosphere which assumes that the temperature of the atmosphere depends only on the vertical coordinate. In other words, we ignore the presence of horizontal temperature gradients on the path of formation of outgoing radiation. Different exponents in equation (10.5.3) take into account the contributions of different elements of the path of formation of radiation to the outgoing radiation of the horizon.

Figure 10.9 shows the behaviour of another characteristic of the formation of outgoing radiation, namely the contribution of the layers of the atmosphere to outgoing radiation which together with the kernels, the weight functions and the variation derivatives are also used for analysis of the problems of remote sounding of the atmosphere. Figure 10.9 shows the contributions to the radiation of layers at different measurement tangent heights for the problem described by the equation (10.5.3) for a spectral interval in the 15 μm CO_2 band. In the problem of thermal sounding discussed here, the contributions of the individual layers of the atmosphere to the outgoing radiation are equal to the product of the Planck function by the difference of the transmittance functions at the boundaries of the layer. They characterise the fraction introduced by the layer of the atmosphere to the outgoing radiation. It is important to note that these weight functions are equal to zero below the measurement tangent height because of the fact that the subjacent layers (at a fixed tangent height) have no effect on the

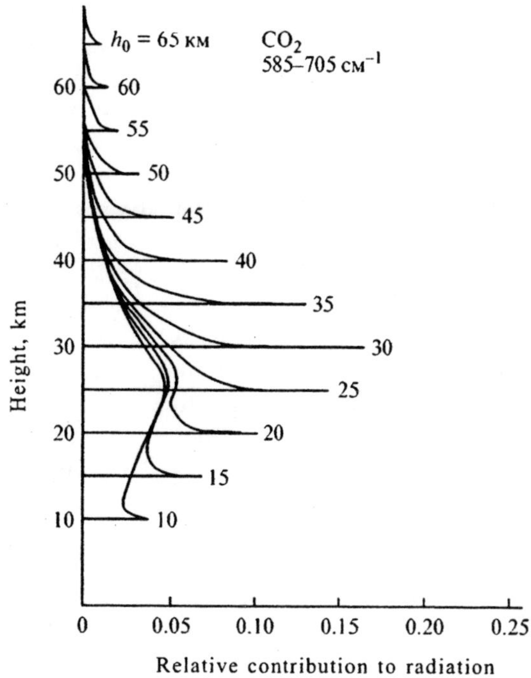

Fig.10.9. Contribution of atmospheric layers at different tangent heights h_0 to the outgoing radiation for the spectral interval in the CO_2 band at 15 μm [35].

formation of outgoing thermal radiation (in the absence of scattering and under LTE conditions). Therefore, the type of functions characterising the contributions to the outgoing radiation of different layers of the atmosphere is slightly different in comparison with the nadir geometry of measurements. These functions show a rapid decrease to zero for the appropriate tangent heights. It is clear that this special feature should result in higher vertical resolution of the discussed remote method of determination of temperature profile in comparison with nadir geometry.

It should be stressed that for the limb geometry of measurements in solving the inverse problems we use the measurements of the angular dependence of outgoing radiation or, which is the same, dependences of the intensity of radiation on the measurement tangent height. This means that to solve this problem it is sufficient to measure the outgoing radiation in one spectral range of the CO_2 or O_2 absorption bands. In a general case, we can use both angular and spectral dependences. Then we can determine in an independent manner the vertical profile of pressure and avoid the assumption on the spherical layer-stratified model of the atmosphere. In this case we can formulate a more complicated inverse problem in respect of the determination of the temperature of the atmosphere as the function of altitudes and horizontal coordinate.

The considered schemes of satellite experiments (nadir and tangential) have been used many times in a large number of space experiments. Nadir thermal sounding of the atmosphere is carried out at present to obtain global information on the three–dimentional temperature field at altitudes from 0 to 40–50 km. The most informative systems of temperature sounding use the measurements of outgoing radiation in all spectral ranges suitable for solving the examined problem, – the absorption bands of carbon dioxide at 4.3 and 15 μm and the microwave band of oxygen. This integrated scheme is especially effective is sounding the cloudy atmosphere because of the fact that microwave radiation is only slightly affected by the clouds. Limb sounding is carried out for scientific research problems and makes it possible to determine the altitude profile of the temperature in the altitude range 10–120 km. This method was used for the first time to obtain climatological information on the temperature condition of the atmosphere of the Earth for the middle atmosphere. Current errors of temperature sounding are ~1–2 K.

Determination of the characteristics of the gas composition of the atmosphere

Knowing (determining) the vertical profile of temperature and using measurements in the absorption bands (radiation bands) of other gas components, we can obtain information on the content of, for example, water vapour, ozone and other gases. When solving these problems, we can used both nadir and limb geometry of measurements. Attention should be given to the fact that the required characteristic, for example, the vertical distribution of the densities of absorbing gases, are located in the index of the exponent of the transmittance function and under two integrals in respect of the spatial variable. Evidently, the smoothing effect of these operators should complicate obtaining the information on these profiles. In addition to this, it may be shown that for the nadir scheme of measurements in the conditions of the isothermal atmosphere and the absolutely black underlying surface the measurements of outgoing thermal radiation do not contain any information on the composition of the atmosphere. This conclusion follows from the definition of the absolutely black radiation as the radiation of an isothermal cavity.

Figure 10.10 shows the kernels of the integral equation for determining the vertical profile of the content of water vapour obtained after linearization of equation (10.5.1) for the nadir scheme of measurements:

$$\delta I(\lambda,\theta) = \int_0^\infty K(\nu,z,\theta)\delta\rho(z)dz. \tag{10.5.4}$$

here $\delta I(\lambda,\theta)$ and $\delta\rho(z)$ are the variations of the intensity of outgoing radiation and of the content of the water vapour in relation to the mean values. The kernels are given for different frequencies in the contour of the spectral line of H_2O in the infrared range of the spectrum at different shifts from the centre of the line (Fig.10.10). The figure shows the specific behaviour of the kernels of the discussed inverse problem for the infrared range of the spectrum – tendency of the kernel to zero in the vicinity of the Earth's surface. This special feature is characteristic for the absolutely black underlying surface in the absence of a sudden change of temperature in the vicinity of the surface.

For the limb geometry of measurements, the behaviour of the kernel of the appropriate integral equation is similar to that in Fig.10.9. This method, as the method used previously for temperature, is characterised by a higher vertical resolution (this is its significant advantage) in comparison with the nadir method. It should be mentioned that the methods of limb sounding have another advantage. Because of the considerably longer (in comparison with nadir measurements) paths of formation of outgoing radiation (tens of times), the limb measurement geometry provides information on the temperature and gas composition of the atmosphere up to considerably higher altitudes.

At present, satellite systems for nadir sounding in the infrared and microwave ranges of the spectrum are used in the determination of the total content of water vapour and ozone. They also provide some information on the content of water vapour in individual layers of the atmosphere. Satellite sounding with limb measurement geometry is used to obtain vertical profiles of many gas components of the atmosphere (H_2O, O_3, CH_4, etc.). The errors of determination of the characteristics of the content of the different

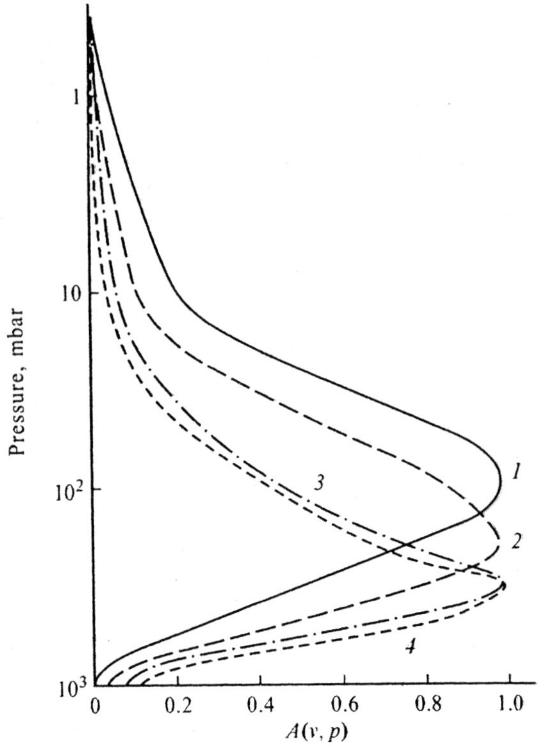

Fig.10.10. Kernels of the integral equation for determining the vertical profile of the content of water vapour (nadir measurement scheme) for different frequencies in the contour of the spectral line of H_2O for different shifts from the centre of the line in fractions of its half widths α_0. 1 – $\alpha_0/10$; 2 – $\alpha_0/5$; 3 – $\alpha_0/2$; 4 – α_0.

gases depend on the given geometry or measurements and the type of gas. For total contents of the water vapour and the ozone for the nadir geometry of measurements these errors equal usually 5–10 %. The vertical profiles of the content of different gases are determined with errors of 10–30%.

Determination of other parameters of the atmosphere and surface

The measurements of thermal radiation are used to determine not only the previously mentioned parameters of the atmosphere and underlying surface – the temperature of the atmosphere and the surface, the characteristics of the gas composition, but also to determine the characteristics of the cloud cover (for example, the water content of clouds) the intensity of precipitation, the properties of the underlying surface (emissivity, moisture content of soil) and the characteristics of atmospheric aerosols. One of the simplest but relatively important parameters of the cloud field – the cloud amount – is determined using devices with high horizontal resolution by analysis of the spatial pattern of the brightness temperature of radiation in different transmittance windows of the atmosphere. The clouds usually located in the troposphere have a considerable lower brightness temperature of the radiation than the temperature of the underlying surface and can be easily seen on its background.

The method has been used for a long time to study different characteristics of the cloud cover of our planet – amount and altitude of clouds, their type. These investigations were used for developing the climatology of the cloud cover of the Earth. The importance of these investigations is associated with the fact that, as mentioned previously, the clouds are one of the most important parameters determining the climate of our planet.

Infrared sounding of non-equilibrium atmosphere

At present, inverse problems for the non-equilibrium atmosphere become more and more important. Conventional and inverse problems of the thermal range of the spectrum are based on an important assumption on local thermodynamic equilibrium (LTE). As shown previously (Chapter 7), this assumption is however not fulfilled for the upper layers of the atmosphere. Recently, special attention has been given to the development of remote methods of sounding the non-equilibrium atmosphere for the limb measurement geometry – measurements of the radiation of the horizon of the Earth.

The inverse problems of the non-equilibrium atmosphere are characterised by a number of special features:

1. The number of parameters, characterising the physical state of the atmosphere, greatly increases because the state of the atmosphere is no longer described by a single kinetic temperature and the total concentration of absorbing molecules. Since the Boltzmann law is not fulfilled, this state is additionally described by the population of the appropriate levels of the molecules and vibrational (electronic, rotational, etc) temperatures. The number of these parameters corresponds to the number of states of the internal energy of the molecules, and the transitions between these states form the field of outgoing radiation of the atmosphere. For example,

for one of the formulations of the inverse problem of determination of the parameters of the non-equilibrium atmosphere from the measured infrared radiation in the absorption bands of CO_2 and O_3 the required values are 40 vertical profiles of the kinetic and vibrational temperatures, the content of O_3 and CO_2 and pressure. In the LTE conditions, this problem is reduced to finding only four profiles.

2. When studying the inverse problems of the non-equilibrium atmosphere there is another new class of parameters (denoted by B), describing the processes of excitation and de-excitation of different levels of the molecules – rates of different collisional processes, chemical reactions, etc. These parameters appear in appropriate kinetic equations describing the population of different excited states.

3. The role of boundary conditions becomes more important. For example, solar radiation becomes a source of excited molecules. The boundary conditions in a general case include new quantities, for example, corpuscular radiation.

As shown in a number of studies, the measurements of non-equilibrium radiation make it possible to determine the population of excited vibrational levels. This is accompanies by difficulties in determining the population of the ground states of the molecules, i.e. the total concentration of absorbing and emitting molecules. For example, in interpretation of the measurements of the radiometer ISAMS on board the UARS satellite, the content of CO in the conditions of a non-equilibrium atmosphere is determined using a new type of *a priori* information – a kinetic model of the population of the first excited vibrational state of the CO molecule, – which enables the function of the source to be calculated and tabulated and to solve the inverse problem. This results in a large increase of the volume of *a priori* information used. In addition to the kinetic model, i.e. the assignment of the physical mechanisms and appropriate constants of the population of excited states, the information on the thermal structure of the troposphere and stratosphere, on the CO content in the lower layers of the atmosphere, on the state of cloud cover, etc. was also used.

For a non-equilibrium atmosphere there are several types of inverse problems:

1) in relation to the physical parameters of the atmosphere –

a) in accordance with the scheme (see section 10.1)

$$J+A+G \rightarrow X, \tag{10.5.5}$$

b) using kinetic models

$$J+A+G+B \rightarrow X; \qquad (10.5.6)$$

2) in respect of the parameters of the processes of excitation and de-excitation – scheme

$$J+A+G+X \rightarrow B. \qquad (10.5.7)$$

The two last approaches use the kinetic equations for describing the populations of the excited states. In the first case (scheme (10.5.6)) we determine different parameters of the physical state of the non-equilibrium atmosphere, in the second case (scheme (10.5.7)) – the parameters of the process of excitation and de-excitation of the molecules. This approach was used to determine the profiles of the content of atomic oxygen and hydrogen in the upper layers of the atmosphere from radiation measurements in the 9.6 μm ozone absorption band using appropriate kinetic relationships. The same measurements were used to determine more accurately the important constant of the rate of quenching of excited molecules of CO_2 by atomic oxygen. Thus, in this case, the combined application of the transfer equation and kinetic equations enables us to determine the physical parameters of the atmosphere which do not have a direct effect on the radiation field but strongly effect the population of the excited states of emitting molecules.

An example of the retrieval of kinetic temperature of the atmosphere in a wide range of altitudes (40–120 km) including heights at which LTE is not established, is shown in Fig.10.11. These data were obtained using the measurements of the radiation spectrum of the horizon of the Earth in the 15 μm band of CO_2 (apparatus CRISTA) and the interpretation method not using kinetic equations. When interpreting these measurements we also retrieved simultaneously the profile of the kinetic and vibrational temperatures of a number of states of the CO_2 molecule, pressure and CO_2 content.

10.6. Remote measurement methods based on recording the scattered and reflected solar radiation

These methods are based on the integral – differential equation of radiation transfer (for example, equation (3.4.35)), and not on its partial cases used previously. The theory and realization of these methods are relatively complicated because it is necessary to take into account the multiple scattering of solar radiation and reflection

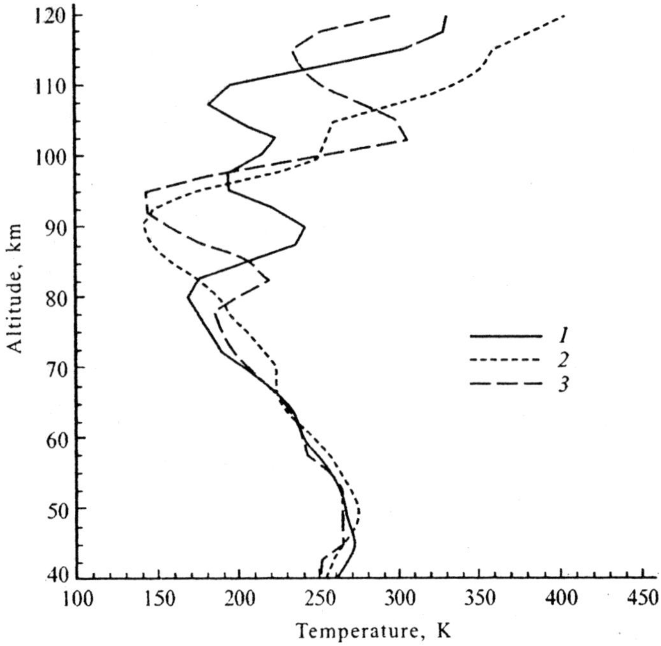

Fig.10.11. Examples of retrieval of the kinetic temperature of the atmosphere in a wide range of altitudes (40–120 km) from measurements of radiation spectra of the horizon of the Earth in the 15 μm band of CO_2 (CRISTA apparatus). 1) 1 – 37° N., 82° W., 2 – 49° S., 172° W., 3 – 19° S., 110° W.

of radiation from the surface. The most frequently used satellite method is the remote determination of the vertical profile and of the total ozone content from measurements of reflected and scattered solar ultraviolet radiation.

Study of the Earth ozonosphere

As mentioned in Chapter 4, the values of the absorption coefficient of ozone in the ultraviolet range are very high. For this reason, ultraviolet radiation with the wave length smaller than 290–300 nm does not reach the surface of the Earth. When observing the Earth from space in this range of the spectrum, the solar radiation scattered by different layers of the atmosphere is registered. In the centre of the Hartley absorption band where the absorption coefficients of ozone are very large, scattering takes place in the upper layers of the atmosphere at altitudes of 40–60 km. In the central part of the band where the absorption coefficients of ozone are not very large, the solar radiation penetrates into the thickness of the atmosphere and is scattered at small altitudes of the 20–40 km. Finally, in the wing of the band at still smaller values of the absorption coefficients of ozone (Huggins band) solar radiation reaches the Earth surface – it is reflected by the surface and also scattered by tropospheric layers. Here we have the same effect as the one discussed in the problem of temperature sounding by the measurements of the spectrum of outgoing thermal radiation in the absorption band of carbon dioxide. Measuring the spectral dependence of the outgoing scattered and reflected solar ultraviolet radiation, we carry out 'vertical' scanning of the atmosphere.

For the quantitative illustration of this class of methods we consider the simplest case. It is assumed that the field of outgoing scattered solar radiation is determined by single molecular scattering in the presence of molecular absorption. This simplified formulation is close to the actual problem of determination of the vertical profile of the ozone content from measurements of scattered solar radiation in the ultraviolet range of the spectrum. For the component of single-scattered solar radiation at angle θ to the nadir we can write

$$I(\lambda,\theta) = \alpha_R P(\theta)\frac{F_0}{4\pi}\int_0^{p_0}\exp\left(-(1+\sec\theta_S)\times\right.$$

$$\left.\times\left(\frac{1}{g}\int_0^{p} k'_\nu(p')q_{O_3}(p')dp' + \tau_R(p)\right)\right)dp, \qquad (10.6.1)$$

where α_R is the Rayleigh scattering coefficient; $P(\theta)$ is the Rayleigh scattering phase function; F_0 is the extra-atmospheric radiation flux; θ_S is the zenith angle of the Sun; q_{O3} (p) is the ozone mixing ratio as a function of pressure; $\tau_R(p)$ is the optical thickness of Rayleigh extinction.

The linearization of equation (10.6.1) reduces the examined inverse problem to the Fredholm integral equation of the first kind. Fig.10.12 shows the relative variational derivatives (the kernels of the linearized equation) of this problem characterising the regions of formation of outgoing scattered solar radiation in different spectral ranges of the Hartree–Huggins ozone absorption band. The derivatives are related to the intensity of radiation in the appropriate channel of measurements.

Figure 10.12 shows that the outgoing radiation in different spectral intervals forms in different layers of the atmosphere: in spectral ranges with high values of the absorption coefficient of

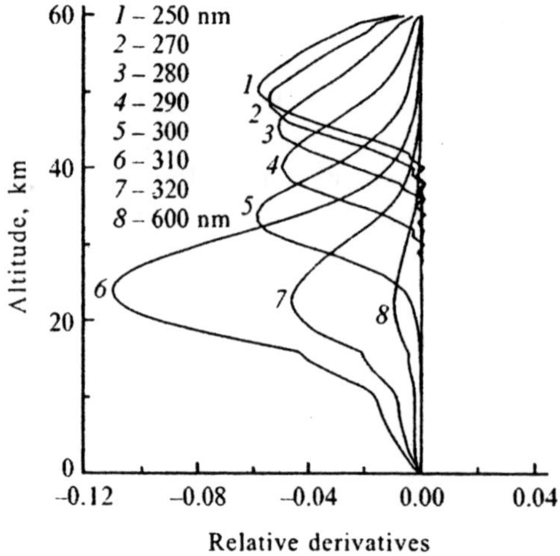

Fig.10.12. Relative variational derivatives (kernels of the linearised equation), characterising the ranges of formation of outgoing scattered solar radiation in different spectral interval of the Hartley–Huggins ozone absorption band.

ozone – in the upper stratosphere, in spectral ranges with small values – in the lower stratosphere and the troposphere. In particular, this special feature of the formation of scattered solar radiation is also a physical basis for the possibility of solving the given inverse problem and obtaining the information on the vertical profiles of the ozone content. This indirect method is used actively for studying the Earth ozonosphere and it makes it possible determine in particular the such the phenomenon as 'ozone holes' above the Antarctica. The same principle is also used for studying the content of other gases in the atmosphere – water vapour, methane, etc.

Space survey

Many remote methods of studying the atmosphere and surface originated in aerospace methods for military applications. However, it is usually the case, there were other fields of application of spy satellites. For example, earth resources satellites were put into operation. The natural resources include usually a wide complex of the characteristics of the underlying surface relating to geology, water resources, or oceanological, fishing industry, forestry, agriculture, etc.

Measuring the outgoing radiation in the 'transmittance' windows, i.e. in spectral ranges with small atmospheric extinction, we 'see' the underlying surface. As discussed in Chapter 6, different surfaces have different spectral optical characteristics. Thus, it is quite easy to solve the problem of identifying the type of underlying surface. Further, the optical characteristics of the surface are dependent on the state – moisture content, vegetation of different types, and so-on. Using these dependences we can determine, for example, the salt and moisture contents of soil, the condition of vegetation, etc.

Remote refractometry

There is also a whole group of inverse problems combined in remote refractometry. To determine different atmospheric parameters we utilize the refraction effects, i.e. distortion of the trajectory of propagation of radiation in the atmosphere as a result of heterogeneities of the refractive index, changes in the phase and the amplitude of electromagnetic radiation. For example, for the microwave range of the spectrum, the refractive index depends on the pressure of the atmosphere, temperature and partial pressure or water vapour. Determining, from space observations, the vertical course of the refractive index and solving the inverse problem using the equation of hydrostatic and the equation of state of the ideal gas, we can formulate and solve the inverse problem of determination of the vertical temperature profile in the stratosphere (where the effect of water vapour on the refractive index is very small) or the inverse problem of determination of the profile of moisture content on the troposphere (for the given temperature profile). Figure 10.13 shows an example of determination of the vertical profile of the atmospheric humidity by this method [104]. In the figure, this profile is compared with the results of independent radiosonde measurements and the results of synoptic analysis of the state of the atmosphere.

10.7. Active remote measurements methods

Radar sounding of atmosphere

The radar methods of studying the atmosphere were developed as a consequence of using military radars after the First World War. The principle of radiolocation is relatively simple. Pulses of electromagnetic radiation of a specific frequency are sent into the atmosphere and special systems record reflected (back-scattered) signals. The intensity of the reflected signal depends on the distance to the 'object' and its properties. Pulse radiolocation

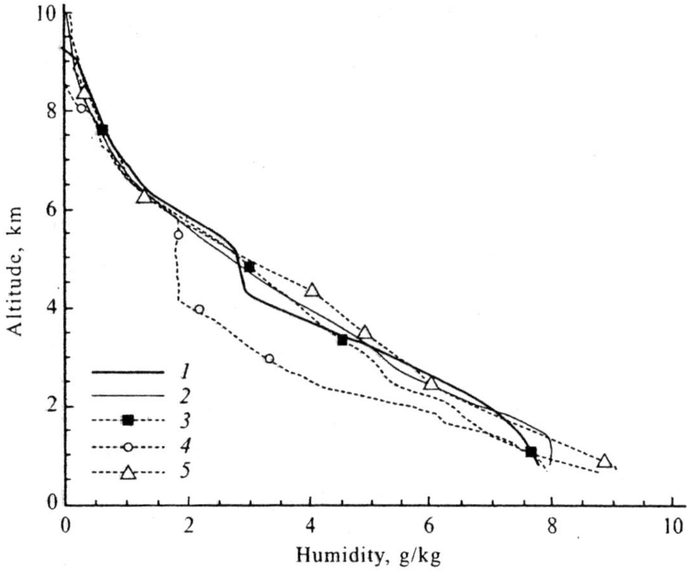

Fig.10.13. Comparison of the vertical profile of atmospheric humidity, determined by refractometry (1), with the results of independent radiosonde measurements (3–5) and synoptic analysis (2) of the state of the atmosphere [104].

enables easy determination of the distance to the object (because the speed of propagation of electromagnetic radiation is known).

The radiolocation methods are most effective is studying clouds and precipitation. They can be used to determine the spatial distribution of clouds and precipitation, the water content of clouds, intensity of precipitation. The physical basis of these methods is the dependence of the coefficients of backscatter of electromagnetic radiation on the number and size of the particles of clouds and precipitation (Chapter 5). At present, there is a large network of meteorological radars, firstly in the system of servicing air transport.

In addition, space radiolocation has been intensely developed in the last decade. In this case, the radars are used not only for examining the atmospheric parameters but also the properties of underlying surfaces. Analysing the signal reflected from the surface, we can determine the type and properties of the surface. This principle is also used for such remote methods such as the determination of the velocity of near-water wind and moisture content of soil.

Lidar sounding

Lidar sounding (LIDAR – Light Detection And Ranging) relates, like radiolocation sounding, to active remote methods of measurements [30, 38, 44]. It may be assumed that lidar sounding is an advanced modification of projector sounding used for examining the atmosphere in the 20s and 30s of the previous century. The physical principles of lidar sounding are identical to radar sounding. The differences are in the wavelengths of the used electromagnetic radiation (visible to long range infrared radiation) and small angle divergence (sharp directionality) of lidar radiation.

The lidar sounding methods are based on the equation of laser location which maybe presented in the following simplified form:

$$E(R) = E_0 \sigma_\pi \exp\left(-2 \int_0^R \alpha(r) dr \right). \qquad (10.7.1)$$

where R is the distance to the sounded volume of the atmosphere. The energy of the recorded 'reflected' signal $E(R)$ is proportional to the initial energy of radiation E_0, directed into the atmosphere, the value of the backscattering coefficient σ_π (the case in which the source and receiver of radiation are situated at the same point – monostatic measurement scenario) and the value of the transmittance function along the path of propagation of radiation from the source to the sounded volume and backward (multiplier 2 in the exponent – transmittance function $P(\theta)$). It should be mentioned that equation (10.7.1) is written in the approximation of single scattering of lidar radiation. In this case, if the effects of multiple scattering (clouds, mist) are important, it is necessary to modify equation (10.7.1) or solve the problem of the propagation of lidar radiation on the basis of the theory of radiation transfer.

Analysis of the equation of lidar sounding shows that the information on the atmosphere is contained in the backscatter coefficient σ_π and in the transmittance function $P(R)$. There are different methods of extracting this information which received full coverage in appropriate monographs [38, 44]. It should only be mentioned that at present the lidar method (mostly ground-based ones) provide a large amount of information on different parameters of the atmosphere and the underlying surface.

An important special feature of the lidar measurement method is its high information content. This is caused by a number of reasons:

1.Use of different types of lasers of pulse and continuous type in a wide range of the spectrum from ultraviolet to far infrared;

2. Presence of spectrally tuneable lasers of different power;

3. A variety of the process of interaction of laser radiation with the atmosphere – molecular and aerosol absorption, scattering of different types – Rayleigh, aerosol, Raman, etc.

4. Use of different measurement scenarios and different carriers – aircraft, satellites, etc.

5. Short duration and high repetition frequency of laser radiation pulses.

6. The possibility of using laser radiation with specific polarization properties.

All these special features make the lidars into a unique tool to examine the atmosphere nd the underlying surface. As an example, we mention the most important parameters of the environment which can be studied using different lidars, and the main mechanisms of interaction of radiation with the medium used for this purpose:

1.Temperature and density of the atmosphere – from the values of molecular (Rayleigh) scattering, Raman scattering, etc.

2. Characteristics of the gas composition – from the values of molecular absorption, Raman and absorption scattering.

3. Characteristics of the aerosols – from the values of aerosol scattering and absorption, polarization characteristics of scattered radiation.

4. The wind field – on the basis of the Doppler shifts of the frequency of scattered radiation.

5. Characteristics of turbulence of the atmosphere – from fluctuations of the characteristics of transmitted and scattered radiation.

6. Different characteristics of the underlying surfaces – from the strength of the reflected signal and its spectral and polarization properties.

It should be mentioned that to solve these problems in a general case we can use the measurements of spectral, time, angular and polarisation characteristics of scattered and reflected laser radiation in a wide spectral range from ultraviolet to far infrared.

APPENDIX

FUNDAMENTAL UNITS IN ATMOSPHERIC OPTICS AND PHYSICS

A.1. Molecular mass of dry and moist air

Molecular mass of a mixture of ideal gases

In atmospheric physics, the air, like all its components, is assumed to be an ideal gas. Consequently, for the air we can write the equation of state of the ideal gas [19, 68]

$$pV = \frac{m}{\mu} RT, \tag{A.1.1}$$

where p is the pressure of air; V is the volume of air; m is the mass of air in volume V; μ is the molecular mass of air; T is the temperature of air; R is the universal gas constant.

$$R = 8.31441 \text{ J·mol}^{-1}\text{·K}^{-1}.$$

However, equation (A.1.1) should be regarded as formal because air is a mixture of different gases and, therefore, the concept of its molecular mass requires definition. The ratio m/μ is the number of mols of air in volume V. According to definition, one mol of any substance contains the same number of molecules equal to the Avogadro number

$$N_A = 6.0221 \cdot 10^{23} \text{ mol}^{-1}$$

Consequently, if N is a number of molecules of air in volume V, then

$$\frac{m}{\mu} = \frac{N}{N_A}, \tag{A.1.2}$$

If the mixture contains K gases, the number of molecules of mixture N is equal to the sum of the molecules of each gas

$N=\sum_{i=1}^{K} N_i$. Substituting this equation into (A.1.2) we obtain

$$\frac{m}{\mu}=\frac{\sum_{i=1}^{K} N_i}{N_A}=\sum_{i=1}^{K}\frac{m_i}{\mu_i},$$

where m_i is the mass of the i-th gas in the volume V; μ_i is the molecular mass of the i-th gas; m/μ is the number of moles of the i-th gas. Consequently,

$$\mu=\frac{1}{\sum_{i=1}^{K}\frac{m_i}{m}\frac{1}{\mu_i}}. \tag{A.1.3}$$

Molecular mass of dry air

Equation (A.1.3) gives the required definition of the molecular mass of a mixture of gases, in particular, air. In the atmosphere, the composition of the main gases is almost constant (up to an altitude of 90 km), see section 2.1. The variable components are only trace gases but their concentrations (m_i/m) are so low that their effect on the molecular mass of air is always ignorable. The only exception is made for the water vapour. The variation of the content of water vapour in the lower troposphere may change the value of μ by several percent. Therefore, as the universal constant we use the molecular mass of dry air μ_0, i.e the molecular mass of air which does not contain the water vapour [19,68]. It is equal to

$$\mu_0 = 28.9645\ \text{g}\cdot\text{mol}^{-1}.$$

It should be mentioned that μ_0 is constant only up to an altitude of 90 km (see Table 2.2).

Dalton's law

We introduce the partial pressure of a gas in the mixture p_i defining it as the pressure of the given gas if at temperature T this gas would occupy the same volume V as the mixture of gases.

According to the definition

$$p_i V = \frac{m_i}{\mu_i} RT. \tag{A.1.4}$$

Summing up all these relationship and taking into account (A.1.3) and (A.1.1) we obtain Dalton's law: the pressure of a gas mixture is equal to the sum of the partial pressures of gases forming the mixture:

$$p = \sum_{i=1}^{K} p_i.$$

Molecular mass of moist air

From equation (A.1.3) we obtain

$$\frac{1}{\mu} = \frac{m_0}{m\mu_0} + \frac{m_w}{m\mu_w},$$

where m_0 is the mass of the fraction of dry air in volume V; m_w is the mass of the water vapour in V; μ_w is the molecular mass of water vapour. Because of the additivity of the mass $m_0 = m - m_w$ we obtain

$$\frac{1}{\mu} = \left(1 - \frac{m_w}{m}\right)\frac{1}{\mu_0} + \frac{m_w}{m}\frac{1}{\mu_w},$$

from which

$$\mu = \frac{\mu_0}{1 + \frac{m_w}{m}\left(\frac{\mu_0}{\mu_w} - 1\right)}. \tag{A.1.5}$$

It should be mentioned that the molecular mass of moist air is always lower than that of dry air (since $\mu_0/\mu_w > 1$) and, consequently, with other conditions being equal, equation (A.1.1) shows that the mass m and the density m/V of moist air are also lower. This means that moist air is lighter than dry air (a paradoxical conclusion because it would appear that the reverse should be the case). The mass of water vapour m_w is not suitable for measurements. It is very simple to measure the partial pressure of water vapour denoted by e. Therefore, writing (A.1.4) separately for dry air and water vapour, we obtain a system of equations which is in fact the equation of state of the system 'dry air plus water vapour' [43]:

$$(p-e)V = \frac{m-m_w}{\mu_0}RT.$$

$$eV = \frac{m_w}{\mu_w}RT. \qquad (A.1.6)$$

Separating the first equation by the second one gives

$$\frac{p-e}{e} = \left(\frac{m}{m_w}-1\right)\frac{\mu_w}{\mu_o},$$

from which

$$\frac{m_w}{m} = 1+\frac{\mu_0}{\mu_w}\frac{p-e}{e}$$

or

$$\frac{m_w}{m} = \frac{e}{\frac{\mu_0}{\mu_w}p-\left(\frac{\mu_0}{\mu_w}-1\right)e}. \qquad (A.1.7)$$

Usually in calculations the value $\frac{\mu_0}{\mu_w} = \frac{28.9645}{18.015} = 1.608$ is immediately substituted into (A.1.7) and therefore (A.1.7) has the form

$$\frac{m_w}{m} = \frac{e}{1.608p-0.608e}.$$

Substituting (A.1.7) into (A.1.5) gives the equation for calculating the molecular mass of air on the basis of its pressure and the partial pressure of water vapour:

$$\mu = \mu_0\left(1-\left(1-\frac{\mu_w}{\mu_o}\right)\frac{e}{p}\right) \qquad (A.1.8)$$

or with the substitution of the value μ_w/μ_0

$$\mu = \mu_0 \left(1 - 0.378\frac{e}{p}\right).$$

Now, taking into account (A.1.8) we can use for all atmospheric gases, with the exception of water vapour, the equation of state (A.1.1) and all consequences from it. For water vapour, it is necessary to use the equation of state (A.1.6). However, the consequences from (A.1.1) in which there is no dependence on μ, are also valid for the water vapour. In addition to this, if the partial pressure of the water vapour e is low, we can ignore the dependence of μ on e and use in calculations the molecular mass of dry air; here everything depends on the required accuracy of calculations.

A.2. Units of measurement of temperature, air pressure and gas composition of the atmosphere

Units of temperature measurements

In atmospheric physics, we use the Kelvin and Celsius scales. The relationship between them is:

$$T[\mathrm{K}]=T[\mathrm{C}]+T_0,$$

$$T[\mathrm{C}]=T[\mathrm{K}]-T_0,$$

where T[K] is the temperature in Kelvin; T[C] is the temperature in degrees of Celsius; T_0 is a universal constant (T_0 is the temperature of absolute zero) whose value is equal to [13, 68]

$$T_0=273.15\mathrm{K}$$

In Great Britain, USA and a number of other countries, the Fahrenheit scale is used but it is found very rarely in the scientific literature. The ratio of the Fahrenheit and Celsius scales

$$T[C] = \frac{5}{9}(T[F]-32),$$

$$T[F] = 32 + \frac{9}{5}T[C],$$

where T[F] is the temperature in degrees of Fahrenheit.

Physical quantities, determining the concentration of gases

The concentration of individual gases in the composition of air in atmospheric physics is characterised using not only different units of measurements but also different physical quantities [25]: density, partial pressure, volume, mass and number concentrations. The relationship between them is determined using the equation of state (A.1.1) and may include temperature, pressure and molecular mass of air. Therefore, we shall discuss also the units of measurements of physical quantities determining the concentration of the gas and the relationship between these quantities. In addition to this, taking into account the special equation of state for water vapour (A.1.6), for H_2O there are special measurements units.

Pressure of air and partial pressure of gas

The main unit of pressure used in atmospheric physics is millibar (mbar) or hectapascal (GPa). the derivatives of the millibar, used usually for measuring the partial pressures of trace gases, are: pascal (Pa), i.e. (N·m^{-2}), millipascal (mPa), micropascal (μPa). An important constant in atmospheric physics is the physical atmosphere or bar [19, 68], equal to

$$p_0=1013.23 \text{ mbar}.$$

In Russia, the unit of measurements of atmospheric pressure is the millimetre of the mercury column (mm Hg) for which p_0=760 mm Hg, which determines its relationship with the milllibar.

A chain for converting the units of measurement of (partial) pressure is:

$$p[\text{mbar}]=1013.25\ p[\text{bar}]=1.33322\ p[\text{mm Hg}]=$$

$$=0.01\ p[\text{Pa}]=10^{-5}\ p[\text{mPa}]=10^{-8}\ p[\mu\text{Pa}],$$

where $p[...]$ is the value of pressure in appropriate units.

Partial density

The partial density of a gas ρ_i is its density in the conventional meaning of the word, i.e. the mass of the gas in the unit volume of air. Dividing both parts of (A.1.4) by V we have

$$\rho_i = \frac{p_i \mu_i}{RT}, \tag{A.2.1}$$

which gives the relationship of partial density and partial pressure at the temperature of air T. Equation (A.2.1) also holds for the density of air as a whole if the quantities in this equation are written without the index i.

The set of the units of partial density varies greatly depending on the selection of the mass and volume units. In atmospheric physics gram per cubic meter is used in most cases ($g \cdot m^{-3}$). The chain for converting the units of measurement of partial density is:

$$\rho[g \cdot m^{-3}] = 10^6 \rho[g \cdot cm^{-3}] = 10^{-3} \rho[mg \cdot m^{-3}] = 10^{-6} \rho[\mu kg \cdot m^{-3}],$$

where $\rho[...]$ is density in appropriate units.

When substituting the quantities in specific units of measurements, equation (A.2.1) has the form

$$\rho[g \cdot m^{-3}] = 100 \frac{p_i[\text{mbar}]\, \mu_i[r \cdot \text{mol}^{-1}]}{R[J \cdot \text{mol}^{-1} \cdot K^{-1}] T[K]}.$$

Volume (molar) concentration (fraction by volume)

The volume concentration of a gas q_i is the ratio of the partial volume V_i – the volume, which the gas would occupy under temperature and pressure of the gas mixture (air), – to the volume of the gas mixture (air). In some cases the volume concentration is also referred to as the volume mixing ratio. According to the definitions of partial volume

$$pV_i = \frac{m_i}{\mu_i} RT. \tag{A.2.2}$$

Dividing both parts by V, we obtain the relationship of volume concentration and partial density

$$q_i = \frac{\rho_i}{\mu_i} \frac{RT}{p} \tag{A.2.3}$$

at the given temperature T and air pressure p. Expressing ρ_i from (A.2.1), we obtain a simple relationship of volume concentration with partial pressure

$$q_i = \frac{p_i}{p}. \tag{A.2.4}$$

The molar concentration of a gas is the ratio of the number of molecules of the gas in the unit volume to the number of molecules in the gas mixture (air). The number of molecules of the gas is m_i/μ_i, for air – m/μ, and the division of (A.1.4) by (A.1.1.) gives the molar concentration p_i/p, but this according to (A.2.4) is q_i. This means that the molar and volume concentrations of the gas are equal. It should be mentioned that this holds only for the ideal gas but, for example, not for a mixture of liquids.

Volume concentration is a dimensionless quantity but this does not mean that it has no measurement unit: these units are the fractions of the volume. The main unit of the volume concentration of the gases in atmospheric physics is the millionth fraction in respect of volume – mln^{-1} (ppm V, sometimes written simply as ppm, indicating that we are discussing volume fractions). Other widely used units are: the volume mixing ratio, i.e. the direct (without conversion) fraction of the volume (vmr); volume percent (vol%); volume promille – prom; the billionth part by volume (ppbV), the trillionth fraction by volume (pptV). The conversion chain of the units of measurement of volume concentration is:

$$q[\text{ppmV}]=10^{6}\,q[\text{vmr}]=10^{4}\,q[\text{vol\%}]=10^{3}\,q[\text{prom}]=$$

$$=10^{-3}\,q[\text{ppbV}]=10^{-6}\,q[\text{pptV}],$$

where $q[\ldots]$ is the concentration in appropriate units.

When substituting the values in specific units of measurement, equation (A.2.3) has the form

$$q_i[\text{mln}^{-1}]=10^{4}\,\frac{\rho_i[\text{g}\cdot\text{m}^{-3}]}{\mu_i[\text{g}\cdot\text{mol}^{-1}]}\,\frac{R[\text{J}\cdot\text{mol}^{-1}\cdot\text{K}^{-1}]T[K]}{p[\text{mbar}]}.$$

Mass concentration (fraction by mass)

The mass fraction (concentration) of the gas w_i is the ratio of the mass of the given gas to the mass of the gas mixture (air) in the given volume. Dividing (A.1.4) by (A.1.1) gives $w_i=\frac{\mu_i}{\mu}\frac{p_i}{p}$ and taking into account (A.2.4) the simple relationship of the mass fraction and volume concentration, we have:

$$w_i=\frac{\mu_i}{\mu}q_i. \tag{A.2.5}$$

It should be mentioned that equation (A.2.5) cannot be used for water vapour because the value of μ for the latter depends on w_i. Equation (A.2.5) shows that for the gases in which the molecular mass is higher than the mass of air, the mass concentration is larger than volume concentration; for gases in which the molecular mass is smaller than the mass of air, the mass concentration is also smaller than volume concentration. The relationship of the mass fraction with partial pressure and density of the given temperature and pressure of air, according to (A.2.3)–(A.2.5) is

$$w_i = \frac{\mu_i}{\mu}\frac{p_i}{p} = \frac{\rho_i}{\mu}\frac{RT}{p}. \qquad (A.2.6)$$

Mass concentration is a dimensionless quantity and is expressed by different fractions of the ratio. The main unit of the mass fraction in atmospheric physics is gram per gram, g/g (kilogram per kilogram, and so on). This unit shows how many grams of gas are present in the gram of air. Other units include: gram per kilogram, g/kg (milligram per gram mg/g, promille by mass – prom); per cent by mass – %; milligram per kilogram, mg/kg (millionth fraction by mass – mln^{-1} (ppm); (microgram per kilogram μg/kg (ppb)) the trillionth fraction by mass, (ppt). The chain of conversion of the units of measurement of the mass fraction of the gases:

$$w[\mathrm{g/g}]=10^{-3}\, w[\mathrm{g/kg}]=10^{-2}\, w[\%]=10^{6}\, w[\mathrm{mg/kg}]=$$

$$=10^{-9}\, w[\mu\mathrm{g/kg}] = 10^{-12}\, w[\mathrm{ppt}^{-1}],$$

where $w[\ldots]$ is the fraction in the appropriate units.

When substituting the quantities in specific measurement units, the second of equations (A.2.6) has the form:

$$w_i[\mathrm{g/g}] = 10^{-2}\frac{\rho_i[\mathrm{g\cdot m^{-3}}]}{\mu[\mathrm{g\cdot mol^{-1}}]}\frac{R[\mathrm{J\cdot mol^{-1}\cdot K^{-1}}]T[K]}{p[\mathrm{mbar}]}.$$

Number concentration

The number concentration of a gas n_i is the number of molecules of the gas in the unit volume. Writing (A.1.4) through the number of gas molecules N_i and dividing by volume, according to the definition we obtain $n_i=p_iN_A/RT$. The constant $k_B=R/N_A$ is the Boltzmann constant – one of the fundamental physical constants [19, 68]

$$k_B=1.380662\cdot 10^{-23}\ \mathrm{J\cdot K^{-1}}.$$

Finally we obtain the relationship of n_i with the partial pressure of the gas and the temperature of air:

$$n_i = \frac{p_i}{k_B T}. \tag{A.2.7}$$

Formula (A.2.7) can be used for air if we set $p_i=p$. Taking into account (A.2.7) from (A.2.1), (A.2.4), (A.2.6) we obtain the relationship of the number concentration with the partial density, volume and mass concentrations at the given temperature T and pressure P of air

$$n_i = N_A \frac{\rho_i}{\mu_i}, n_i = q_i \frac{p}{k_B T}, n_i = w_i \frac{\mu}{\mu_i}\frac{p}{k_B T}. \tag{A.2.8}$$

The third of the equations (A.2.8) cannot be used for water vapour.

It should be mentioned that number concentration is important in atmospheric optics because it is essential for calculating the volume coefficient of molecular absorption from the available absorption cross-section of the molecule (see section 3.3). Therefore, in optics it is often simply referred to as 'concentration' without the word 'number'. The measurement unit for number concentration is represented in most cases by the inverse cubic centimetre ($\mathrm{cm^{-3}}$), i.e. concentration is expressed by the number of particles in the cubic centimetre. Other units are usually not used for number concentration.

When substituting the quantities in specific measurement units, equation (A.2.7) and (A.2.8) have the forms

$$n_i[\mathrm{cm^{-3}}] = 10^{-4}\frac{p_i[\mathrm{mbar}]}{k_B[\mathrm{J\cdot K^{-1}}]T[K]},$$

$$n_i[\mathrm{cm^{-3}}] = 10^{-6} N_A[\mathrm{mol^{-1}}]\frac{\rho_i[\mathrm{g\cdot m^{-3}}]}{\mu_i[\mathrm{g\cdot mol^{-1}}]},$$

(A.2.9).

$$n_i[\mathrm{cm^{-3}}] = 10^{-10} q_i[\mathrm{ppm\ V}]\frac{p[\mathrm{mbar}]}{k_B[\mathrm{J\cdot K^{-1}}]T[K]},$$

$$n_i[\mathrm{cm^{-3}}] = 10^{-4} w_i[\mathrm{g/g}]\frac{\mu}{\mu_i}\frac{p[\mathrm{mbar}]}{k_B[\mathrm{J\cdot K^{-1}}]T[K]}.$$

A.3. Units of measurement of the concentration of water vapour

Specific forms of expressing the concentration of water vapour

For the concentration of water vapour, we use both the previously mentioned five types of units (partial pressure and density, volume, mass and number concentrations) and also additional specific types of units [25,43].

Partial pressure of water vapour

Using the partial pressure of H_2O denoted by e, we can express other concentrations of H_2O. In some cases e is also referred to as the elasticity of water vapour. The units of measurements of this parameter are the conventional units of partial pressure (see A.2).

Partial density of water vapour
Absolute humidity of air

The partial density of water vapour is determined using the same procedure as the conventional partial density (see A.2). It is described by the relationship (A.2.1) in which p_i=e. The synonym of the partial density of.water vapour is the absolute humidity of air. Absolute humidity is the mass of liquid water which could be expressed from the unit volume of air. The standard unit of absolute humidity a is g·m^{-3}. In substitution μ_i into (A.2.1) for water we obtain

$$a[\mathrm{g{\cdot}m^{-3}}] = 217\frac{e[\mathrm{mbar}]}{T[K]}. \qquad (A.3.1)$$

Volume concentration of water vapour

This concentration does not differ at all from the volume concentration of other gases and is expressed in the same units. From (A.2.4) we have

$$q_i = \frac{e}{p}. \qquad (A.3.2)$$

Mass fraction of water vapour
The mixing ratio of of water vapour

In old terminology, the mass fraction of water vapour is often referred to as the specific humidity of air [43]. Its definition and the units of measurement are identical with the mass fraction of any gas but the equations for expressing it should be corrected taking into account the dependence of the molecular mass of air on the value e. The formula of the relationship of $w=\frac{m_w}{m}$, where w is in g/g, with the partial pressure is (A.1.7). This equation gives the required expression for the partial pressure of water vapour through its mass fraction at the given pressure of air p:

$$e = p\frac{\frac{\mu_0}{\mu_w}w}{1+\left(\frac{\mu_0}{\mu_w}-1\right)w}. \qquad (A.3.3)$$

With numerical substitutions

$$e = p\frac{1.608w[g/g]}{1+0.608w[g/g]}.$$

Using (A.3.3) to express e through w from (A.2.1) (or (A.3.1)) and (A.2.4) or (A.3.2), we obtain the relationship of partial density and the volume concentration of the water vapour with its mass fraction. Taking into account numerical substitutions we obtain

$$a[g{\cdot}m^{-3}] = 349\frac{w[g/g]}{1+0.608w[g/g]}\frac{p[mbar]}{T[K]},$$

$$q_i[ppm] = 1.608\cdot 10^6\frac{w[g/g]}{1+0.608w[g/g]}.$$

In addition to the conventional definition of the mass fraction of water vapour we introduce also the mixing ratio s – the ratio of the mass of water vapour in the unit volume to the mass of dry air in the same volume. According to the definition $s = m_w/m_0 = m_w/(m-m_w)$, from which by dividing by m we obtain the relationship of the ratio of the mixture and the mass fraction of water vapour

$$s = \frac{w}{1-w}, w = \frac{s}{1+s},$$

where w, s are in g/g.

Number concentration of water vapour

This concentration is defined as the number concentration of any gas and is measured in cm^{-3}. The relationship of the number concentration with partial pressure, density and volume concentration is given by equation (A.2.7) and the two first equations of (A.2.8). To determine the expression for the number concentration through the mass fraction of the water vapour we substitute (A.3.3) into (A.2.7) and, taking into account numerical values, we obtain

$$n\,[\mathrm{cm}^{-3}] = 1.165\cdot 10^{19}\,\frac{p[\mathrm{mbar}]}{T[\mathrm{K}]}\,\frac{w[\mathrm{g/g}]}{1+0.608w[\mathrm{g/g}]}.$$

Relative humidity of air

The partial pressure of water vapour cannot be higher than some value – saturation pressure E which depends on the temperature of air, T, and as T increases the value of E also increases. For the function $E(T)$ we have theoretical expressions but they all were derived using some approximations and are fulfilled with insufficient accuracy. Therefore, for $E(T)$ it is necessary to use usually semi-empirical relationships obtained by correction of the theoretical equations for better agreement with the experimental data. In particular, the WMO (World Meteorological Organisation) recommends the equations [69, 70]

$$\lg(E) = 10.79574(1-T_0/T) - 5.028\lg(T/T_0) + $$
$$+0.42873\cdot 10^{-3}(10^{4.76955(T/T_0-1)} - 1) + 0.78614, \qquad \text{at } T \geq T_0,$$
$$\lg(E) = -9.09685(T_0/T-1) - 3.56654\lg(T_0/T) + \qquad \text{(A.3.4)}$$
$$+0.87682(1 - T/T_0) + 0.78614, \qquad \text{at } T < T_0.$$

where T is the temperature of air in K; E is the saturation pressure, mbar; T_0 = 273 K.

The relative humidity of air (u) is the ratio of the partial pressure of water vapour to saturation pressure:

$$u = \frac{e}{E(T)}. \qquad \text{(A.3.5)}$$

The meaning of relative humidity is the capacity of liquid water for evaporation: as the value of u increases the rate of this process decreases, at u close to unity evaporation cannot take place and the inverse process of condensation of water from the atmosphere (dew and frost phenomena, formation of fog and cloud) starts. The relationship of the relative humidity with other units of the concentration of water vapour can be easily determined expressing partial pressure e from (A.3.5) $e=uE(T)$ at the given temperature of air. The relative humidity of air is expressed in percent and, less frequently, in fractions.

The characteristic of the concentration of water vapour, similar to the relative humidity is also the deficit of pressure – the difference between saturation pressure and the partial pressure of water vapour $E(T) - e$.

Dew point

Dew point t is the temperature at which the partial pressure of the water vapour present in the atmosphere becomes equal to saturation pressure, this means that

$$e = E(t). \tag{A.3.6}$$

It should be mentioned that the dew point is the unit of the concentration of water vapour, although it is expressed in temperature. To determine other concentration units from the known value of the dew point it is sufficient to use the partial pressure of water vapour determined from (A.3.6). To determine the dew point from other units it is necessary to find the value of e and then use the function $t = E^{-1}(e)$ inverse to (A.3.4). The inverse function $T(E)$ exists because of the monotonicity of $E(T)$. Equation $E(t) = e$ is easily solved in a computer (for example, by the halving method). The dew point is usually measured in degrees Centigrade. The dew point deficit – the difference between the temperatures of air and the dew point $T-t$ is also examined.

In night cooling of the Earth's surface and air (see paragraph 2.3) when the temperature decreases below the dew point, the relative humidity of the air reaches 100% and water vapour condenses. If the dew point temperature is higher than 0°C, dew appears, if $t<0$°C frost forms (in this case, the freezing point is considered). Therefore, the determination of the dew point temperature represents the basis of predicting night (morning) light frost. Because of the high heat capacity of water after the formation of dew further cooling slows down; consequently, as the air humidity increases, i.e. the dew point deficit decreases, the formation of light frost becomes less likely.

A.4. Gas content and units of measurement

Total gas content in the column of the atmosphere

The vertical profile of the gas concentration is the function of the height also and for the general estimation of the gas content of the atmosphere, i.e. the determination of whether the content is high or low, a scalar quantity is more suitable. This quantity is represented by the total content of the gas in the column of the atmosphere (briefly – the total gas content) determined by the formula

$$W = \int_0^\infty Q(z)dz, \tag{A.4.1}$$

where W is the total gas content; $Q(z)$ is the vertical profile of the concentration, and function $Q(z)$ may be represented by the partial pressure, partial density or number concentration. The meaning of the total content is the amount of the gas (for example, in grams or molecules) in the column of the atmosphere above the area with unit surface. It can be mentioned that integration in (A.4.1) should be continued not up to the upper boundary of the atmosphere but to some height z for determining the total content in the part of the column of the atmosphere (for example, in the troposphere). In the general form of the Bouguer's law (Chapter 3) integration for determining the gas content is carried out along the length of the optical path.

Atmosphere–centimetre

When using the partial pressure of the gas as the concentration $Q(z)$ the total content is expressed in the atmospheres (physical), and the altitude in centimetres. Correspondingly, the unit of measurement of the total concentration is the atmosphere–centimetre, atm·cm. Equation (A.4.1) for standard initial dimensions has the form

$$W[\mathrm{atm \cdot cm}] = 98.69 = \int_0^\infty \rho_i[\mathrm{mbar}](z[\mathrm{km}])dz.$$

Dobson unit

The special unit for measurement of the total concentration used for ozone is the Dobson unit. One Dobson unit is 10^{-3} atm·cm.

The total content according to partial density

When using the partial density of gas as $Q(z)$ it is expressed in $g{\cdot}cm^{-3}$, and the altitude – in cm. Correspondingly, the unit of measurement of the total content is gram per square centimetre ($g{\cdot}cm^{-2}$). For standard initial dimensions of density and height equation (A.4.1) has the following form

$$W[g{\cdot}cm^{-2}] = 0.1\int_0^{z} \rho_i[g\cdot cm^{-3}](z[km])dz.$$

The height of the column of precipitated water

For water vapour the unit of the total content is represented by centimetres of precipitated water, i.e. W is reduced to the height of the column of liquid water deposited from the vapour onto the unit of the surface area [25]. The equation for calculating this height is obvious

$$h[cm] = \frac{W[g{\cdot}cm^{-2}]}{\rho_w[g{\cdot}cm^{-3}]},$$

where ρ_w is the density of water. Assuming that $\rho_w = 1\ g{\cdot}cm^{-3}$ we obtain numerically equality between W and h.

Total content according to number concentration

If $Q(z)$ is represented by number concentration, the unit of measurement of the total content is the molecule per square centimetre (cm^{-2}) and formula (A.4.1) has the form

$$W[cm^{-2}] = 10^5\int_0^{z} n_i[cm^{-3}](z[km])dz.$$

Relationship of different expression for total content

Since the partial pressure is linked with the partial density and number concentration through temperature (see A.2), which depends on height, there is no unambiguous relationship between the total content, expressed in atm·cm and cm^{-2}, and also expressed in atm·cm^{-2}. The relationship between the total content in $g{\cdot}cm^{-2}$ and cm^{-2} according to (A.2.8) is given by the relationship

$$W[cm^{-2}] = \frac{N_A[mol^{-1}]}{\mu_i[g\cdot mol^{-1}]}W[g{\cdot}cm^{-2}].$$

A.5. Units of measurement of spectral intensities in radiation fluxes

Planck formula in different units

Expressions for spectral dependences

Spectral dependences in atmospheric physics are expressed in the wavelength, wave number and frequency of electromagnetic waves (the latter usually only in the microwave range).

Units of measurement of wavelength

The universal unit for measuring the wavelength used in all ranges of the spectrum is the micrometer (μm). In the range of short wavelengths (UV, visible and near infrared ranges) nanometer (nm) and angström (Å) are also used. In the microwave range, we use the centimetre (cm) and millimetre (mm). The chain of conversion of the units of wavelength is:

$$\lambda[\mu m] = 10^{-3}\,\lambda\,[nm] = 10^{-4}\,\lambda[Å] = 10^{4}\,\lambda[cm] = 10^{3}\,\lambda[mm],$$

where λ[...] is the value of the wavelength in corresponding units.

Wave number and units of measurement

The scanning of the spectrum in spectral devices in inversely proportional to the wavelength of light. Therefore, to ensure that the scale of the device is linear, it should be calibrated not in the wavelength but in inverse values, i.e. wave numbers. The expression for the spectral dependences in wave numbers is traditional in optics and spectroscopy. The inverse centimetre (cm^{-1}) is the generally accepted unit of measurement of the wave number.

Frequency measurement units

The frequency of electromagnetic waves (in the microwave range) is expressed in gigahertz (GHz), megahertz (MHz) and kilohertz (kHz). The chain of conversion of the frequency units is:

$$\nu[GHz]=10^{-3}\,\nu[MHz] = 10^{-6}\,\nu[kHz],$$

where ν[...] is the frequency in appropriate units.

Relationship between units of measurement of wavelength, wave number and frequency

This relationship is expressed by the formula [19, 68]

$$\lambda[\mu m] = \frac{10^4[\mu m \cdot cm^{-1}]}{\tilde{\nu}[cm^{-1}]} = \frac{10^{-5}c[cm \cdot s^{-1}]}{\nu[GHz]}, \qquad (A.5.1)$$

where λ is wavelength, $\tilde{\nu}$ is wave number, ν is frequency, c is the velocity of light in vacuum – the fundamental physical constant

$$c = 29979245800 \text{ cm}\cdot\text{s}^{-1}.$$

It should be mentioned that (A.5.1) gives the wavelengths in vacuum. In other cases, they must be divided by the refractive index of the medium. For air this difference in the wavelengths (and wave number) is also ignored.

Units of measurement of spectral fluxes and intensities

The dimension of the spectral fluxes is the power per area, per spectral range, the intensity – power per area per spectral range per steradian. To simplify this, we shall discuss only the fluxes, and the units of measurement of the intensities are obtained by automatic division of the flux units by steradian. The units of measurement of spectral fluxes differ greatly depending on the selection of the units of power, area and spectral range. The fluxes per unit wavelength are usually measured in watts per square meter per micrometer ($W/m^2 \cdot \mu m$), milliwatts per square meter per micrometer ($mW/m^2 \cdot \mu m$), milliwatts per square centimetre per micrometer ($mW/cm^2 \cdot \mu m$) and so on. The conversion chain is:

$$F\left[\frac{W}{m^2 \cdot \mu m}\right] = 10^{-3}F\left[\frac{mW}{m^2 \cdot \mu m}\right] = 10F\left[\frac{mW}{cm^2 \cdot \mu m}\right],$$

where $F[\ldots]$ is the value of the flux in appropriate units.

Identical units are used for the fluxes per unit wavelength and frequency (for example, $W/m^2 \cdot cm^{-1}$, $W/m^2 \cdot GHz$ and so on).

To determine the relationship between three groups of units it is necessary to differentiate (A.5.1)

$$d\lambda = \frac{d\tilde{\nu}}{\tilde{\nu}^2} = \frac{c}{\nu^2} = d\nu \qquad (A.5.2)$$

(the minus sign has been omitted). Substituting (A.5.2) into the definition of the fluxes (intensities) (3.2.3) and taking into account (A.5.1) gives the following chain:

$$F_\lambda\left[\frac{\mathrm{W}}{\mathrm{m}^2\cdot\mu\mathrm{m}}\right]=\frac{10^4}{\lambda^2[\mu\mathrm{m}]}F_{\tilde{\nu}}\left[\frac{\mathrm{W}}{\mathrm{m}^2\cdot\mathrm{cm}^{-1}}\right]=$$

$$=\frac{10^{-5}c[\mathrm{cm{\cdot}s^{-1}}]}{\lambda^2[\mu\mathrm{m}]}F_\nu\left[\frac{\mathrm{W}}{\mathrm{m^2{\cdot}GHz}}\right]. \qquad (\mathrm{A.5.3})$$

Units of measurement of integral fluxes

Integral radiation fluxes (see Chapter 7) are the integrals of spectral fluxes over the wavelength and, consequently, their measurement units are determined automatically from the previously examined units of the spectral fluxes by excluding the units of wavelength (wave number, frequency) from their denominator.

Planck law in different spectral units

The intensity of thermal radiation is expressed by Planck's function (3.4.19) which, taking into account (A.5.2), may also be written in different spectral units. The initial equation is (3.4.19) which, taking into account dimensions, has the form

$$B_\nu\left[\frac{\mathrm{W}}{\mathrm{m}^2\cdot\mathrm{GHz}}\right]=10^{22}\frac{2h[\mathrm{J{\cdot}s}]\nu^3[\mathrm{GHz}]}{c^2[\mathrm{cm}\cdot\mathrm{s}^{-1}]}\times$$

$$\times\frac{1}{\exp\left(10^9\dfrac{h[\mathrm{J}\cdot\mathrm{s}]\nu[\mathrm{GHz}]}{k_B[\mathrm{J}\cdot\mathrm{K}^{-1}]T[K]]}\right)-1},$$

where h is Planck's constant [19, 68]

$$h = 6.6254\cdot10^{-34}\ \mathrm{J{\cdot}s}$$

Now, taking into account (A.5.1), (A.5.3), we obtain an equation for intensity in the unit wavelength interval [19, 68]

$$B_\lambda\left[\frac{\mathrm{W}}{\mathrm{m}^2\cdot\mu\mathrm{m}}\right]=10^{20}\frac{2h[\mathrm{J}\cdot\mathrm{s}]c_2[\mathrm{cm}\cdot\mathrm{s}^{-1}]}{\lambda^5[\mu\mathrm{m}]}\frac{1}{\exp\left(10^4\dfrac{c_2[\mathrm{cm}\cdot\mathrm{K}]}{\lambda[\mu\mathrm{m}]T[K]}\right)-1},$$

where $c_2 = hc/k_B$ is the second radiation constant (see (3.4.21)):

$$c_2 = 1.43879 \text{ cm}\cdot\text{K}.$$

Finally, for the intensity in the units of the wavelength and wave number, we have

$$B_\nu\left[\frac{\text{W}}{\text{m}^2\cdot\text{cm}^{-1}}\right] = 10^{-4}\frac{2h[\text{J}\cdot\text{s}]c_2[\text{cm}\cdot\text{s}^{-1}]}{\tilde{\nu}^3\,[\text{cm}^{-1}]}\frac{1}{\exp\left(\dfrac{c_2[\text{cm}\cdot\text{K}]}{\tilde{\nu}[\text{cm}^{-1}]T[K]}\right)-1}.$$

A.6. Units of measurement of the coefficients of molecular scattering and absorption

Volume coefficients

The volume coefficients of molecular scattering and absorption (and also of aerosol scattering and absorption) have the dimensions of the reciprocal length and are almost always expressed only in inversed kilometres (km^{-1}). Taking into account the standard measurement of wavelength in μm and molecule concentration in cm^{-3}, this gives the conversion factor of 10^{21} in the equation for the volume coefficient of molecular scattering (5.1.14). The cross-sections of gas absorption are usually measured in square centimetres per molecule (cm^2). Taking into account the previously mentioned units of measurement of the concentration and volume coefficient of absorption, the factor 10^5 appears in equation (3.3.22).

Different forms for representing coefficients of molecular absorption

The dependence of the volume coefficient of absorption on the number of molecules of matter enables us to introduce different forms of the absorption coefficient. Actually, we write formally Bouguer's law (3.4.5) in the absence of scattering in the form

$$dI = -kIdl = -k\frac{q}{q}Idl = k_q qIdl, \qquad \text{(A.6.1)}$$

where q is the concentration of attenuating substance; $k_q = kq$ is its coefficient of absorption. Since product kdl is a dimensionless quantity, the dimension of k is inversed in relation to the dimension dl. Selecting various types of measurement units for concentration (see A.2), we obtain absorption coefficients determined variously. Usually, four forms of the expression for the molecular absorption coefficient are used.

– the volume coefficient of molecular absorption k_v (if $q \equiv 1$ in equation (A.6.1)), which characterises this absorption per unit length of the optical path,

– the mass coefficient of absorption k_m (if q in equation (A.6.1) is partial density), characterising absorption per unit mass of gas.

– absorption cross-section of the molecule C_a (if q in equation (A.6.1) is number concentration), characterising the absorption by one molecule.

– Absorption coefficient per centimetre under standard conditions k_{vs}. This is the volume coefficient of absorption determined (usually from laboratory measurements) for some standard pressures and temperatures. It should be mentioned that the term 'standard conditions' is not suitable because it differs with different authors and it is therefore better to regard the coefficient k_{vs} as the volume coefficient of absorption for determined strictly fixed parameters of the state of the gas.

Relationship between the volume coefficients of molecular absorption

We present Table A.6.1 for the conversion of the volume coefficient of molecular absorption from one type to another. It should be noted that the Table gives the most frequently used measurement units.

It should be mentioned that the relationship of transition from the coefficients in the standard conditions to the coefficients in specific conditions should be regarded as approximate because (see Chapter 4) the dependence of the coefficients of molecular absorption on temperature and pressure is complicated.

A.7. Units of measurement of the concentration of aerosols and volume coefficients of aerosol extinction

Forms for expressing the concentration of aerosol particles

The content of aerosol particles in the atmosphere is usually expressed in the form of number, mass or volume concentration [23].

The number concentration of aerosol particles n is determined as in the case of gases: it is the number of particles in the unit volume of air. It is usually expressed in particles per cubic centimetre (cm^{-3}) and less frequently in particles per litre (l^{-3}). The relationship between the units is

$$n[cm^{-3}] = 10^{-3}\, n[l^{-3}],$$

where $n[...]$ is the concentration in appropriate units.

The mass concentration of aerosol particles m is equivalent to the partial density of gases: it is the mass of the aerosol particles in the unit volume of air. The measurement units are the same as for the gases (see A.3). For liquid droplet clouds, the mass concentration has the special name – the water content of the cloud – and is determined as the mass of liquid water in the unit volume of the cloud [51]. For crystal clouds, the water content is sometimes replaced by the term iciness of the cloud.

The volume concentration of the aerosol particles v is the total volume of the aerosol particles in the unit volume of air. This concentration has no analogues amongst the gas concentrations. It is expressed in cubic micrometers per cubic meter ($\mu m^3 \cdot m^{-3}$) or other similar units (volume ratios).

Relationships between the units of concentration of aerosol particles

The relationships between the concentration of aerosol particles are not unambiguous because the aerosol particles have different

Table A.6.1. Units of measurement and conversion multipliers for extinction coefficients

Name	k_v	k_m	C_a	k_{vs}
Specific name	Volume	Mass	Extinction cross-section	Per 1 cm under standard conditions
Measurement units	km^{-1}	$g^{-1} \cdot cm^2$	cm^2	cm^{-1}
k_v	1	$10^5\, \rho$	$10^5\, n$	$10^5\, n/n_s$
k_m	$10^{-5}\, \rho^{-1}$	1	$N_A \mu$	ρ_s^{-1}
C_a	$10^{-5}\, n^{-1}$	$\mu^{-1} N_A^{-1}$	1	n_s^{-1}
k_{vs}	$10^{-5}\, n_s/n$	ρ_s	n_s	1

Comment. ρ – partial density of the absorbing gas, $g \cdot cm^{-3}$; n – the number concentration of the absorbing gas, cm^{-3}; μ – the molecular mass of the absorbing gas, g/mol; ρ_s – the partial density of the absorbing gas in the standard conditions, $g \cdot cm^{-3}$; n_s – number concentration of the absorbing gas in the standard conditions, cm^{-3}; N_A – Avogadro number, mol^{-1}.

dimensions. To convert different concentrations, it is necessary to know the size distribution function of the particles $f(r)$ (Chapter 2). It is also taken into account that the radius of the aerosol particles is usually measured in micrometers (μm), consequently, according to (2.4.4), $f(r)$ has the dimension μm^{-1}. We use the approximation of homogeneous spherical particles. Consequently, determining the volume of the aerosol particles in the unit volume of air we obtain

$$V = n\int_0^\infty \frac{4}{3}\pi r^3 f(r)dr = \frac{4}{3}\pi n\int_0^\infty r^3 f(r)dr.$$

Similarly, for the mass concentration

$$m = \frac{4}{3}\pi n\rho\int_0^\infty r^3 f(r)dr,$$

where ρ is the density of aerosol substance which is assumed to be the same for all particles. Taking into account the dimensions, these equations have the form:

$$V[\mu m^3 \cdot m^{-3}] = 10^{-6}\frac{4}{3}\pi n[cm^{-3}]\int_0^\infty r^3[\mu m]f(r)dr,$$

$$m[g \cdot m^{-3}] = 10\frac{4}{3}\pi n[cm^{-3}]\rho[g \cdot cm^{-3}]\int_0^\infty r^3[\mu m]f(r)dr.$$

Volume coefficients of aerosol scattering, absorption and extinction

The volume coefficients of aerosol extinction, scattering and absorption are measured in inverse kilometres (km^{-1}). The cross-section of aerosol extinction, scattering and absorption are measured in square centimetres (cm^2). Taking into account the standard measurement of the particle radii in μm and its number concentration in cm^{-3} the equations (3.3.21), (3.3.22) contain the conversion factor 10^5 and the equations (5.3.2), (5.3.3) the factor 10^{-3}.

It should be mentioned that for aerosol extinction we can formally use the general form of Bouguer's law and, consequently, consider different units of the aerosol content on the optical path and different expressions of its absorption coefficient. However, usually, in addition to the sections and volume coefficients no other types of characteristics are used for aerosol extinction. The mass coefficient of aerosol extinction, associated with the volume extinction and the cross-section through the density of the aerosol substance (see Table A.6.1) is used only seldom.

References

1. Atlas of clouds, ed. by Khrgian A.Kh. and Novozhilov N., Gidrometeoizdat, Leningrad, 1978.
2. Atmosphere (handbook data, models), a textbook, Gidrometeoizdat, Leningrad, 1991.
3. Bauer Z., Physics of atmospheres of planets, Mir, Moscow, 1976.
4. Belov N.N., in: Physics and chemistry of atmospheric aerosols. Problems of atmospheric physics, No. 20, ed. by L.S. Ivlev, St Petersburg State University, 1997, 209–215.
5. Bogorodskii V.V. and Kozlov A.I., Microwave radiometry of earth covers, Gidrometeoizdat, Leningrad, 1985.
6. Bohren C.F, Huffman R.K., Absorption and scattering of light by small particles, Wiley, New York, 1983.
7. Born M. and Wolf E. [Russian translation], Fundamentals of optics, Nauka, Moscow, 1973.
8. Brasseur H. and Solomon S., Aeronomy of the middle atmosphere [Russian translation], Gidrometeoizdat, Leningrad, 1987.
9. Van de Hulst, Scattering of light by small particles [Russian translation], IL, Moscow, 1961.
10. Vasil'ev A.V., Vestnik SPbGU, ser. 4, Fizika, Khimiya, 1996, No. 4, 25, 3–11.
11. Vasil'ev A.V. and Ivlev, L.S., Izvestiya, Atmospheric and Oceanic Optics, 1997, **10**, No. 8, 534–540.
12. Vasil'ev A.V., Vestnik SPbGU, ser. 4, Fizika, Khimiya. 1997, No. 1, 25, 14–24.
13. Vasil;ev A.V., et al., Issledovanie Zemli iz Kozmosa. 1997, No. 3, 25–31.
14. Vasil'ev A.V., et al., Issledovanie Zemli iz Kozmosa, 1997, No. 4, 43–51.
15. Vasil'ev, A.V., et al., Issledovanie Zemli iz Kosmosa, 1997, No. 5, 25–32.
16. Vorontsov-Vel'yaminov B.A., Astrōnomy, Prosveshchenie, Moscow, 1976.
17. Givishvili G.V., et al., Izvestiya, Atmospheric and Oceanic Physics, 1996, **32**, No. 3, 303–313.
18. Grechko G.M., et al., Optical phenomena in sunrise and sunset. Temat. vyp. GOI, No. 205, Leningrad, 1989.
19. GSSSD 1-76 , Fundamentals physical constants, Publishing House of Standards, Moscow, 1976.
20. Goody R.M., Atmospheric radiation, 1. Fundamentals of Theory [Russian translation], Mir, Moscow, 1966.
21. Gurvich A.S., et al., Izv AN Seria: Fizika Atmosfery i Okeana, 1995, **31**, No.6, 774–781.
22. Deirmendijan D., Scattering of electromagnetic radiation by spherical polydisperse particles [Russian translation], Mir, Moscow, 1971.
23. Dovgalyuk Yu.A. and Ivlev, L.S., Physics of water and other atmospheric aerosols, St Petersburg State University, 1998.
24. Zuev V.E., Propagation of visible and infrared waves in the atmosphere, Sovetskoe radio, Moscow, 1970.
25. Zuev V.E. and Komarov V.S.. Statistical models of temperature and gas components of the atmosphere, Gidrometeoizdat, Moscow, 1986.

26. Zuev V.E. and Krekov G.M., Optical models of the atmosphere, Gidrometeoizdat, Leningrad, 1986.
27. Zuev V.E., et al., Spectroscopy of the atmosphere, Gidrometeoizdat, Leningrad, 1987.
28. Zuev V.E., et al., Nonlinear optics of the atmosphere, Gidrometeoizdat, Leningrad, 1987.
29. Zuev V.E. and Naats I. E., Inverse problems of atmospheric optics. Current problems of atmospheric optics, vol. 7, Gidrometeoizdat, Leningrad, 1990.
30. Karol' I.L., et al., Gas impurities in the atmosphere, Gidrometeoizdat, Leningrad, 1983.
31. Kondrat'ev K.Ya., Meteorological investigations by rockets and satellites, Gidrometeoizdat, Leningrad, 1962.
32. Kondrat'ev K.Ya., Actinometry, Gidrometeoizdat, Leningrad, 1965.
33. Kondrat'ev K.Ya., Radiation transfer in the atmosphere, Gidrometeoizdat, Leningrad, 1972.
34. Kondrat'ev K.Ya. and Timofeev Yu.M., Thermal sounding of atmosphere from satellites, Gidrometeoizdat, Leningrad, 1970.
35. Kondrat'ev K.Ya. and Timofeev Yu.M., Meteorological sounding of the atmosphere from space, Gidrometeoizdat, Leningrad, 1978.
36. Krinov E.L., Spectral reflectivity of natural formations, Academy of Sciences of the USSR, Leningrad–Moscow, 1947.
37. Liou Kuo-Nan, Fundamentals of radiation processes in the atmosphere, Gidrometeoizdat, Leningrad, 1984.
38. Hinckley E.D., (editor), Laser monitoring of the atmosphere [Russian translation], Mir, Moscow, 1979.
39. Landsberg G.S., Optics, Publishing House of Technical and Theoretical Literature, Moscow, 1957.
40. Levich V.G., Course in theoretical physics, vol. 1, Fizmatgiz, Moscow 1962.
41. McCartney E., Atmospheric optics [Russian translation], Mir, Moscow, 1979.
42. Malkevich M.S., Optical investigations of the atmosphere from satellites, Nauka, Moscow, 1973.
43. Matveev L.T., Fundamentals of general meteorology: Atmospheric physics, Gidrometeoizdat., Moscow, 1965.
44. Measuris R., Remote laser sounding [Russian translation], Mir, Moscow 1987.
45. Kondrat'ev K.Ya. (editor), Meteorological sounding of the underlying surface from space, Gidrometeoizdat, Leningrad, 1979.
46. Marchuk G.I. (editor), The Monte Carlo method in atmospheric optics, Nauka, Novosibirsk, 1976.
47. Minin I.N., Theory of radiation transfer in the atmosphere of planets, Nauka, Moscow 1988.
48. Mitra S.K., The upper atmosphere [Russian translation], IL, Moscow, 1955.
49. Monin A.S., The solar cycle, Gidrometeoizdat, Leningrad, 1980.
50. Moroz V.I., Physics of planets, Nauka, Moscow, 1977.
51. Mazin I.P. and Khrgian A.Kh. (editors), Clouds and cloudy atmosphere, a handbook, Gidrometeoizdat, Leningrad, 1989.
52. Penner S.S., Quantitative molecular spectroscopy and emissivity of gases [Russian translation], IL, Moscow, 1963.
53. Lenoble J. (editor), Transfer of radiation in scattering and absorbing at-

mospheres. Standard calculation methods, Gidrometeoizdat, Leningrad, 1990.
54. Perov S.P. and Khrgian A.Kh. (editors), Current problems of atmospheric ozone, Gidrometeoizdat, Leningrad, 1989.
55. Poberovskii A.V., et al., Izvestiya, Atmospheric and Oceanic Physics, 1999, **35**, No. 3, 282–290.
56. Polyakov A.V., et al., Izvestiya, Atmospheric and Oceanic Physics, 1999, **35**, No. 3, 291–298.
57. Polyakov A.V., et al., Atmospheric and Oceanic Optics, 1997, **10**, No. 9, 613–616.
58. Polyakov A.V., et al., Izvestiya, Atmospheric and Oceanic Physics, 1998, **34**, No.3, 318–327.
59. Polyakov A.V., et al., Izvestiya, Atmospheric and Oceanic Physics, 1998, **34**, No. 3, 328-333.
60. Polyakov A.V., et al., Atmospheric and Oceanic Optics, 1997, **10**, No. 2, 97-99
61. Rozenberg GV, Izvestiya, Atmospheric and Oceanic Physics, 1998, **34**, No. 3, 328–333.
62. Sivukhin D.V., General course in physics. Optics, Nauka, Moscow, 1980.
63. Smirnov V.I., Course in higher mathematics, Vol. 2, Part 2, Nauka, Moscow, 1974.
64. Sobolev V.V., Course in theoretical astrophysics, Publishing House of Physical and Mathematical Literature, Nauka, Moscow, 1967.
65. Sobolev V.V., Scattering of light in the atmosphere of planets, Nauka, Moscow, 1972.
66. Sokolov A.A., et al., Quantum mechanics. State Teaching and Pedagogical Publishing House of the Ministry of Education of Russian Federation, Moscow, 1962.
67. Stepanov N.N., Spherical trigonometry, Gostekhhizdat, Moscow–Leningrad, 1948.
68. Kikoin I.K. (editor), Tables of physical quantities, a handbook, Atomizdat, Moscow, 1976.
69. Technical regulations, 2nd edition, WMO, Appendix 2, Geneva, 1963.
70. Technical regulations, Vol. 1 (general part), 2nd edition, WMO, No. 49, OD2, Geneva, 1959.
71. Timofeev Yu.M., Izvestiya, Atmospheric and Oceanic Physics, 1998, **34**, No.6, 712–717.
72. Timofeev Yu.A., and Polyakov A.V., Mathematical aspects of solving inverse problems of atmospheric physics, a textbook, St Petersburg State University, 2001.
73. Timofeev Yu.M., Remote (indirect) methods of measuring atmospheric parameters. Methodical instructions for the lectures "Indirect methods for studying planet atmospheres", Leningrad, 1988.
74. Timofeev Yu. M., Izv AN SSSR, Fizika Atmosfery i Okeana, 1989, **25**, No.5, 451–472.
75. Turchin V.F., et al., Uspekhi Fysicheskikh Nauk, 1970, **102**, 3, 345–387.
76. Farafonov V.G., Differentsial'nye uravneniya, 1983, **19**, No. 10, 1765–1777.
77. Fleagle R. and Businger D., Introduction into atmospheric physics [Russian translation], Mir, Moscow, 1965.
78. Hippel A.R., Dielectrics and waves [Russian translation], IL, Moscow 1960.
79. Khrgian A.Kh., Atmospheric physics, Moscow State University, Moscow 1986.
80. Sian Sue-sen, Physical mechanics [Russian translation], Mir, Moscow, 1965.

81. Chamberlain J., Theory of the atmosphere of planets. Introduction to their physics and chemistry [Russian translation], Mir, Moscow, 1985.
82. Chamberlain J., Physics of polar glow and atmospheric emission [Russian translation], IL, Moscow, 1963.
83. Shifrin K.S., Introduction into the ocean physics, Gidrometeoizdat, Leningrad, 1983.
84. Shifrin K.S., Scattering of light in a turbid medium, Gostekhizdat, Moscow, 1951.
85. Shutko A.M., Microwave radiometry of water surface and soils, Nauka, Moscow, 1986.
86. Angstrom A. Techniques of determining turbidity of the atmosphere, Tel-lus, 1961. Vol. **13**, No 1, p. 214.
87. Asano S. and Yamamoto G., Applied Opt., 1975, **14**, No 1, 29–49.
88. Bojkov R.D., The changing ozone layer, New York, WMO, UNEP, 1995.
89. Elachi C., Introduction to the physics and techniques of remote sensing. John Wiley & Sons. New York, Chichester, Brisbane, Toronto, Singapore, 1987.
90. Frohlich C. and Lean J., Geophys. Res. Lett., 1998, **25**, No. 23, 4377–4380.
91. Goody R.M. and Yung Y.L., Atmospheric radiation. Theoretical basis. Oxford University Press, New York, 1989.
92. Goody R.M., Principles of atmospheric physics and chemistry. Oxford University Press, New York, 1995.
93. Hollweg H.D., et al., Interaction at mm and optical frequencies. Part I: Current problems in radiative transfer simulations, Final Report ESA ESTEC Contract N 1 0603/93/NL/NB, Berichte aus dem Zentrum fur Meeres- und Klimaforschung. Reihe A: Meteorologie No.17, Zentrum fur Meeres– und Klimaforschung der Universitat Hamburg Meteo-rologisches Institut, 1995.
94. Hanel R.A., et al., Exploration of the solar system by infrared remote sensing. Cambridge University Press, Cambridge, New York, Port Chester, Melbourne, Sydney, 1992.
95. Harries I.E., Quart. J. Royal Meteor. Soc., 1996, **122**, No. 532, 799-818.
96. Heney L. and Greenstain J., Astrophys. J., 1941, **93**, No. 1, 70–83.
97. Hollweg H.D., et al., Interaction at mm and optical frequencies. Part II: Specific atmospheric absorption and emission featu-res: Investigation and modelling, Final Report ESA ESTEE Contract N 0603/931NL/NB, Berichte aus dem Zentrum fur Meeres- und Klimatforsc-hung. Reihe A: Meteorologie N. 18, Zentrum fur Meeres- und Klimaforsc-hung der Universitat Hamburg Meteorologisches Institut, 1995.
98. Huffman R.E., Atmospheric ultraviolet sensing. Academic Press Inc., Boston, San Diego, New York, London, Sydney, Tokoy, Toronto, 1992.
99. Intergovernment Panel on Climate Change (IPCC) Climate Change 1994: Radiative Forcing of Climate Change and an Evaluation of the IPCC IS 92 Emission Scenarios. Cambridge University Press. Cambridge, UK.
100. Jones P.D., et al., Reviews Geophysics, **37**, No. 2, 1999, 173–199.
101. Kidder S.Q. and Vonder Haar T.H., Satellite meteorology. An Introduction. Academic Press, San Diego, New York, Boston, London, Sydney, Tokyo, Toronto, 1995.
102. Kondrat'ev K.Ya, Radiation in the atmosphere, Academic Press, New York, 1969.
103. Liou Kuo-Nan, Radiation and cloud Processes in the atmosphere. Theory,

observation and modeling. Oxford Monographs on Ecology and Geophy-sics, 20, 1992.

104. Kursinsky E.R., et al. J. Geophys. Res. 2001, **106**, No DI, 1113–1123.
105. Li J., J. Appl. Meteorol., 2000, **39**, No.8, 1248–1268.
106. Mlawer E.J., et al., J. Geophys. Res., 1997, **102**, 16,663-16,682.
107. Manuilova R.O., et al., J. Quant. Spectros Radiat. Trans., 1998, **59**, No. 3–5, 405–422.
108. Nalli N.R., et al., Applied Optics., 2001, **40**, No. 9, 1343–1353.
109. Nardi B., et al., Geophys. Res. Lett., 1999, **26**, No. 6, 723–726.
110. Parki J.H., et al., Atlas of absorption lines from 0 to 17 900 cm^{-1}. NASA Reference Publication 1188, Sept. 1987.
111. Frohlich C. And London J. (editors), Revised instruction manual on radiation instruments and measurements. WMO, WCRP Publications series N 7, WMO/TD-N 149, October 1986.
112. Seinfeld J.H. and Pandis S.N., Atmospheric chemistry and physics. From air pollution to climate change. A Wiley-Interscience Publication, J. Wi-ley & Sons, Inc., 1998.
113. Shved G.M., et al., J. Atmosp. and Solar-Terrestr. Phys. 1998, **60**, No. 3, 283–314.
114. Solanki S.R. and Fligge M., Geophys. Res. Lett., 1999, **26**, No. 16. 2465–2468.
115. Stephens G.L., Remote sensing of the lower atmosphere. An introduction, New York, Oxford, Oxford University Press, 1994.
116. Stewart R.H., Methods of satellite oceanography. University of California Press. Berkeley/Los Angeles/London, 1985.
117. Thomalla E. and Quenzel H., Appl. Optics, 1982, **21**, No 17, 3170–3177.
118. Tonkov M.V., et al., J. Q. S. R. T., 1996, **56**, No. 5, 783–795.
119. Ulaby F.T., et al., Microwave remote sensing. Active and passive. Vol. III, From theory to applications, Artech House, 1986, 1065–2162.
120. Waters J.W, Absorption and emission by atmospheric gases, Methods of experimental physics. 1976, Vol. 12, Pt. B. Academic Press, New York, 142–176.
121. Wayne R.P., Chemistry of atmosphere, Oxford University Press, Oxford, 1985.

Index

A

B

C

D

E